SEED CONDITIONING

Volume One

MANAGEMENT

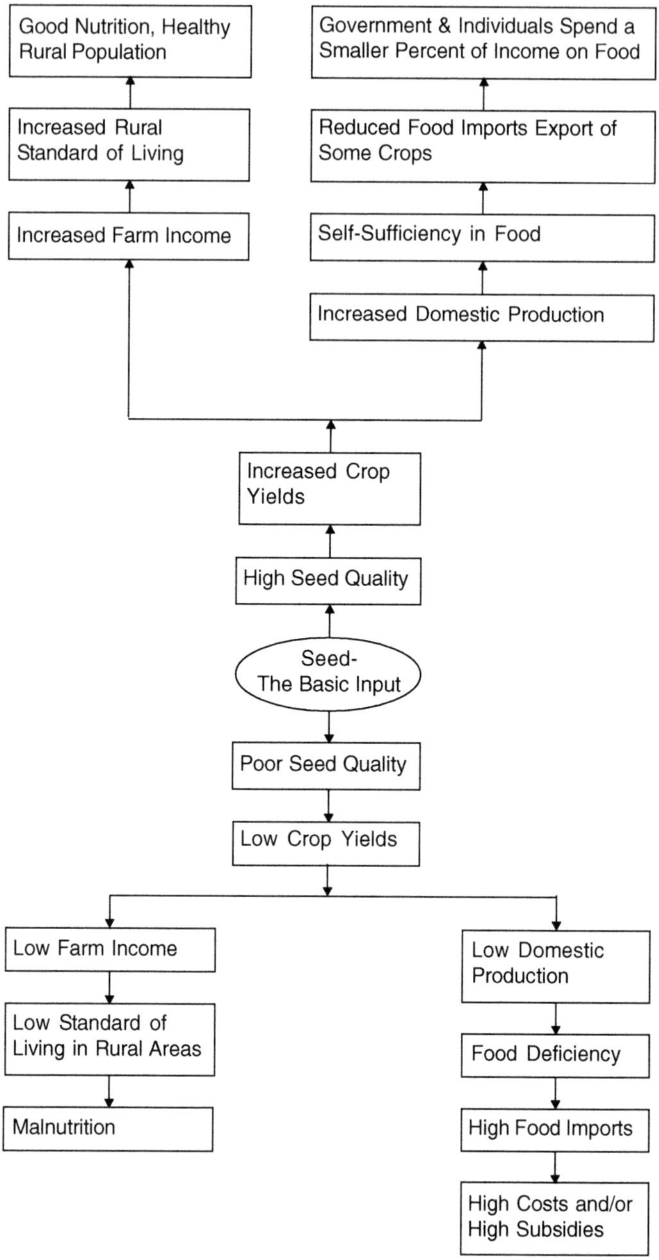

Frontispiece: The chain of effects resulting from farmers using good seed or poor seed. The influence extends to and affects the national economy and the general standard of living of the people, both rural and urban.

SEED CONDITIONING

Volume One

MANAGEMENT

A Practical Advanced-level Guide

Bill (B.R.) Gregg, Ph.D.
Seed Industry Development & Conditioning Specialist

Gary L. Billups
Seed Conditioning Specialist & President
Crippen International, Inc.

CRC Press
Taylor & Francis Group
Boca Raton London New York

CRC Press is an imprint of the
Taylor & Francis Group, an **informa** business
A SCIENCE PUBLISHERS BOOK

Science Publishers *www.scipub.net*
234 May Street
Post Office Box 699
Enfield, New Hampshire 03748
United States of America

General enquiries : *info@scipub.net*
Editorial enquiries : *editor@scipub.net*
Sales enquiries : *sales@scipub.net*

Published by Science Publishers, Enfield, NH, USA
An imprint of Edenbridge Ltd., British Channel Islands

ISBN 978-1-57808-572-9

Library of Congress Cataloging-in-Publication Data

Gregg, B. R.
Seed conditioning. Volume one, Management : a practical advanced-level guide to help managers, technical specialists, professionals, to achieve better seed in less time and cost / Bill (B.R.) Gregg & Gary L. Billups.
--1st ed.
p. cm.
Includes bibliographical references and index.
ISBN 978-1-57808-572-9 (hardcover)
1. Seed technology. I. Billups, Gary L. II. Title.
SB113.7.G74 2009
631.5'21--dc22
2009000072

Dedication

Seed conditioning is essential to providing high-quality seed, which ultimately benefits farmers, rural people and urban populations.

This guide and compendium of information is respectfully dedicated to the pioneering leaders who recognized the role of seed and conditioning—and then proceeded to do something about it! These included:

Dr. Louis N. Wise
Gus Stricker
Lakshmi Sagar
Oliver Steele
Jim Henderson
Jesse Harmond
Virgil Frevert
Lewis M. Carter
Ronald Banton

Their work is still with us!

Foreword

Welcome to *Seed Conditioning!*

Seed conditioning processes raw harvested seed into pure crop seed which is free of undesirable materials, protected from pests and diseases, and can be planted properly to give the farmer a good stand of healthy plants of the desired crop.

As industry and technology have developed, seed conditioning has also developed. The seed industry went from being a local industry to a global powerhouse in agricultural economics affecting the world's ability to feed and equip mankind. Today, seed conditioning is an essential tool for ensuring good planting seed at reasonable cost so as to enable farmers to produce crops more efficiently, and sell the produce at costs which permit consumers to live better. It is the major capital investment and process operation in the seed industry, which is a business in itself which exists only to serve farmers who plant seed to produce the food, feed and fiber needed to maintain life in a healthy condition. Seed conditioning involves machines, process engineering, operations, biology, physics, plant physiology and pathology, science, and business. It must be done accurately, carefully, economically and practically. The successful seed conditioner must efficiently use science and engineering, operate cost-effectively and with good business practices.

The ability to do this does not come without considerable effort. To make seed conditioners more efficient with less loss of seed/time/investment, this *Seed Conditioning* compendium was prepared. It covers a wide range of topics that are both useful and essential for the beginner and experienced seedsmen. It is based a lifelong collection of information and real-life experiences from a broad geographic and cultural field. The trials and learning and efforts of many seed conditioners are compiled and distilled into a brief but hopefully easy-to-understand form.

Seed Conditioning is divided into three volumes: *Volume I, Management,* covers essential aspects of organizing and managing a seed conditioning operation. *Volume 2, Technology,* covers the equipment and processes used in seed conditioning. *Volume 3, Crop Seed Conditioning,* compiles all available information on experiences in conditioning the seed of different crops.

Many basic topics are covered but also included are secrets of the trade that are important for all members of the seed industry to be aware of so they can see how their contribution fits into the entire program.

We hope that the many decades of experience, by the authors and many seedsmen, that went into making this publication can truly serve as a useful tool and guide in helping you make a significant contribution to your own operating organization and the global seed industry and agriculture.

THE AUTHORS

Preface

A first requirement to improve rural and urban living standards requires plentiful food supplies at reasonable cost. The first step toward this is to provide farmers with high-quality seed of higher-yielding varieties. A critical element in providing better seed is seed conditioning.

Seed has been an important crop production input since the first crop plant was domesticated by prehistoric man. For thousands of years, man cleaned the seed of his food crops by winnowing. This is still an important process in some countries, but is no longer adequate to supply the kind of seed needed by the modern farmer who must produce food for more than just his own family.

Seed conditioning is a vital part of the total technology needed to move higher-yielding, cost-effective improved genetic materials from the plant breeder to the farmers who must produce food for the rapidly expanding world population. The modern farmer must have seed that is free of inert material, weed seed, other crop seed, and disease to obtain maximum production. He must have crop seed that are sized for accurate planting, high in germination, and free of damage, to minimize planting rate.

The seed industry developed to bridge the gap between the breeder and the farmer, to carry the breeder's developments to the farmer without adulterating it. The seed industry involves practical, economic application of many facets of science, technology, and business. It has become a highly specialized business, essential to efficient and productive farming. No substitute has yet been found for its product—good seed of high genetic, physical, physiological, and phyto-sanitary quality. Even the supposedly "informal seed sector" cannot provide the quality assurance which helps farmers get the most production from the least investment.

In recent years much detailed information has been developed on improving seed through conditioning. Yet with all the knowledge available, seed conditioning still remains the one aspect of the seed industry which is something of an 'art'. This book is a follow-on and major update and expansion of the original work published in India in 1970. It brings together, in a logical sequence, present knowledge—both old and new—of seed conditioning, to help seed managers, specialists, and conditioning operators reduce loss of good seed while improving seed quality and honing the efficiency of their operations.

The Seed Industry Today

Seed conditioning today is as essential as ever in bringing seed to high quality. However, the seed industry has changed remarkably, and has become almost different sciences and industries, under different economic and agricultural situations.

On the one hand, in highly-developed economies, farming has become a business which is carefully managed, technically, operationally and financially.

A great deal of information is available on the management and technology of seed conditioning. However, it is scattered in many publications, many of which are not generally available.

This manual is an effort to comb these many sources, select the most useful information, and organize it into a form which is practical, readily useful to seed conditioners, and is available to them when they need it.

Seed conditioning plants and operations range from small hand-powered machines on farms and in developing areas, to high-capacity automated operations in developed agricultures. An attempt, although admittedly inadequate, was made here to include aspects which will be useful to operations at all levels. If a certain aspect is not useful to you now, at least we want to make you aware of what is possible, and possibly help you as you grow!

It is impossible to make a complete or artificial separation of different aspects of seed conditioning. For this reason, many topics will often be discussed in different places herein. In such instances, what is an apparent duplication is in reality an essential discussion of a different aspect of the same topic.

A generalized list of references utilized is included at the end. These, and many other materials, can be studied for more in-depth information. The operating, maintenance and repair manuals of the specific brands and models of machines installed in the facility should be studied and referred to for valuable in-depth information. And, the seed enterprise manager, conditioning manager and conditioning operator should maintain close working contact with local representatives of the equipment suppliers and manufacturers.

Good seed is essential to agricultural productivity, and conditioning is essential to good seed! We salute you, the seed conditioner, and wish you all the best in improving the efficiency of your operations so that you can be of more help to farmers and the people who depend on them for food.

Machines Described and Illustrated

For most seed conditioning machines, there are a number of models, brand names, and capacities available in different locations.

It was an impossibility to include in this guide for conditioning and management specialists, details of all machines available. Information also was not available on all models and brands; Some companies did not provide materials, while others were quite generous in supporting this work.. For each kind of machine, we tried to select a representative model, and then describe it as well as possible. While they differ in details and in operations, the basic principle of machines in a specific purpose group are similar. In general, the specialist can easily use the information included here as a guide to familiarize himself with, and properly use, any locally-available model of a particular machine. However, herein we are not recommending any specific brands or models, and are not downplaying any brands or models; we have tried to be completely unbiased. All have utility in certain installations.

We have assiduously avoided any bias toward particular models or brands. This has been a characteristic even of the industry specialists called upon for assistance.

Information Included and Organization

Over years of working with seed conditioning, the conditioning managers and operators have emphasized the need for specific information on specific topics. We have tried to cover these topics with concise information on the methods which have seemed to be most efficient, effective and useful.

Each chapter in this book has been included because one or more seed conditioners mentioned the need for information on the chapter's topic. Information is organized so that it will be readily available and "findable" when needed. Conditioners have indicated that "don't worry about duplicating information in different sections; make each section complete, so that we don't have to look back and forth to get complete information". We have tried to do this.

Topics discussed are those which have been identified over the years by seed conditioners. To the extent possible, each discussion is complete, without requiring cross-references to other sections.

Illustrations included play a very important role in presenting information on the machines. However, illustrations were selected so as to fit the entire presentation, rather than a single paragraph or item. Therefore, Figures are not referenced in the text, but the interested specialist should closely study both text and illustrations.

Better conditioning means better business for seed enterprises, and better crops for farmers. We hope this book helps!

Dr. Bill Gregg, Seed Industry Development & Conditioning Specialist
PO Box 1756, Starkville, MS 39762
Email: billgregg1@bellsouth.net, Tel. (662) 323-0035

Mr. Gary Billups, President, Crippen International, Inc.
12237 Montfort Drive, Suite 701, Dallas, TX 75240-5114
Tel. (972) 924-4242, fax (972) 924-446, Email: baryb@crippeninternational.com

Acknowledgements

Information presented herein is not original, but is information on existing commercial and technical operations, on past methods, and known research, presented in a manner intended to assist managers and specialists.

A great deal of information in this management and technology guide came from many people. Grateful acknowledgement is made to all those listed in the References, and to the many friends and co-workers who made useful suggestions on the material included here. Without their encouragement and contributions, this conditioner's helper could never have been completed.

Special thanks is due to ICARDA (International Center for Agricultural Research in the Dry Areas, Aleppo, Syria) for permission to use illustrations in "Seed Processing, ICARDA Training Manual", authored by L. Grass and B. Gregg, published in 2000, and for its support in improving the use and efficiency of seed conditioning.

And, special thanks to the following, for providing information and helpful suggestions. Without such wholehearted support, this could not have been possible.

Chris Ansel, Creamer Metal Products, London, Ohio 43140

Mark Avery, Grain Journal, Decatur, IL 62526

Stew Bass, American Crystal Sugar

Zewdie Bishaw, Seed Unit, ICARDA, Aleppo, Syria

Randolph E. Block, Screw Conveyor Corporation, Hammond, IN 46327

Dr. A.H. Boyd, Professor of Agronomy-Seed Technology (retired), MS State University

Catherine Byrnes, Western Pistachio Association, cbyrnes@westernpistachio.org

Dr. William (Bill) Chancellor, University of California Davis

Don Clark, Continental Eagle Corp., Prattville, AL 36067
Ed Clutes, EDCO Metal Works, Rocky Ford, CO
Mogens Dahl-Nielsen, Westrup A/S, Slagelse, Denmark
John DeWall, Chantlant Material Handling, Humboldt, IA 50548
William J. Donnell, Arrow Corp., Alma, Michigan 48801
Robert Dyar, Petkus Agri Engineering, Des Moines, IA 50321
Mette Eising, DAMAS A/S, DK-5600 Faaborg, Denmark
J.E. Elsworth, Ltd., Norfolk England PE 34 4AS
Pat Fanjoy, Focus Industries, Inc., St. Albert, Alberta, Canada T8N 5C3
Joe Funk, Editor, Seed Today, Decatur, IL 62526
David J. Goldenberg, California Pistachio Board, golden59@pacbell.net
Armand Golian, DEAMCO Corp., City of Commerce, CA 90040
Dr. Lahcen Grass, Syngenta Seeds, Inc., Washington, IA 52353
Mike Hanna, Schutte-Buffalo Hammermills, Hattiesburg, MS
Jens A. Henriksen, DAMAS A/S, Faaborg, Denmark
James B. Henson, American Red Cross, Starkville, MS 39760
H.J. "Hank" Hill, Seed Dynamics, Inc.,
Lyle Hollander, Sudenga Industries, Inc., George, Iowa 51237
Burkhard Issbruecker, Petkus Wutha GmbH, 99848 Wutha-Farnroda, Germany
ISTA and its excellent reports and publications
Kenny Johnson, Grainway LLC, Aplington, IA 50604
Tom Johnson, AEC Companies, 114 E Main, St. Charles, IA 50240
Bennie C. Keith, Miss. Crop Improvement Assn., Miss. State, MS 39762
Dr. John Kern, American Crystal Sugar Co., Moorhead, MN 56560
Dr. Bob Klein, California Pistachio Commission, Fresno, CA 93710-8044
MS Sandra Langlois, Satake USA Inc., Houston, TX
MS Kristi Lawless, Universal Industries, Cedar Falls, IA 50613
Hooman Mahdavinader, AR Machine Co., Tehran, Iran
Buddy Mathis, Lewis M. Carter Co., Donalsonville, GA 31745
Samad Mobasser, Seed & Plant Registration & Certification, Tehran, Iran
David Mrozinski, Continental Eagle Corp., Prattville, AL 36067
MS Beatrice Muehlenbeck, Petkus Wutha GmbH, 99848 Wutha-Farnroda, Germany
Dennis Neal, Enviropac, Inc., Compton, IL 61318
Abdul Aziz Niane, Conditioning Specialist, Seed Unit, ICARDA, Aleppo, Syria

Patric Pike, Satake America, Houston, TX
Jerry Perkins, Essmueller Materials Handling Co., Laurel, MS 39441
Kathy Reading, Seedburo Equipment Co., West Chicago, IL
William A. (Bill) Rice, W.A. Rice Seed Co., Jerseyville, IL 62052
Kyle Rushing, Integrated Technology, Salinas, CA
Ajay Sagar, Indosaw Equipment, Ambala Cantt., Haryana, India
Mark Scholze, Petkus Wutha GmbH, 99848 Wutha-Farnroda, Germany
Claus Simonsen, Westrup A/S, Slagelse, Denmark
Todd Sparrow, Rapat Corp., Hawley, MN 56549
Tom Spitzer, Fischbein-Inglett Co., Augusta, GA 30914
Darin Stutler, Bratney Companies, Des Moines, IA 50322
James Thomas, Biological Engineering, MS State Univ., MS State, MS 39762
Jim Thomas, Oliver Mfg Co., Rocky Ford, CO 81067
James Thompson, Horticulture, University of California-Davis
Robert Trimbee, F.H. Schule Muhlenbau, Alma, AR 72921
Don Uglow, Sortex Inc., Lafayette, IN 47909
Marcos Valdez, Valdez Engr. Co., and Three Star USA, Conroe, TX 77302
Randy Vaughan, Foundation Seed Program, MS State Univ., MS State, MS 39762
Thomas Warne, Schutte-Buffalo Hammermills, Buffalo, NY 14240-0546
MS Pamela Weber, W.A. Rice Seed Co., Jerseyville, IL 62052
Kurt Weieneth, AEC Enterprises, New Virginia, IA 50210
Jay Williams, Extension Engineer, Peanut Mechanization Specialist, Univ. of Georgia

A "special special" acknowledgement must be made to:

1. Hossam Lasheen, Computer Design Specialist, Cairo, Egypt, for all his work which made this book possible. He computerized the majority of the illustrations in this book, including the designs of Mr. Billups.
2. S.S. Rekhi, retired FAO Vegetable Seed Specialist, New Delhi, India, who reviewed and edited the chapter on vegetable seed extraction.
3. Darin J. Stutler, Vice President Sales, Bratney Companies, Des Moines, IA 50322, who provided many pictures, information and evaluation of the materials.

Contents

List of Illustrations

Chapter 18: Mechanical Injury to Seed

Chapter 19: Waste Products

Chapter 20: Dust Control

Chapter 22: Internal Quality Control (IQC)

Chapter 23: Sampling

Chapter 24: Testing Seed

Chapter 26: Labels and Seed Identification

Chapter 27: Blending

List of Tables

Chapter 1

Business Laws

In most places, a number of laws and regulations affect the establishment and operation of any business enterprise. It is advisable to obtain the services of a specialist to identify all relevant legal requirements, and then make and follow a checklist of everything that has to be done and the legal/technical assistance required.

Business Operating License

Most countries require that every commercial business have a business operating license before the business can be established, and the license must be periodically renewed for the business to continue. These may be called by different names, and there may be different types of licenses, with different fees.

Types of Businesses

Basic Legal Forms

There are four major legal enterprise structure frameworks for private persons who form a business for profit:

sole proprietorship,
partnership,
limited liability company, or
cooperative.

A sole proprietorship with "one boss or one owner" may be legally organized as a "sole proprietor" (*i.e.,* self employed) or as a "limited company". When several "bosses or owners" are involved, the form could be a "partnership", "limited company" or "cooperative".

Sole Proprietor

This is probably the most common form when a new small enterprise is started. The owner may trade under his own name or a business name; the name usually does not have to be registered with the agency implementing the Company Law. If a business name is used, the owner's name should appear on all letterheads, etc. The owner of the business solely holds all profits (and losses), and is fully and entirely responsible for the business. He may decide to employ people to work for him or to manage the business for him, but they will remain employees and he has final authority and responsibility.

Adequate finance must be provided to start and operate the business. If your own funding is not sufficient, you must borrow money. Banks and other financial institutions usually require detailed plans of the proposed business operations (Business Plan). If the business runs into financial problems, the sole proprietor is personally responsible for paying the debts, even if this means selling his personal assets including house, car, etc.

A sole proprietor has few legal requirements to fulfill, as he is personally responsible for all obligations of the proprietorship. Generally, taxes are paid only on profit. The sole proprietor is interested in the amount of profit the enterprise makes in the year, but there is no legal requirement to produce a statement showing this. If the value of taxable assets and goods (per annum) is over a certain sum, in most countries, the enterprise must be registered with the Tax Authorities (depending on the tax laws of the country). It then is a legal requirement to maintain accurate records. However, in the proprietor's best interests, to manage the enterprise effectively and ensure that the correct tax is paid, proper accounting records must be kept from the beginning of the enterprise.

Partnership

A partnership is two or more people conducting a business for profit. Partnerships are only suitable for a relatively small number of persons who know and trust each other. There is a limit to the number of partners; usually, it should not exceed 20, but this depends on legislation in each country. The partners are all part-owners of the enterprise, and they jointly make decisions on running the enterprise. Alternatively, they may appoint one partner, or hire another person, as the manager. They may operate under a business name, which does not have to be registered. Depending on the laws in your country, the names of the partners may have to appear on the enterprise's stationery.

Partnership law is based on contract laws. In addition, your country may have a Partnership Act to specify the following:

1. Proper accounting records must be kept.
2. Capital must be distinguished from profits and losses.
3. A record must be kept of profit shares and withdrawals by each partner.

Partnerships must render true accounts and full information of all things affecting the partnership, only to any partner or his legal representative. This is not public information.

Partners normally have an attorney-at-law to draw up a legal agreement (contract) showing the proportions in which they share profit or loss, and other aspects of their operations. This agreement is the basis for the existence of the partnership.

It is easier for a partnership to raise funds to start an enterprise, as all the partners share the investment. At the same time, all partners share any profit or loss. In case of a loss, they also personally share responsibility.

If different partners invest different amounts of funds (capital), they can choose to either:

1. Apportion the profits in the same ratio as the amounts invested, or
2. Allow enough time, say 6 months, for all partners to invest more, so that their investments are equal.

In either case, voting rights of each partner should be equal to his individual investment amount.

Partners are personally responsible for the debts of the business. If one partner disappears, the remaining partners are liable for all debts. Requirements of tax and financial statements for partnerships are similar to those for sole proprietors.

Limited Liability Company

Unlike a sole trader, a limited liability company is a legal entity in its own right. Because of this, a number of legal requirements must be met by a limited company. No one should set up a company without understanding all implications and getting professional advice.

The most important aspect is that only the company is responsible for debts or liabilities it incurs. If the company is unable to pay its debts, the company can be sued in its own name. The owners (shareholders) are responsible only up to the amount of money they invested in the company, usually the value of their shares of stock in the company. Shareholders' liability is limited to that amount; their personal assets are not subject to takeover. If the company is unable to pay its debts, then it may go into liquidation.

Shareholders and directors may change, but the company continues to exist until it is legally disbanded.

A company must have at least one director and a company secretary who could be a second director, another shareholder, an accountant, or an attorney. To form a company, certain documents (articles or memorandum of association) and the name of the company (no two companies can have the same name) must be registered with the Registrar of Companies or similar agency in your country.

Companies limited or owned by shares or stock shares are the most important form of business organization, not because of their number, but because of their size. Companies whose shares can be offered to the public are known as public limited companies (PLC) and PLC must appear as part of the name of the company. There are also private limited companies whose shares are not be offered to the public; in most cases, the founders and owners hold all shares. Legally, the word "limited" must appear as part of the company's name. A public limited company can offer its shares to the public through the stock exchange; a private limited company cannot do this.

Each year, a limited company must send certain financial information to all shareholders and to the government's Registrar of Companies. As any person can have access to the files of the Registrar, the data submitted by the company become public documents. The legal reporting requirements, included in the regulations of the Company Act of most countries, are complex. While these vary according to the type of company and its size, general requirements are:

1. Accounting records must be kept to show and explain the company's transactions.
2. At the end of each financial year, accounts must be prepared and include:

   A. Profit and loss account
   B. Balance sheet
   C. Auditor's report
   D. Director's report

These accounts are given or made available to each shareholder, debenture holder, and any other person entitled to attend the Annual General Meeting. These are also filed with the Registrar of Companies. In this way they become public documents; for a small fee, anyone can obtain a copy of them.

Certain accounting standards (national and international) are guidelines to be followed by the limited company. These are issued by the professional accounting body in your country. These do not have the force of law, but are recommendations as to how certain accounting matters should be handled by limited companies.

Cooperative

A cooperative or "worker's cooperative" is a specialized form of limited company. They are operated and managed in the same manner as any other small business with managers or supervisors. However, the difference is that the enterprise is owned by everyone who works in it, and/or is served by it, and decisions are made democratically. The same laws and regulations as for a limited company apply to a cooperative.

Factors Affecting the Choice of Legal Form

Each of the different legal forms of business enterprise is most appropriate in different circumstances. The form you select, and the kind of seed enterprise you establish, is determined by:

1. The kind of activities you will conduct.
2. The legal requirements and liabilities of your operating business environment.
3. The scope of your operations and the adequacy of your financial means.
4. How much liability/risk you are willing to take, and are willing to risk your family's personal assets on.
5. Your tax environment, and the amount of tax liability incurred by different legal forms of enterprises.

Choosing the Best Legal Form

Choosing the most suitable legal framework for a small-scale seed enterprise depends on several factors, primarily:

1. The amount of capital needed to start the enterprise: Adequate finance must be available and secured to allow the required investment in equipment and infrastructure, and working capital required by different sizes of enterprise.
2. The amount of financial risk involved: If a venture is risky, it is much safer to spread the risk in case of financial failure.

Taxes

Every government levies taxes, and enterprises with seed conditioning also have tax liabilities. Check with the local authorities, register and obtain a tax payer number, and become thoroughly familiar with taxes which must be paid (often including profit tax, operating tax, personnel social security tax, real estate tax, and others). Learn when and how taxes must be paid, and either have a qualified accountant and tax specialist in your staff or contract the services of an accountant with tax qualifications.

Accounting Systems and Financial Reports

The type of accounting system you use, and the kind and frequency of financial reports you prepare and file may be specified in local laws.

Legal Liability

Every enterprise faces risks and is legally liable for certain actions, such as injury to personnel, injury to "outside" persons on the enterprise's property, damage to neighbors due to conditions such as dust on the seed enterprise property, losses caused by poor seed quality, etc. Some governments require liability insurance to cover certain aspects. These requirements should be determined from local authorities. As a matter of good management, the seed enterprise should be ensured to cover all possible legal liabilities which may result in damage claims.

Personnel

Responsibilities to enterprise personnel are generally spelled out in local laws. These range from hiring and firing, working hours, medical care, withholding and payment of employee income tax, social benefits payments, to retirement. Check local laws on personnel.

Some governments require periodic inspection of the enterprise to ensure that safe working conditions are maintained.

Health Conditions

Some governments require certain standards of healthy working conditions and amenities for personnel. In terms of a seed enterprise, this includes such factors as control of dust in the air, exhausting fumes from the treater, facilities for fumigation, bath/toilet facilities of specified standards, etc. Periodic inspections are usually made to ensure continued compliance.

Environmental Effects

Some governments require evaluation of environmental impact before the enterprise can be established, and may also require periodic inspections and re-approval. This includes such aspects as disposal of liquid waste, handling and disposal of dry solid wastes such as screenings, preventing environmental contamination by treatment materials, avoiding creating sources of rats and insects which may affect neighbors, etc.

Zoning and Location

Most areas are zoned and only certain types of structures are allowed in specified areas. Types include single-family homes, several-family homes, manufacturing, business, retail, etc.

The location of a seed enterprise is important in terms of easy access for farmer-customers, and for trucks carrying seed. Also, be sure to check the zoning of the proposed site, and obtain a legal building permit.

Fire Safety

Some governments require periodic inspections of enterprises for fire safety, and adherence to specified requirements. When deficiencies are found, these must be immediately corrected, or the government may shut down the enterprise.

Vehicles

Licensing of vehicles, and often insurance, is controlled by law. Check and abide by local laws, including not only "over the road" vehicles such as cars and trucks, but also for "in plant" vehicles such as tractors and forklifts.

Chapter 2

Seed Law

Farmer Trust in the Seed

It is almost impossible to determine seed quality by looking at the seed. By the time the farmer knows if his seed is good, it is usually too late to find new seed and replant. Because most farmers have little risk capital and any problem with seed usually cannot be corrected and thus affects the entire crop, farmers have been described as being "a suspicious lot". In fact, farmers will not accept seed that they cannot or do not trust.

The successful seedsman knows that his most valuable asset is the ability to ensure that each and every bag of seed is of high quality, and builds and holds the trust of farmers. Without this, there is no market and his seed will never sell.

This requires some means of identifying seed as of specified high quality, so that farmers can identify the seed and trust the seed and its supplier.

Seed Movement

Many areas are not suited to production of seed for needed crops. Producing small quantities of seed increases the per-kg cost of the seed; it is far more per-kg cost-efficient to produce larger quantities of fewer varieties and then market them over wider areas. This reduces the cost of seed to farmers while maintaining reasonable profit for the seedsman, and also results in more farmers using improved seed.

Another factor in supplying seed over wide areas is that often there are crop failures or seed shortages, creating situations where farmers require more seed than is available locally-produced.

There is thus a constant need to move seed over wider areas, to reach and serve more farmers while getting maximum use from economic resources and ensuring markets which in turn ensure the development of good seed suppliers. It is to the advantage of farmers, seedsmen, and the area's economy to be able to move seed quickly from the production area to many using areas, and ensure that the seed is high quality and that farmers will buy it.

A problem in many developing economies is that there are no specific legal measures to ensure unimpeded movement of seed into needy areas. As a result, local police check posts stop, hinder, and delay movement of badly-needed seed, and even extort significant bribes to allow it to pass.

For seed to move freely within and between states, countries and other political zones, there must be some means of ensuring seed quality in a way which farmers can identify and trust. And, equally important, government administrative procedures and systems must not interfere with movement of good seed. Instead, government systems must promote the movement of seed and farmer acceptance and use of it.

Government Seed Quality and Seed Enterprise Regulation

Government has the responsibility for improving general social and economic conditions for all citizens, and establishing systems which protect citizens from harmful conditions while encouraging investment, business development, and productivity.

A major role of government is to guide and support farmers in supply and use of inputs which improve their productivity and income. This includes promoting the production, supply and use of improved higher-yielding seed in a manner which ensures a reliable, dependable and ongoing production and supply system, with effective extension educational promotion to get farmers to understand and use improved seed which costs more.

Government regulations include a special Seed Law(s), controlling the minimum quality of seed offered for sale, and regulating the agencies which produce, market and distribute seed subject to the Seed Law.

Achieving General Control of Seed Quality

To achieve the required supply and general use of improved seed, governments have established minimum standards to ensure that all seed meet or exceed known minimum quality levels. To ensure unbiased and widespread implementation of these standards, governments have established Seed Laws and regulations, and created agencies to inspect and test seed quality so that farmers have access to good seed of known quality.

Seed Law

Minimum standards of quality have been established for most crops grown by farmers. Quality factors established under Seed Laws normally cover physical quality (percentage content of pure seed of the specified crop/ variety, other crop seed, weed seed, freedom from noxious weeds or diseases) and physiological quality (germination percentage).

Enforcement of seed quality minimum standards under the Seed Law is implemented by the Seed Law Implementing Agency. This is usually a division under the Ministry of Agriculture of the country, with divisions in each state and operations in most local areas.

It should be noted that in most cases, a Seed Law applies only to seed which are marketed in a business-like manner. Seed exchanged between farmers (farmer to farmer seed exchange) is usually exempt from provisions of the Seed Law. However, if the selling farmer advertises the seed he has for sale, this is usually considered as a business and becomes subject to the Seed Law.

There is no practical means of monitoring and controlling the quality of seed exchanged "informally" between and among farmers on a local basis. Thus, seed quality control essentially applies only to seed moving in the "formal" seed trade involving registered producers, distributors, and marketers. In many developing economies, the greater part of seed comes from informal sources, and consists of grain grown by farmers and used for seed without quality improvement or control.

Eliminating Low-Quality Seed Supply

Seed Laws protect farmers by ensuring that licensed seed dealers supply seed of known minimum quality, as shown on the required labels. However, Seed Laws also protect reliable seedsmen by helping eliminate or minimize unfair competition from unscrupulous dealers who sell poor seed and falsely-labeled seed.

Regulations Under the Seed Law

As the economy and technology develop, conditions change. A law is difficult to change quickly enough to keep up with changing conditions, especially progress and development of seed supply. Often, years are required to change a law; by this time, a problem with crop varieties or seed may have had disastrous effects.

To offset this and keep the laws current of changing conditions, the Law generally embodies policies and sets general requirements and directions. Detailed requirements and procedures are then enumerated in official "Regulations Under the Seed Law", which actually control details of implementation. Such Regulations can usually be changed rather rapidly through Ministerial Decree, public hearings to allow interested parties to express their viewpoints, and public notices.

Registering and Obtaining a License to Operate a Seed Enterprise

Before establishing a seed enterprise, it is necessary to register with the appropriate government agency, pay the licensing fee, and obtain a license to do business in seed. The license is usually for a one-year period, and must be renewed each year.

Seed Laws and Seed Quality

A Seed Law generally specifies the following, and requires certain aspects relating to the seed and its identification:

1. *Labeling the seed*: Each container of seed must be labeled in a certain manner.
2. *Label content*: The nature and content of the label is specified.
3. *Components of the labeled seed lot*: The components of the seed lot must be specified, along with the percentage of each. In some cases, the name of contaminant seed kinds must also be listed.
4. *Definition of components of the seed lot*: What constitutes each component of the seed lot (*e.g.*, pure seed of the labeled kind, noxious weeds, etc.) may be specifically defined.
5. *Date of testing*: The date when the seed were officially tested (for germination) must be specified. The test date must be within a certain time limit before the date of sale of the seed.
6. *Name and address of the producer/seller* must be shown on the label.
7. *Disclaimers*: some governments prohibit or restrict disclaimers.
8. *Minimum quality standards*: Each lot and bag of seed must meet or exceed the minimum quality standards established in Regulations under the Seed Law.
9. *Year of production* sometimes must be shown on the label.
10. *Seed treatment*: if seed are treated with chemical or other protectants, these must be shown on a label, along with "poison" warnings and antidotes.

Seed Laws and Seed Conditioning

Seed conditioning is a major part of the seed supply process, and has a major impact on seed quality. Seed Law and Regulations requirements thus have a major impact on seed conditioning.

The Seed Law usually requires specific facilities, operations and management of the seed conditioner, to help ensure the purity and quality of seed. These include:

1. *Operating/approval license*: The conditioning plant must have an official license to operate. Its facilities, equipment and records are periodically inspected and approved or improvements required, to renew the license. The operating license is valid for a specified period, usually one year.
2. *Facilities*: The plant must have facilities which can handle seed without adulteration or mixing, or quality deterioration.
3. *Conditioning equipment*: The equipment must be able to turn out pure seed of the crops handled, by separating commonly-encountered contaminant seed and materials, and be in good maintenance.
4. *Elevators and conveyors*: The seed handling system must be able to handle seed without causing undue mechanical damage or mixing, and not cause adulteration.
5. *Clean-up*: The plant must be designed and constructed so as to minimize adulteration with undesirable seed. There must be adequate clean-up equipment (vacuum, compressed air, etc.) and there must be a good clean-up procedure and operations.
6. *Records*: The record system must be adequate to record accurately and completely all operations on every seed lot, and the current status and condition of the lot.
7. *Storage*: Seed storages must be able to protect the seed from insects, pests and moisture, and maintain lots separately and completely identified.
8. *Seed identity system*: The plant must have a system which ensures complete and correct identity of every stack, bulk container, lot, and bag of seed.
9. *Access for sampling and inspection*: Seed Law inspectors regularly inspect and sample seed for official testing to ensure that the seed meet minimum standards. Storing and stacking must make seed readily accessible for sampling, and each lot must be well-identified.

Chapter 3

Seed Enterprise Structure and Organization

Most modern seed enterprises and/or programs are vertically-integrated, including production (owned and/or contracted), conditioning, storage, and marketing. Thus, seed conditioning is a major part of complete seed supply operations conducted by the enterprise. However, there are several different operating approaches.

Types of Seed Enterprises

Seed conditioning involves drying the seed to safe moisture contents, cleaning the seed to remove undesirable materials, sometimes treating the seed with a protective chemical, and packaging the seed in the package sizes needed by the using farmers. Classified according to the level of operations, a seed enterprise may be of the following categories.

"Informal" Farmer Seed Producer

If the small informal farmer-seed producer sells seed to his neighbors, he needs to clean out undesirable impurities, to at least a minimum degree. This is usually done by hand picking, winnowing in the wind, or hand sieving/sifting/shaking. The informal seed producer usually cleans only the seed he produces for his own use and for sale/barter to other farmers. Note that, in most legal environments, if the farmer advertises his seed for sale, he is considered to be part of the "formal" rather than "informal" un-regulated seed sector.

Cooperative

Often, a farmer cooperative will maintain a small seed conditioning facility. This may be used to clean seed produced for the cooperative, and/or to provide seed cleaning services for members and others who produce seed.

"Custom" Seed Conditioner

"Custom" seed conditioners have seed cleaning equipment and facilities (often including maize sheller, seed dryer, etc., as well as cleaning and bagging equipment) to clean seed for seed producers, farmers, etc. They charge a fee per kg or ton of seed cleaned.

Custom seed conditioning is economically feasible only in specific local situations such as where seed companies and enterprises are not developed well enough to have their own seed cleaning facilities, but farmers want clean and high-quality seed. In this case, the extra cost of cleaning is worthwhile and there is enough cleaning business for small local seed producers to justify a "custom" seed cleaning enterprise but each producer cannot justify his own facility.

The custom seed conditioner often buys seed from farmers, cleans and bags it, and re-sells it for his own account. In many cases, the custom conditioner contracts with farmers to produce seed for him.

Vertically-Integrated Seed Company

The vertically-integrated seed company is by far the most economically feasible seed operation, once local farmer demand for good seed develops sufficiently. In many cases where the owner's finances permit or sale of other inputs maintains cash flow, the vertically-integrated seed company actually works to develop sufficient farmer seed market to justify the facility.

These companies usually condition only their own seed. They may condition seed for other companies or growers and then buy the seed, or they may condition seed for others on a fee basis.

Seed Supply Work and Activities

Different kinds of work activities involved in producing and supplying high-yielding seed including (Figure 3.1):

1. Management and planning.
2. Finance and accounting management and reporting.
3. Seed production and harvest.
4. Seed drying, conditioning, treating and bagging/packaging.

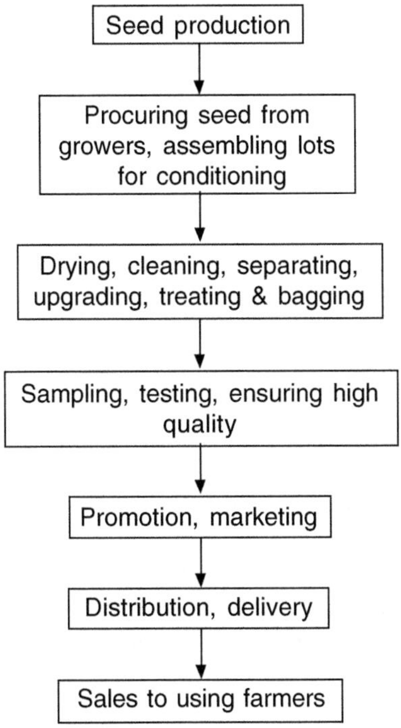

Figure 3.1 Sequence of handling operations in a vertically-integrated seed enterprise, required to make higher-yielding seed available to farmers.

5. Seed storage.
6. Seed promotion, distribution, marketing and sales.
7. Quality control.

Types of Responsibility

Some work activities have direct responsibility for accomplishing necessary operations. The jobs which carry these responsibilities are called "line" or "line of command or responsibility" positions.

In other cases, work activities are not directly involved in producing the goods sold, but are essential to support managers or staff in their decisions and operations. These support activities are carried out by persons in "staff" positions. Staff positions report to one (or more) managers, who use the staff-provided information to guide their decisions and work.

In a small operation, the same person usually carries out all, or several, activities. In a larger operation, organizational structure includes "line" positions of responsibility, and "staff" positions which provide specific support to managers as they make decisions and carry out operations.

Line positions of responsibility in a seed enterprise include managers, foremen and staff who manage and conduct operations of:

1. Production
2. Conditioning
3. Storage
4. Marketing

Staff units and positions which provide supporting activities and information to managers include:

1. Quality control
2. Finance and accounting
3. Personnel
4. Secretarial and filing work

Organize to Handle Work Efficiently

Each staff and work category includes many specific jobs. In a small enterprise, one person may carry out all jobs in several of these activity categories. In a large enterprise, one or more persons may be required to handle each activity in each category.

Regardless of enterprise size, the work must be organized and conducted efficiently, to minimize costs and get the most done in the least possible time. In a small enterprise or even a "one man" operation, the work is organized so that one (or a few) person can handle it logically. In a large enterprise, work is organized so that different persons can handle different aspects most efficiently.

Efficient Seed Enterprise Organization

Experience has shown that the most cost-and-quality-efficient organizational structure for a seed enterprise is to organize according to category of work. This simply means that all work which is logically similar, or related to one activity, is organized together into a single unit. In this manner, a single manager or person can direct and coordinate all related activities so they are all completed as required, and in the most efficient manner. Avoid organizational structure according to crop seed; this has proven to be duplicative and more costly.

In the continuing effort to gain improved efficiency, organization of a seed enterprise has evolved into the basic structure, with variations, of the general structure shown in Figure 3.2.

Staff operations are (1) quality control, (2) finance, and (3) personnel. Finance and personnel are often combined as sections in a single

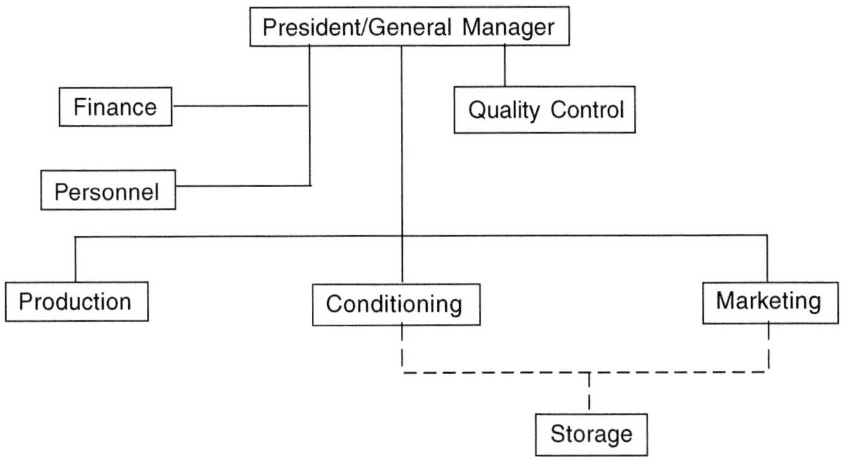

Figure 3.2 Organizational structure of a modern seed enterprise.

Administrative unit. Line responsibility operations are (1) production, (2) conditioning, and (3) marketing.

Each line function includes many activities; in a small operation, one or a few persons can conduct these; in a larger organization, one or a few persons may be assigned to each activity. In a larger organization, several sub-divisions or sub-units may be organized under each of these major units. For example, the "Conditioning" unit may contain sub-units for drying seed, conditioning seed, repair and maintenance of equipment and buildings, etc.

Production Activities and Responsibilities

Production is responsible for seed production, managing contract growers, supervising harvest and on-farm handling and drying, delivering produced seed to the conditioning plant, and related activities. During field production, this department ensures that quality control—internal and external (Certification and Seed Law) inspect fields at the required stages to ensure that seed meets standards and to minimize losses.

Responsibilities of this organizational unit include:

1. Obtaining stock seed for planting seed fields.
2. Selecting fields and contracting growers.
3. Crop production operations, done by the enterprise or supervised if done by contract farmers.
4. Roguing and inspection of fields and equipment, to ensure that all seed are of high quality and no contamination or losses occur.

5. Ensure that Internal Quality Control adequately inspects all fields and operations, and that Certification inspections are made as required.
6. Ensure safe harvest, on-farm drying and bagging, and delivery of the seed to conditioning.

Conditioning Activities and Responsibilities

Conditioning handles all operations of getting the harvested seed ready for marketing. Responsibilities of this organizational unit include:

1. Receive raw (uncleaned) seed from growers and the Production Unit.
2. Keep each seed lot separate, pure and uncontaminated.
3. Operate a labeling and identification system which maintains identity, separateness, and purity of all seed lots.
4. Have IQC examine incoming seed, organize operational lots of uniform quality, and reject unacceptable lots.
5. Dry seed as required.
6. Control insects and pests as required.
7. Clean the seed to remove all undesirable materials.
8. Apply chemical treatments as required.
9. Package the seed in the desired size and kind of bags/packages, and apply appropriate tags and labels after testing and inspection procedures are complete.
10. Ensure sampling and testing of each lot, as required by the enterprise and seed law and/or certification.
11. Deliver cleaned, treated, packaged, labeled seed, which meets all standards, to storage.
12. Maintain equipment and facilities in good condition, so they can operate efficiently, clean seed properly, and keep it pure.

Marketing Activities and Responsibilities

Marketing determines the amounts and kinds of seed to produce, the amounts and kinds of other inputs to purchase, and handles advertising, promotion, sales, and deliveries.

Responsibilities of this organizational unit include:

1. Conduct market surveys and determine the amount and kind of each seed to produce, and of each fertilizer, chemical and equipment item to purchase. This must be done at least one year in advance, to allow production and/or procurement of the required seed and inputs.
2. In more efficient seed operations, conditioned/packaged seed is turned over to the Marketing Unit, which manages the storages.

3. Deliver seed and other inputs to retailers who purchase from the enterprise.
4. Enlist and manage dealer-distributors.
5. Sell seed and inputs and remit the proceeds to the Finance Unit.
6. Advertise and promote the seed and inputs.
7. Provide educational/promotional information to farmers.
8. Provide after-sales service to farmer and dealer customers.
9. Compile feed-back information to guide current and future operations.
10. Conduct demonstrations, programs, and other promotional and sales activities.

Seed Storage

Seed storage is often included as part of the Conditioning Section, because of the close relationship between conditioning and handling activities. However, perhaps equally often, storage of cleaned seed is included under the Marketing Section, to enable marketing to keep track of its stocks and seed movement.

The individual situation determines which is the best structure. In any case, there must be close working relations between Conditioning and Marketing, not only in handling seed stocks but also in scheduling conditioning so the seed is ready when needed for marketing.

Administrative Activities and Responsibilities

Staff units/activities work directly under the Manager, and reports directly to him. The Administrative Unit has persons or sections to handle different areas of responsibility.

The **Finance Section** keeps prepares necessary documents and payments, and records of the financial operations of the enterprise by maintaining cash book, sales book, ledgers, and reports. Through these records, the accounts of the enterprise are prepared at the end of each quarter and financial year. Records of seed stocks and assets of the enterprise are also handled by this unit.

The **Personnel Section** manages, and maintains records of, the human resources of the enterprise. Qualified personnel must be recruited for all positions, and must be regularly motivated and trained/advised on enterprise operations. Seed and input supply is a highly technical venture and requires specialists with special training.

Internal Quality Control Activities and Responsibilities

This staff unit works directly under the Manager. It acts as the quality and technical control arm of management. The IQC can enter, inspect, and supervise any operation of the enterprise. It is the only unit, short of the General Manager, which can stop and change any operation in order to achieve the desired seed quality or operating efficiency. Its aim is to ensure that all seed meets the required standards, and to prevent losses and excessive costs.

Responsibilities of this organizational staff unit include:

1. Formalize the standards, procedures and conditions required by the market, and as determined by management and external quality control systems such as Certification and Seed Law.
2. Set up, operate and maintain adequate testing facilities of lab and equipment.
3. Regularly observe each operation of the enterprise, evaluate the operation, sample and test seed, and stop, change, eliminate or adjust the operation until it fits the desired requirements.
4. Provide quality control supervision for all operations of the enterprise and all contracted growers, dealers, and others.

Sub-Dividing Responsibilities as the Enterprise Grows

As a seed enterprise grows, more people will be needed to carry out the essential activities efficiently. Regardless of size, the enterprise must be managed so that all personnel are well-informed of overall activities and how their individual responsibilities fit in and must be done. And, activities of a like nature or objective must continue to be focused in one manager.

As an example, Marketing organization can be further sub-divided into two areas of different activities, as shown in Figure 3.3.

Inputs Supply Work and Activities

The typical small-medium seed enterprise markets not only seed, but also sells the entire line of inputs and goods needed by farmers. This (1) helps get more farmers as customers for the enterprise's seed and (2) evens out the cash flow, as seed sales are seasonal. The enterprise does not manufacture fertilizers, chemicals or equipment; these are purchased from manufacturers/suppliers and resold. Thus, the only operations involved in handling non-seed inputs are financing, stocking, marketing, sales and delivery.

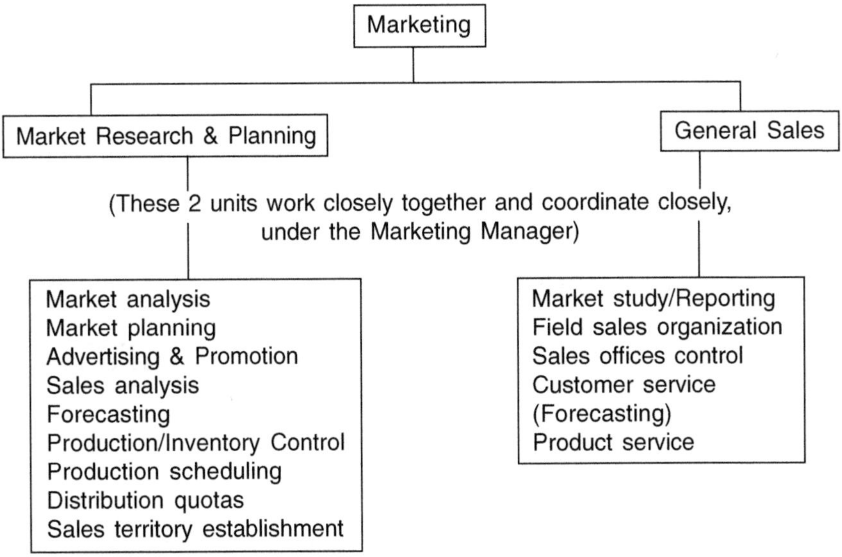

Figure 3.3 Example of separate units established within a division of a seed enterprise, to focus responsibilities of staff on specific activities. Source: Gregg, Seed Marketing.

With additional training, the staff which handles these activities for seed can also handle other inputs. No additional organizational structure is needed, unless the enterprise is a very large company.

Chapter 4

Enterprise and Conditioning Planning

Planning is critical to success of the seed enterprise. Conditioning the seed to prepare it for marketing, is a critical part of the seed enterprise's operations, and must be included in the overall planning. And, the conditioning operations themselves must be carefully and completely planned.

Management Tools

Careful personnel, financial, operating and technical management is essential to the profitability of a seed enterprise. Planning is a major aspect of good management. The areas which must be covered by management and management tools include:

1. Market Survey
2. Marketing Plan
3. Business Plan
4. Financial, Production, and Conditioning Plans
5. Internal Quality Control (IQC)
6. Assigning (Delegating) Authority
7. Records
8. Reports
9. Staff Meetings
10. Individual Conferences/Discussions
11. Public Meetings and Groups
12. Trade/Professional Associations
13. Extension and Research
14. Advertising and Promotion
15. Farmer/Customer Contacts

16. Field Days, Demonstrations, Open House
17. High Employee Morale and Loyalty
18. Follow-up

Enterprise Planning

Planning by enterprise management is essential. Planning is not only an analysis, but is also a guide for operations, and must include details of all operations. It must tie them together in a sequence which fits the local agriculture, and is efficient and cost-effective. Management guides always describe planning and business plans as a map or guide to direct the operations. Questions which should be answered in plans include: Where are you going? Can you profit from it (should you go)? How will you get there? How far have you gone? Are you stumbling around blindly, or do you have a realistic plan that lets you measure your progress?

Planning requires investigation, study, evaluation, compilation, and other analyses to determine if the planned business will be successful. Successful seed enterprise management requires that managers at all levels:

1. Plan ahead.
2. Identify what you want to do.
3. See if you can do it profitably.
4. See how, and if, you can do it.
5. Be able to measure your progress.
6. See how you can adjust to meet changing conditions.

These are then incorporated into formal plans. The plans must be detailed, realistic, and specific. At the same time, they must allow flexibility to meet changing conditions.

Plans Needed

The results of studying the potential for the seed enterprise are compiled into "plans" which outline and detail the proposed activities. These are compiled into the basic planning document, the BUSINESS PLAN. The business plan itself includes, and is based on, a MARKETING PLAN. The marketing plan is based on, and supported by, (1) a FINANCIAL PLAN which evaluates the financial aspects of the planned work, and (2) an OPERATING PLAN which outlines the operating approach, (3) a PRODUCTION PLAN which defines how the required seed will be produced, (4) a CONDITIONING PLAN which shows how/when/where the seed will be prepared for marketing, and (5) a TRANSPORT-DISTRIBUTION PLAN spelling out how the seed will be moved to retail markets and farmer customers.

Planning Process

The planning process must include a complete evaluation of the potential for a successful seed enterprise operation. This includes:

1. Objectives of the enterprise's operations.
2. Evaluation of the current situation and farmer needs, including:
   National and agricultural economy
   Agriculture and agricultural regions
   Farmer operations and income
   Rural and farmer marketing conditions
   Farmer credit and financing
   Level of farmer development and assistance
3. Potential for business for a seed enterprise, including:
   Best location for a seed enterprise
   Analysis of competition and market requirements for an enterprise
4. Potential seed marketing by the enterprise, including:
   Sources of seed and other inputs to sell
   Initial and planned operations
   Marketing Plan
5. Structure of the enterprise.
6. Financial Plan, and how the enterprise can be financed.
7. Production, Conditioning, and Operating Plans.
8. Business Plan for the enterprise.

General planning sequence is shown in Figure 4.1.

Marketing Plan

The ultimate goal of a seed enterprise is to sell the seed it produces, conditions, and stores. The first step in planning seed enterprise operations is to conduct a detailed, realistic, "in-the-field" analysis of the potential market for seed. This must accurately answer the question—Can we sell enough to make a business profitable? The only way to determine this is to conduct a realistic Market Survey, and canvass farmers to find out what they will buy, how much they will buy, what price they are willing to pay and where they are.

The results of the Market Survey are the first and key management effort in preparing the Marketing Plan. The Marketing Plan:

1. Guides all operations; there is no profit in producing and conditioning seed if they cannot be sold.
2. Advises what kind, how much and when to produce seed.
3. Advises Conditioning how much of what seed to have ready to sell by what dates.

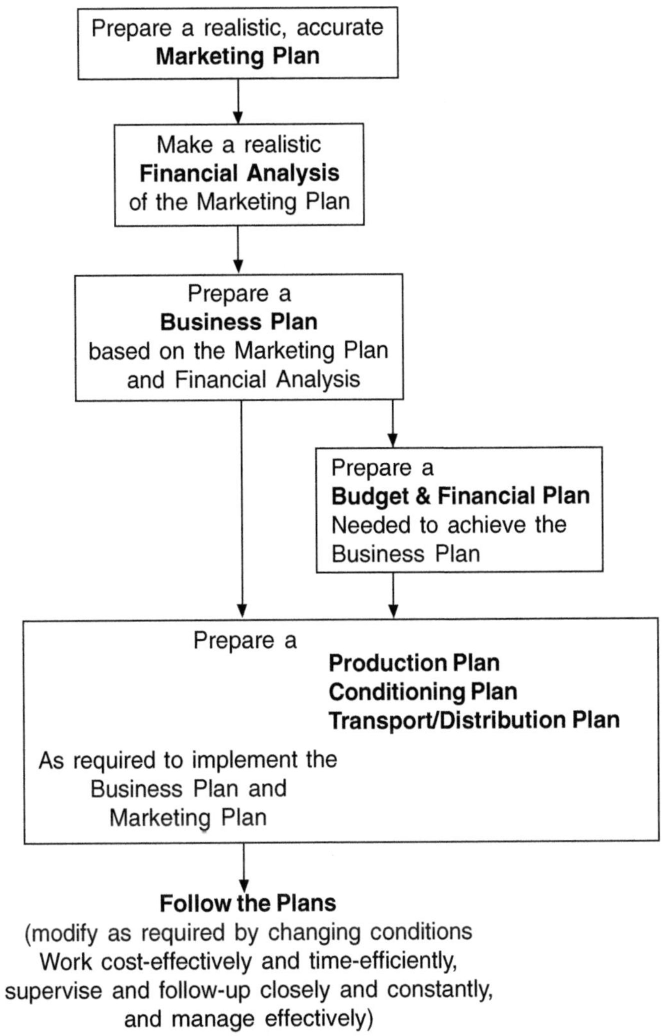

Figure 4.1 Sequence and steps of the planning process to ensure successful operations of a seed enterprise. Source: Gregg *et al.*, Business Plans.

4. Tells Management what can be done, so management and operations can be cost-effective and efficient.
5. Tells sales and income to expect, to prepare a Financial Plan and budget, establish seed prices, and finance operations.
6. Tells Marketing what/where/when/how to sell, how to plan and prepare to market seed.

To do this, the Marketing Plan shows:

1. What seed crops and varieties can be sold.
2. How much of each can be sold.
3. Where each can be sold.
4. How each can be sold (dealers, sales trucks, directly, etc.).
5. What promotion and sales efforts and costs will be required.
6. When each seed kind can be sold.
7. What price can be charged (farmers will pay!) for each kind.

Forming Business Operating Plans

The Marketing Plan shows what could be sold in the enterprise's operating area. Then, to convert this data to operating plans, prepare:

1. Financial Plan to show financial needs, uses, sources, etc., required to achieve the Business Plan (which is based on the Marketing Plan).
2. Business Plan, detailing all aspects of the enterprise.
3. Production Plan, of what is required/how to achieve the Business Plan.
4. Conditioning Plan, of what is required/how to achieve the Business Plan and Marketing Plan.
5. Transport/Distribution Plan, to detail how the seed will be moved to the market so that the farmer customers can buy it.

Financial Plan

The Financial Plan shows whether or not a profit can be made by doing the business shown in the Marketing Plan. It must be complete and in detail, showing:

1. funds available.
2. funds needed.
3. income expected.
4. sources and amounts of required funding and their timing.

The Financial Plan should include:

1. Basic financial statements.
2. Timings of expenses and income (cash flow analysis).
3. If the Plan allows you to earn the necessary margin.
4. If farmers/customers will pay the prices shown in the Financial Plan.
5. Kind of accounting system and records used, and their suitability for tax reporting.

Business Plan

All this information is combined into a cohesive, complete, detailed Business Plan. This gives a complete summary of the planned operations and their realistic chances for success. The business Plan is often called the "key to successful business management. If you need a loan, a Business Plan is usually the first requirement. The Business Plan helps to:

1. Know from the beginning if your intended operations will be profitable.
2. Know what you should do, when you should do it, how you should do it.
3. Compare actual operations/costs/income with what was intended, to measure progress and identify/correct problems.

A complete Business Plan is based on the other detailed plans, and includes:

1. Brief executive summary.
2. The seed industry and agriculture in which the enterprise operates.
3. The enterprise and its structure, ownership, recent performance and history, management, strengths, personnel, facilities and capabilities.
4. Marketing survey, plan, methods, approaches and conditions.
5. Operations for production, conditioning, storage and distribution facilities, plans and methods.
6. Financial status of the enterprise, its recent operations, and the planned operations. This includes projected cash flow and relevant financial statements.
7. Sensitivity or "What if" analysis and plans to adjust to unexpected conditions.

Operating Plans

After the Financial Plan shows that a profit can be made on the sales indicated by the Marketing Plan, plans for efficient operations must be made. These include all aspects, production, conditioning, and transport/distribution of the seed during the sales periods (marketing has already been covered in detail, in the Marketing Plan on which all activities are based).

Production Plan

The Production Plan details all activities: contracting growers, getting stock seed, land preparation, fertilizer, irrigation, roguing, isolation, inspection and supervision, certification, harvest, bagging, etc. It also includes time schedules of all activities, and labor/personnel schedule to conduct/

supervise all activities, and when needed. The estimated costs and expenses are included, showing both amounts and dates needed, which are essential in planning the operating cash flow.

Conditioning Plan

The Conditioning Plan details all activities: receiving, fumigation, drying, shelling/threshing, cleaning, treating, bagging, storage, transport, etc. The following are some aspects which must be carefully planned in advance, in the Conditioning Plan:

1. When will seed be received, and processing must be complete? A detailed time schedule should be planned.
2. What are the facilities for receiving/holding incoming raw seed? What is the maximum amount that can be handled per day? Can seed be safely held on farms for some time?
3. Is drying required? What is the capacity of drying facilities? How many different lots/varieties can be handled at the same time? What is the weather when seed are received?
4. When will how much of each kind of seed be required for marketing? How much time do we have for conditioning?
5. What is the capacity of conditioning? How much time must be allowed for clean-out between varieties? Are two operating shifts required? How much time will be needed for conditioning?
6. How much storage is required, and is it available where it is needed?
7. What are labor requirements? When, how many of what kind of workers, for how long?
8. What supplies are required? Electricity and fuel? Operating expenses?

Transport and Distribution Plan

Seed must be delivered to locations which are readily accessible for the potential farmer customers. It must be within their reach, at a price they consider reasonable, and at the time they need to obtain seed. This requires careful development and planning of the transport and distribution of seed, in addition to marketing and promotion methods. Planning must consider:

1. Storage conditions required, locations, amounts, dates, periods.
2. Transport arrangements must include dates, amounts, locations, conditions required, vehicles, maintenance/backup, etc.
3. Distribution system and components must include a network of dealers, support operations, etc., to get the seed to the places where it can be sold.

Chapter 5

Financing a Seed Enterprise and Conditioning

From beginning to end of all operations, the seed enterprise must plan for and calculate costs and income. Finance impacts every step of planning, managing and operating every aspect. Financial analysis must focus on the entire enterprise and cover production, conditioning, storage, marketing and administration.

Financial Analysis

After the Marketing Survey and Plan shows projected and planned sales, you must determine:

1. What will be the cost of each operation, and the total cost of all operations?
2. What will be the income from probable sales?
3. Will the planned operations be profitable?

The Marketing Plan (assuming that it has been realistically developed) shows how much can be sold, at what prices. Then, other operations must work to create the seed required to implement the Marketing Plan. Production, conditioning, quality control, marketing, finance and other units all must be aimed at providing high-quality seed and selling them to farmers; this is the enterprise's primary source of income.

Financial analysis must include details of estimated costs of seed production, conditioning, supply/transport/storage, administration, management, and marketing. Questions which should be considered include:

1. Are operations most cost-effective, while maintaining high seed quality?
2. What are overhead and other costs?
3. Can any costs be reduced, and still maintain high seed quality?

A first step is to analyze current seed prices and farmer willingness to pay:

1. What are farmers now paying for seed?
2. How much (more) could farmers be convinced to pay?
3. What prices are charged for seed by other dealers?
4. What is the difference between cost and price of their sales?
5. Will selling prices cover costs, dealer discounts, overhead, etc., and still give a reasonable profit?

Volume and profitability analysis pro-rates all costs to get the cost per bag and answer the critical questions of:

1. Can we sell enough to get the income and profit needed?
2. How much do we have to sell to make the required profit?
3. Is that the amount shown by the Marketing Plan? Does the Marketing Plan show we can sell enough to profit?
4. Will we lose money? Should we be in the seed business?

Costs

Financial analysis includes a detailed analysis of how much money is needed for costs of investments, operations, inventory, and overhead. Lists of costs which must be considered in setting up and operating the enterprise, especially conditioning and storage facilities, are shown in the chapters on Cost and Income. These include:

Capital investment:

1. Land
2. Land improvements
3. Buildings
4. Access, road, parking, utility connections
5. Equipment
6. Vehicles
7. Furniture

Overhead costs:

1. Utilities
2. Salaries of staff
3. Labor wages

4. Owner/manager's salary and return on investment
5. Mortgage servicing, interest, etc., costs for capital investment
6. Social costs of employees
7. Legal and licensing costs

Operating costs:

1. Inventory of seed, and its production/purchase, is the major operating cost.
2. Operating expenses and supplies
3. Maintenance
4. Promotion, advertising, demonstrations, etc.
5. Transport, storage, handling
6. Servicing dealers.

Income

Income derived from operations is largely income from sales of seed; if other inputs (fertilizers, pesticides, tools, etc.) are sold, this adds to total income.

Funding Needed

Realistic financial analysis can give a good estimate of the funded needed, funding available, and how much and for how long loans may be needed.

Timing of Expenses and Income

Operating costs are incurred on a regular and continuing basis. However, seed sales and income are seasonal. Funds must be available when needed, in amounts needed, to pay for expenses in a timely manner as they come due. Seed must usually be paid for before they are sold to farmers, so advance funding is necessary to pay monthly operating costs. This requires planning and arranging for sufficient funds to cover expenses incurred each month. A Pro Forma Cash Flow analysis (Table 5.1) is the best management tool for this.

Sources and Needs for Financing

Ultimately, the source of funds should be income from sales. However, under some conditions, this is not adequate. A new enterprise sometimes requires 3–5 years to turn a profit, although sometimes with good Business and

Marketing Plans, a small profit may even be achieved in the first operating year. In all cases, funds will be required to carry out operations before seed are sold, and when the enterprise requires new capital investment or seeks to increase operating volume, more funds are needed before additional income is received (Table 5.2).

This usually requires some funding from other sources to cover the operating period before sales income is adequate. This usually requires some financing, either short-term for operations or long-term for capital investment. Such financing may come from your own or family personal funds, from debt financing, or from equity sales.

Raising funds by equity financing involves selling a share in the ownership of the enterprise. This is done in different ways, according to the legal structure of the enterprise. A Sole Proprietor can raise equity capital by taking in a partner. Partnerships can take in new partners or increase the investment of existing partners. A Limited Liability Company can issue or sell more shares of stock.

Debt financing involves borrowing, assuming debts which must be repaid with interest. These loans may be from family, friends, banks, money lenders, venture capitalists, etc. Short-term borrowing is usually to obtain funds for operations, inventory, etc., which will soon generate sales income. Long-term borrowing is usually for capital improvements, equipment, etc., which will be used over a period of years.

Getting Loans

Loans are the usual means of financing a seed enterprise. However, many enterprises lack knowledge of the lending/collateral/management approach required to get the full amount needed, at the best interest rates.

Lenders look for: "Can you earn enough money to repay the loan and interest on time, at minimal risk?" In requesting a loan, the enterprise manager must show that the enterprise can generate enough income to cover its needs and repay the loan principal and interest when it comes due.

The first and most important step in seeking a loan is to prepare detailed information to show what you will do, how you will do it, that you know what is involved and are competent to do it successfully, and how it will generate income. This requires a detailed and realistic Business Plan showing all management, operational and financial implications. The financial aspects should cover costs, income, pro forma cash flow, pro forma profit and loss statement, pro forma balance sheet, etc. The plan should show that the loan can be repaid and that the lender is making a good investment by giving the requested loan. If possible, get the support of government agencies (agricultural research, extension, etc.) and farmers who will be your contract growers and/or customers. If the banker is your friend, get his input in preparing the Business and Financial Plans. See the example formats in Tables 5.3, 5.4, 5.5, and 5.6.

Most lenders (banks, lenders, venture capitalists) require the enterprise owner to invest some of his own money, usually a minimum of 25-50% of the total required. The lender's thinking is "Why should I risk my money on you, if you don't think that your enterprise is good enough to risk your own money in it?"

Venture Capitalists

Venture Capitalists are financiers who make investments usually in new start-up or expanding companies. They may be individuals, banks, organizations, etc., with money to put to work in investments. To deal with a venture capitalist, you must have detailed realistic Business Plan, Financial Plan, and supporting documentation sufficient to justify risking capital investment. Investment may be as a loan, or by taking stock shares in the enterprise.

How Much Interest Can You Afford to Pay?

The margin between cost and selling price of seed is usually small. Interest payments are an added operating expense, and you can afford to pay only so much interest on a loan. There is no benefit to a loan if you pay more interest than the profit you gain from the extra seed production and sale.

To determine how much interest you can afford to pay, have your accountant calculate the IRR—Internal Rate of Return. IRR is essentially the amount of interest you could pay on a loan and still break even (cost = benefit). IRR is a *discount* rate at which the *present value* of investment is equal to the *present value* of the returns on those investments; *i.e.*, break-even rate of return. IRR can be used for a series such as irregular purchases and sales of seed, to show overall rate of return. If IRR is higher than the opportunity cost of capital (the best alternative use for it), the proposed investment is good. Avoid an investment if its IRR is less than the cost of borrowed capital or minimum desired rate of return.

The value of the IRR depends on having good data which shows future conditions accurately. Inaccurate data gives a wrong IRR which can result in serious financial problems. Best for a small seed enterprise is good market analysis, low-cost production, and stable demand for quality seed. No seed enterprise will succeed without enough farmer clients who see the benefit of high-quality seed at an affordable price.

Table 5.1 Projected cash flow by end for year 20__.

Item	J	F	M	A	M	J	J	A	S	O	N	D
BEGINNING CASH												
CASH SALES:												
Fertilizers												
Agro-Chemicals												
Field Crop Seed												
Vegetable Seed												
Farm Equipment												
Total Cash Sales												
DISBURSEMENTS:												
Inventory Purchase												
Salaries and Wages												
Advertising												
Vehicle Expense												
Legal/Accounting												
Office Supplies												
Telephone (office & cell)												
Utilities												
Property Tax												
Rent & Storage												
Loan Principal Payment												
Loan Interest Payment												
Equipment Maintenance/Purchase												

Operating Supplies												
Vehicles												
Other Asset Costs												
Prepaid Income & Business Tax												
Total Disbursements												
NET CASH FLOW												
ENDING CASH												

Table 5.2 Summary of planned sources and uses of funds.

Sources of funds:		
Owner's funds	____	
Other equity	____	
Sales income	____	
Loans	____	
Total Funds		____
Use of Funds:		
Land	____	
Buildings	____	
Equipment	____	
Salaries	____	
Labor wages	____	
Vehicles	____	
Renovations	____	
Promotion/advertising	____	
Inventory	____	
Furniture and fixtures	____	
Taxes	____	
Total Expenditures		____
Cash Balance (working capital):		____

Managing Finances

A key element in having a successful enterprise, and in getting financing at the best rates, is having good financial management.

Some records and plans/projections which are essential to good financial management include Sources and Uses of Funds, Cash Flow Analysis, Profit and Loss Statement, and Balance Sheet. These are described in the chapter on records and reports. Use Pro Forma estimates to plan finances for the coming period; use actual records in the same format to record/report/analyze financial results.

Example of Planning Finance for an Enterprise

An over-simplification of the steps in planning and managing the finance of a seed enterprise is:

1. Market survey and production information show that you can buy or produce maize seed at L$100 per bag, and sell 300,000 bags at L$175 per bag.
2. Total inventory cost would be L$30,000,000, and total income would be L$52,500,000, for a gross profit of L$22,500,000.

Table 5.3 Projected cash flow statement for the coming year of 20__.

Description	Amount	
Beginning Cash		____
Cash Sales:		
Fertilizers	____	
Agro-Chemicals	____	
Field Crop Seed	____	
Vegetable Seed	____	
Farm Equipment	____	
Total Cash Sales		____
Disbursements:		
Inventory purchase	____	
Salaries and Wages	____	
Advertising	____	
Vehicle expense	____	
Legal/accounting	____	
Office supplies	____	
Telephone (office & cell)	____	
Utilities	____	
Property Tax	____	
Rent and Storage	____	
Loan principal payment	____	
Loan interest payment	____	
Equipment maintenance & purchase	____	
Operating supplies	____	
Vehicles	____	
Other asset costs	____	
Prepaid income & business tax	____	
Total Disbursements		____
Net Cash Flow		____
Ending Cash		____

3. Overhead and operating costs for the year are calculated at L$4,800,000.
4. You can buy buildings and equipment on a 10-year loan, paying back the loan at L$50,000 per month (L$600,000 per year).
5. You need L$30,000,000 to buy the seed inventory of 300,000 bags.
6. From the gross projected profit of L$22,500,000, subtract the overhead and operating costs (L$4,800,000) and annual capital investment loan cost (L$600,000), and you still have a profit of L$17,100,000. Thus, the business seems quite profitable.

Table 5.4 Projected profit and loss statement for the period ____ to ____.

Description	($)	($)
Revenue:		
Net Sales		____
Cost of Goods Sold:		
Beginning inventory	____	
Purchases	____	
Goods available for sale	____	
Less ending inventory	____	
Total Cost of Goods Sold		____
Sales, General and Administrative Expenses:		
Salaries & Wages	____	
Advertising	____	
Vehicle expense	____	
Legal & accounting	____	
Office Supplies	____	
Telephone (office & Cell)	____	
Utilities	____	
Property Tax	____	
Rent & Storage	____	
Loan interest payment	____	
Equipment maintenance	____	
Operating supplies	____	
Depreciation	____	
Total S, G & A Expense		____
Estimated Tax:		
Income Tax	____	
Total Taxes		____
Net Income:		____

7. Total expenditures for the year are L$30,000,000 for inventory, L$600,000 capital investment cost, and L$4,800,000 for overhead and operations, or a total of L$35,400,000 for the year.
8. For your own investment, you have L$4,000,000.
9. You prepare a Business Plan, and go to the bank for a loan of L$32,000,000 for operations and inventory.
10. Your banker points out some weaknesses in the preparation of your Business Plan, and would prefer a larger investment on your part. So, he refuses to give you a loan.
11. You re-work your Business Plan to improve the weak points noted by your banker and to justify the high percentage of the loan to costs, and go to another bank to request a loan.

Table 5.5 Projected end-of-year balance sheet, December 31, 200–.

ASSETS:		
Current Assets:		
Cash	____	
Ending inventory	____	
Accounts receivable	____	
Total current assets		____
Fixed assets:		
Land	____	
Buildings	____	
Equipment	____	
Renovations	____	
Total fixed assets	____	
Less accumulated depreciation	____	
Net fixed assets		____
Other Assets:		
Vehicles	____	
Less accumulated depreciation	____	
Net other assets		____
Total assets		____
LIABILITIES:		
Current Liabilities:		
Notes payable	____	
Current part of long	____	
Total current liabilities		____
Long-term liabilities:		
Loans payable	____	
Less current part	____	
Total long-term liabilities		____
Total liabilities		____
EQUITY:		
Equity:		
Owners' funds	____	
Retained earnings	____	
Total equity		____
Total liability and equity		____

12. This banker gives you the desired loan of L32,000,000 at an interest rate of 10%.
13. The loan includes a provision that you can repay all or part of the loan before the end of the year, and not be charged full interest.
14. You calculate that you can pay this operating loan within 6 months (thus half the interest cost, or 5% of L$32,000,000 or L$1,600,000).

Table 5.6 Projected income statement for the period December 31, 20__.

Item	Amount	Amount
Sales	xxxxxx	
Other Revenues	xxxxxx	
Total		**xxxxxx**
Less Cost of Goods Sold		
Opening Inventory, Jan. 1, 20__	xxxxxx	
Goods Produced	xxxxxx	
Goods Purchased	xxxxxx	
Total Cost of Goods Available for Sale	**xxxxxx**	
Less Closing Stock, Dec. 31, 20__	xxxxxx	
Total Cost of Goods Sold		**xxxxxx**
Less: Marketing and Distribution Expenses	xxxxxx	
Administrative Expenses	xxxxxx	
Total Selling Expenses		**xxxxxx**
Operating Profit		xxxxxx
Less Taxes		xxxxxx
Net Profit After Taxes		**xxxxxx**
Appropriation		
Interim Dividends Paid	xxxxxx	
Final Dividends Proposed	xxxxxx	
Transfer to General Reserve	xxxxxx	
Total Appropriations		**xxxxxx**
Retained Earnings as of 12/31/__		**xxxxxx**

15. You subtract the loan cost from your net projected profit (above) of L$17,100,000, and find that you have a final net projected profit of L$15,500,000. This is a profitable investment, given good management, efficient operations and high-quality seed.

Chapter 6

Income from Conditioning

Income

Although the primary output of seed conditioning is cleaned, high-quality seed, income is or may be generated by the activities shown below.

Income	Unit	Unit Price	Units Sold	Total Income
1. Sale of seed	____	____	____	____
2. Sale of deteriorated seed	____	____	____	____
3. Sale of waste products	____	____	____	____
4. Custom seed conditioning fees	____	____	____	____
5. Custom seed treating fees	____	____	____	____
6. Seed delivery charges	____	____	____	____
7. Seed storage charges	____	____	____	____
	____	____	____	____
Total Income	____	____	____	____

Maximizing Income

Good management can minimize costs and maximize income. Suggestions for maximizing income include:

1. Minimizing costs is the best way to maximize income and profit! Constantly analyze and improve all operations, to minimize or eliminate cost factors.
2. The most profitable product is good cleaned seed. All separators and cleaners should be adjusted and monitored so that the least possible amount of good seed is lost with waste products in the cleaning processes.

3. Ensuring that all seed are cleaned to meet or exceed the required standards, ensures that the most possible good seed are obtained.
4. Use waste products in animal feed. Do not include treated seed (there should be no treated waste products). Waste products containing weed seed should be treated in some manner to destroy germination, or should be burned.
5. Ensure that storage conditions and procedures are efficient and maintain seed quality, to prevent losses.
6. Have, and use to the fullest extent, an Internal Quality Control (IQC) system which monitors all operations and tests all seed and separated fractions.
7. Keep all equipment well-maintained and serviced, and ensure that operators work safely and efficiently.
8. Build strong loyalty and enthusiasm in all staff, so they operate efficiently.

Chapter 7

Costs and Cost Control

Seed conditioning is a business operation as well as a technical activity, and must be conducted and managed so that it is profitable, efficient, income is maximized, and costs are minimized.

Identifying Costs

Seed conditioning involves a range of cost items including those of operations, overhead, maintenance, equipment and facilities, etc. Good management requires identification and monitoring of all cost factors, and managing them efficiently.

Minimizing Costs

One of the major ways to improve profit is to minimize expenses. Sometimes, major changes and upgrades are necessary to make a seed conditioning operation efficient and economically feasible and market-competitive. However, the best means of reducing expenses are usually relatively simple procedures to:

1. Reduce time required for an operation.
2. Reduce labor requirements.
3. Eliminate or combine steps in the operation.
4. Replace an inefficient component with a more modern, efficient unit.
5. Increase capacity of a unit which is a limiting factor—a machine, bin, elevator, etc.
6. Improve efficiency of a separation process.
7. Train, and constantly upgrade, all employees including laborers, so they do their jobs more efficiently. To take advantage of their training, use permanent-hire laborers.

8. Reduce cost of the operation or a part of it.
9. Reduce losses of good seed.
10. Plan in advance, to avoid wasted time and start-up problems.
11. Reduce handling required, etc.

Constant Evaluation and Improvement

An effective means of continuous improvement is having a system of receiving suggestions for improvements from staff who constantly work with the operations, and know them in great detail. The system should include worthy rewards for suggested improvements which are adopted.

Regularly, each operation and expense item should be monitored and evaluated, and possible means of reducing its cost—without affecting operating efficiency or seed quality—should be considered.

As a general rule, mechanizing an operation to reduce labor requirements is cost-effective in the long run. Operating costs are usually lower than labor costs, and cost of equipment can be depreciated over a period of time.

Cost Categories and Cost-Saving

Following is an analysis of different categories of expenses in a typical seed conditioning operation. These are arranged in a tabular form useful to a seed enterprise, to show type of cost, unit cost, units used, and total cost as must be calculated in the enterprise's financial records. Some suggestions for improving cost efficiency are also included; detailed study of the enterprise's operations should reveal many more ways to improve cost- and operating-efficiency.

Minimizing Administrative Costs

1. Computerize forms and reports. To the extent possible, use available commercial software systems.
2. Connect all departments/offices with a computer network so that forms/reports filled out by one are automatically available to all others.
3. Make reports and records simple (but complete, with all information!) so they require minimum time to prepare.
4. Design records and reports, and use good computer software, so reports can be generated directly from records.
5. Design records/reports so a minimum of work is required. For example, a single "seed lot record" can be used to follow the lot from the field to final sale.

Overhead Costs—1 Administrative/Support Costs	Unit	Unit Cost	Units Used	Total Costs
1. Secretaries, clerks, accountant, etc.	____	____	____	____
A. Office staff salaries	____	____	____	____
B. Office staff social benefits	____	____	____	____
2. Office space	____	____	____	____
A. Office cost/rent	____	____	____	____
B. Office electricity	____	____	____	____
C. Office water	____	____	____	____
D. Office heating & air conditioning	____	____	____	____
3. Office supplies	____	____	____	____
4. Office records forms, folders, etc.	____	____	____	____
5. Office equipment	____	____	____	____
A. Office equipment depreciation/cost	____	____	____	____
B. Office equipment maintenance	____	____	____	____
C. Office equipment supplies	____	____	____	____
6. Communications	____	____	____	____
A. Telephone	____	____	____	____
B. Fax	____	____	____	____
C. Mail, postage	____	____	____	____
D. Telegrams, telex	____	____	____	____
Other	____	____	____	____
Total Cost	____	____	____	____

6. Hire a short-term specialist to design the overall record and report system so that it is complete yet simple and efficient. (a complete example record/report system for a seed enterprise has been published by Gregg *et al.*).
7. Hire competent staff who turn out good work in the shortest time, with the fewest mistakes. Sometimes, one competent (higher-paid) employee can do more than one kind of work, and reduce the number of employees.
8. Maintain all facilities and equipment in good condition, with regular maintenance, using the enterprise's mechanic/electrician/carpenter.
9. Use standard office efficiency and control methods to minimize costs for electricity, supplies, filing, communications, etc.

Overhead Costs—2 Legal Costs	Unit	Unit Cost	Units Used	Total Costs
1. Seed Law permits, licenses	_____	_____	_____	_____
2. Legal counsel, advisor	_____	_____	_____	_____
3. Certification, etc., fees	_____	_____	_____	_____
4. Business licenses, fees, permits, costs	_____	_____	_____	_____
5. Seed Law compliance, quality, etc., costs	_____	_____	_____	_____
Other	_____	_____	_____	_____
Total Cost	_____	_____	_____	_____

Minimizing Legal Costs

1. Sell only seed of assured high quality, properly labeled, sealed to prevent tampering, so there are few chances for claims of poor quality and lawsuits for damages!
2. Ensure that IQC actively ensures seed quality and compliance with seed law and certification. This minimizes penalties, lost seed, and operating costs of correcting problems.
3. Obtain permits and licenses when due, to avoid penalties.
4. Arrange for legal counsel on a retainer fee which minimizes total costs.
5. Make regular professional safety inspections of all operations, and upgrade them to minimize risks and liabilities. Check with local fire and insurance agencies about getting inspections, and then ensure that improvements are reflected in reduced insurance premiums.

Overhead Costs—3 Tax Costs	Unit	Unit Cost	Units Used	Total Costs
1. Business taxes, costs	_____	_____	_____	_____
2. Income taxes, costs	_____	_____	_____	_____
3. Real estate taxes, costs	_____	_____	_____	_____
4. Customs taxes, duties, costs	_____	_____	_____	_____
5. Costs of collecting/paying sales taxes	_____	_____	_____	_____
6. Levies, fees for special uses on seed sold or business	_____	_____	_____	_____
Other	_____	_____	_____	_____
Total Cost	_____	_____	_____	_____

Minimizing Tax Costs

1. Get good legal advice on how to minimize taxes.
2. Pay taxes on time, to avoid penalties. Ensure cash flow adequate to meet timely payments!
3. Have a good computer software program to record, compile and prepare sales tax payments as they are due.

Delivering Seed to Conditioning Costs

	Unit	Unit Cost	Units Used	Total Costs
1. Labor for loading seed onto vehicle	_____	_____	_____	_____
2. Transport/vehicle costs	_____	_____	_____	_____
A. Vehicle cost/depreciation/rental	_____	_____	_____	_____
B. Vehicle maintenance	_____	_____	_____	_____
C. Vehicle fuel	_____	_____	_____	_____
D. Driver salaries	_____	_____	_____	_____
E. Driver travel allowances	_____	_____	_____	_____
F. Vehicle tag and taxes	_____	_____	_____	_____
Other	_____	_____	_____	_____
Total Cost	_____	_____	_____	_____

Minimizing Costs of Delivering Raw Seed to Conditioning

1. To the extent possible and compatible with maintaining high seed quality, arrange for growers to deliver seed to conditioning on a schedule which brings in only one kind of seed at a time. This avoids clean-out between seed kinds (if seed quality is uniformly good, no clean-out should be required between lots of the same variety), and reduces labor costs and down time. If necessary, receive one variety today, clean out tonight, receive another variety tomorrow, etc.
2. If growers can deliver the seed, have them do so.
3. If possible, handle seed in bulk to minimize bag handling. Before use, clean out bulk trucks, then cover seed with a tarpaulin, etc., to prevent loss or admixture. Bulk forklift toteboxes are very cost-efficient, as they can be filled on the farm, transported to conditioning, and then used as storage bins for the seed until it is conditioned.
4. Use trucks which haul economical loads, for local conditions.
5. Keep trucks and forklifts properly serviced and maintained.
6. If bagged seed are brought in, put pallets on the truck and have bags loaded directly onto pallets at the farm(s).

Receiving & Pre-Conditioning Handling Costs

	Unit	Unit Cost	Units Used	Total Costs
1. Receiving/unloading bulk seed at plant				
A. Labor				
B. Equipment				
i. Forklifts, conveyors, etc. cost/ depreciation				
ii. Forklifts/equipment fuel				
iii. Forklifts/equipment maintenance				
iv. Forklifts/equipment operators salaries				
v. Drivers/operators social costs				
C. Pallets/storage frames				
i. Pallets/storage frames cost/ depreciation				
ii. Pallets/storage frames maintenance				
2. Unloading seed bags when received				
A. Labor				
B. Equipment				
i. Equipment cost/rental/ depreciation				
ii. Equipment maintenance				
iii. Equipment fuel				
3. Bag repair/replacement				
4. Emptying bags into bulk storage/ stacking bags into bag storage				
A. Labor				
B. Equipment				
i. Equipment cost/rental				
ii. Equipment maintenance				
iii. Equipment fuel				
5. Storage warehouse/bins/facilities				
A. Cost/rent/depreciation				
B. Maintenance and repairs				
C. Utilities (electricity, fire protection, etc.)				
6. Supervisory staff for receiving				
A. Staff salaries				
B. Staff social benefits				
7. Receiving supplies				
A. Stack cards				
B. Labels, tags				
C. Markers, chalks, other supplies				
Other				
Total Cost				

Minimizing Costs of Receiving and Pre-Conditioning

1. Design receiving facilities to receive seed and get it into the operating system with minimum time, labor, and risk of mixture.
2. Use forklift(s) to unload and handle pallets, loaded on the farm, with bagged seed.
3. For bulk seed, have truck dump pits, elevators, conveyors, etc., to receive bulk seed and take it to bins. Have all pits, conveyors, elevators, etc., designed and installed so they are self-cleaning. Have easy access to all parts of all equipment, for inspection and clean-out. Have all floors smooth-finished concrete.
4. IQC should sample and quick-test each incoming lot or load. Determine its drying requirements, and schedule it immediately, to avoid loss. If a seed lot cannot economically be cleaned to meet standards, reject it immediately before any costs are expended on it.
5. Use hydraulic truck dump, etc., so that a minimum number of persons and laborers are required to receive bulk seed.
6. Place a blackboard/chalkboard at the receiving point, with handling and storage instructions for each kind of seed. This saves time, and prevents mistakes.

Conditioning Separating/Cleaning Costs

	Unit	Unit Cost	Units Used	Total Costs
1. Equipment	_____	_____	_____	_____
A. Equipment cost/depreciation/ rent				
B. Maintenance and spare parts	_____	_____	_____	_____
C. Equipment fuel/power/ electricity	_____	_____	_____	_____
D. Installation/stand cost/ maintenance	_____	_____	_____	_____
2. Staff and laborers	_____	_____	_____	_____
A. Supervisor	_____	_____	_____	_____
B. Operators	_____	_____	_____	_____
C. Mechanic	_____	_____	_____	_____
D. Laborers (permanent; no temporary)	_____	_____	_____	_____
E. Records/reports clerk	_____	_____	_____	_____
3. Conditioning buildings	_____	_____	_____	_____
A. Cost/depreciation/rental	_____	_____	_____	_____
B. Maintenance and repairs	_____	_____	_____	_____
C. Power, electricity, fire protection, etc.	_____	_____	_____	_____
D. Security, watchmen, etc.	_____	_____	_____	_____
Other	_____	_____	_____	_____
Total Cost	_____	_____	_____	_____

7. Train staff in clean-out, and provide adequate clean-out tools, so they can clean-out the receiving facilities quickly.
8. Keep pallets in good condition, with no loose nails which can tear bags.
9. Keep the receiving area (which should have a smooth-finished reinforced concrete floor) constantly swept clean, free of loose seed, nails, gravel, trash, etc., which could cause damage or contaminate seed.

Minimizing Cleaning Costs

1. Place blackboards/chalkboards in the conditioning receiving and bagging areas, with full information on the seed being conditioned. This prevents mixing seed, loss of identity, mis-handling, etc. On the chalkboard at bagging, show the lot number to put on the bag, bag weight, etc.
2. Allow only one lot/variety in the conditioning area at the same time, to prevent mixing, mistakes, etc.
3. Design and install cleaning and handling equipment so that seed flow mechanically and continuously through the entire operation, without labor handling or constant management.
4. Spend a little more initially, and buy equipment with slightly more capacity (and room for growth in business) and is better-built, so it requires less maintenance and lasts longer.
5. Plan the sequence of conditioning of seed lots and kinds so that all (or as many as the marketing schedule for other varieties will permit) lots of the same variety/class are cleaned in an uninterrupted sequence, to minimize the number of clean-outs between lots/kinds.
6. In advance, prepare a schedule of all conditioning and clean-outs, with all details of each lot, so that all staff know what is to be done. Give all supervisors a printed copy, post this on the blackboard; be sure that even laborers know fully what is to be done.
7. Have self-cleaning elevators, conveyors, spouting, etc., to bring bulk seed from storage to the conditioning intake.
8. Avoid bag handling as much as possible. If bagged raw seed are conditioned, install the receiving elevator in a floor pit (with a grate cover for safety) so workers can empty bags at floor level directly off the pallet, without lifting.
9. Have all required clean-seed bags (with proper lot numbers), tags, and supplies (thread, etc.) ready, and easily moved into place for use.
10. Have the first bin in conditioning (usually, over the air-screen cleaner) large enough to hold all seed which can be cleaned in a few hours. Have a high-capacity elevator feeding this bin, so workers can fill the

bin in a short time. Then, while this bin-load of seed is being cleaned, the workers can do other jobs—bagging cleaned seed, etc. This avoids tying down a crew of workers fulltime to feed seed into conditioning.

11. Have a vibrating (or other type) conveyor to receive all waste products from all cleaners and separators, and deliver them to an elevator feeding a waste product bin. This eliminates labor and delays in bagging and handling waste products.
12. Have easy, safe access to all parts of all machines for adjustment, clean-out, and maintenance.
13. Install two-way or three-way valves on all elevator discharges, so each machine can be used or bypassed as required.
14. Avoid mechanical seed damage and loss, by using proper impact absorption and let-down systems in all bins, downspouts, etc.
15. Have easy, safe access to all bin tops, elevator heads, etc., for sampling and clean-out.
16. Design and construct all bins, spouts, etc., so they are self-cleaning, without places where seed can lodge.
17. Keep detailed records of each machine's adjustment on each lot. This provides approximate cleaner settings when a similar lot is to be conditioned. Equip IQC with a full set of hand screens, small-scale cleaners, etc. In advance, have IQC determine approximate adjustments and settings for each cleaner and separator. DO NOT start with "guess" adjustments, and then have to stop machines and re-adjust them.
18. A trained operator should constantly move among the cleaning machines, and monitor their operation. Seed lot composition may change, feed rate may change, etc., so separation efficiency changes. This is especially important with the gravity separator, length separator, etc. Constant monitoring catches such changes immediately, so small adjustments can be made quickly to avoid non-uniform seed lots or poor separation.
19. Adjust the gravity separator, before operation, to the approximate setting used previously on similar lots (as shown in cleaning records of previous lots). Even then, it takes a few minutes to adjust the gravity exactly; seed discharged during the adjustment period should be re-cleaned on the gravity. Install a short elevator to take all seed discharging from the gravity during adjustment, back into the bin over the gravity. As soon as adjustment is precise, change discharge spouts so clean seed and waste products move to their proper places. The "adjustment return" elevator can continue to be used, to take the middling product back to the gravity bin for re-cleaning with the seed flow.
20. Before each operating season, re-train all staff and workers so they understand their work and how to do it most efficiently. Bring in outside specialists for special training, and have good operating/maintenance

manuals readily available to all, especially so they can read them during off-seasons.

21. IQC should constantly monitor seed quality and operating efficiency, and stop or change any substandard operation. Clean seed should be analyzed for quality, and waste products analyzed for excessive good seed, at least twice daily.
22. Have one laborer constantly sweep up loose seed, and keep the conditioning area clean. It is especially important to collect loose bag thread, etc., which may get into the seed flow and plug the feed hopper of a seed separator.
23. Prolong equipment operating life, by complete overhaul and maintenance in the off-season; during operation, watch for minor problems and correct them immediately.
24. Keep birds out of conditioning (with transparent door "flaps" and screened windows), and keep poisoned baits in place to control rats and mice.

Treating Costs

	Unit	Unit Price	Units Sold	Total Income
1. Equipment	____	____	____	____
A. Equipment cost/depreciation/rent	____	____	____	____
B. Equipment maintenance	____	____	____	____
C. Equipment fuel/power	____	____	____	____
D. Installation/stand cost/ maintenance	____	____	____	____
2. Equipment operator and laborers	____	____	____	____
3. Chemicals (treatment, surfactant, spreader, etc.)	____	____	____	____
4. Water & water supply	____	____	____	____
Other	____	____	____	____
Total Cost	____	____	____	____

Minimizing Treating Costs

1. Install the treater so water and chemicals can be easily measured and fed to the treater. A good method is to install the main chemical tank on the floor so it can be easily checked, filled, and cleaned, with a pump to carry chemicals to the treater. Minimize the time and cost of keeping the treater supplied with chemicals.
2. Have a chart, posted at the treater, showing treater adjustments (seed weight, chemical cup size, etc.) for each seed kind and treating material,

so no time is wasted in adjusting the treater to deliver the proper dosage. Calculate these adjustments well in advance.

3. If slurry treatments are used, have water readily available for mixing solutions. But, be sure that loose water does not affect the seed or operations, and does not provide a source of water for rats.
4. Adjust the treater in advance, while other machines are being cleaned out. Have it ready to begin operating properly as soon as seed are cleaned.
5. Regularly check the treatment tank, flow hoses, and treater chemical measuring system to be sure there is adequate chemical, and nothing is plugged up. Avoid having to bring seed back and re-treat it!
6. Do not treat seed until sales are more or less assured. However, avoid having to store untreated seed, then treat it in a separate operation; this greatly increases the cost of conditioning and treating. Plan sales in advance, so seed can be conditioned, treated and bagged in a single operating sequence. If time allows, store "excess" seed uncleaned, and then condition/treat/bag it in a single later operation.
7. Have facilities so the treater can be easily washed out after use. Be sure that washing water, etc., can be safely drained away; a good method is to install the treater in a large watertight sheet metal pan, with a drain to take wash-out water safely to the main drain.
8. Do not leave chemicals in the treater overnight or over weekends. Liquid treatments may be corrosive, and slurry treatments may settle out and cake up in the treater measuring cups or pipes.
9. Prolong the treater's working life by carefully washing it out when it will not be used for several days. Cover all exposed metal parts with a light oil to prevent rust.
10. Arrange purchase of chemicals at wholesale prices, to save in purchase costs. Buy larger quantities for a quantity discount (but be sure you will use all the chemical before it loses its strength!).

Minimizing Weighing and Bagging Costs

1. Use semi-automatic (or automatic, if possible) bagger-weighers which fill and weigh the bag in one operation.
2. Using valve-pack bags eliminates the need for a bag sewing machine. In most plants, the sewing machine is the most difficult machine to keep in good operating condition.
3. Have the bin over the bagger-weigher large enough to hold a few hours output of cleaned seed. This enables intermittent bagging and weighing, instead of a fulltime crew for bagging and weighing. This allows the same work crew to feed in raw seed and also bag cleaned seed.

Weighing/Bagging Clean Seed Costs				
	Unit	Unit Price	Units Sold	Total Income
1. Equipment	_____	_____	_____	_____
A. Equipment cost/depreciation/ rent	_____	_____	_____	_____
B. Equipment maintenance	_____	_____	_____	_____
C. Equipment fuel/power	_____	_____	_____	_____
D. Installation/stand cost/ maintenance	_____	_____	_____	_____
2. Operators and laborers	_____	_____	_____	_____
A. Supervisor/check	_____	_____	_____	_____
B. Equipment operators	_____	_____	_____	_____
C. Laborers	_____	_____	_____	_____
3. Tags and labels	_____	_____	_____	_____
4. Bags	_____	_____	_____	_____
5. Supplies (thread, stencils/ink, etc.)	_____	_____	_____	_____
A. Pallet cost	_____	_____	_____	_____
B. Pallet maintenance labor	_____	_____	_____	_____
C. Pallet maintenance supplies	_____	_____	_____	_____
Other	_____	_____	_____	_____
Total Cost	_____	_____	_____	_____

4. Pre-adjust the bagger-weigher, to give exact weight as soon as bagging starts. During operation, check-weigh one bag in every 15–20 bags, to ensure that weight does not change.
5. Use a sewing conveyor to move filled bags through the bag sewing machine at the proper speed, without requiring a fulltime operator.
6. Adjust tension on the bag sewer's thread feed so thread moves at the proper tension. Putting some light sewing machine oil on the thread on the spool(s) helps.
7. Load bags directly onto pallets, and move the loaded pallets into storage by forklift.
8. Show on the bagging blackboard the exact storage location where the lot is to be stored, to avoid confusion.

Minimizing Costs of Moving Clean Seed Into Storage

1. Mark off the floor of the storage into numbered rows, to identify the exact area in storage where each lot is stored.
2. Separate storage from conditioning with a fireproof wall and a fire door. However, minimize the distance that seed must be moved from

Moving Clean Seed Into Storage Costs				
	Unit Price	Unit Sold	Units Income	Total
1. Forklifts and other equipment	_____	_____	_____	_____
A. Equipment cost/depreciation/rent	_____	_____	_____	_____
B. Equipment maintenance	_____	_____	_____	_____
C. Equipment fuel/power	_____	_____	_____	_____
2. Staff and laborers	_____	_____	_____	_____
A. Staff salaries and overhead	_____	_____	_____	_____
B. Staff overhead/social costs	_____	_____	_____	_____
C. Laborers	_____	_____	_____	_____
3. Storage buildings	_____	_____	_____	_____
A. Cost/depreciation/rental	_____	_____	_____	_____
B. Maintenance and repairs	_____	_____	_____	_____
C. Power, electricity, fire protection, etc.	_____	_____	_____	_____
D. Security, watchmen, etc.	_____	_____	_____	_____
4. Labeling/identifying stored seed	_____	_____	_____	_____
A. Supplies	_____	_____	_____	_____
B. Supervisor	_____	_____	_____	_____
C. Workers	_____	_____	_____	_____
Other	_____	_____	_____	_____
Total Cost	_____	_____	_____	_____

bagging to storage. Design building and operating layout to minimize distances of all seed movements.

3. Load seed bags onto pallets as the bags are filled. Carry the loaded pallets into storage, and stack them on pallets without re-handling the bags. Avoid stacking bags in piles in storage.
4. Have adequate aisles in storage, so movement is unimpeded and existing stacks are not bumped or damaged.
5. Train forklift operators to move loaded pallets safely, without dropping or damaging bags.

Minimizing IQC Costs

1. Test quality of incoming lots, reject poor lots before wasting time and money trying to clean them up to standards.
2. Install equipment so that materials going into all discharge spouts (clean seed and waste products) from all machines can be easily sampled.
3. Install all elevators and bins so that seed going into, and coming out of, them are easily accessible for sampling seed.

Internal Quality Control Costs				
	Unit	Unit Price	Units Sold	Total Income
1. Sampling when seed is received	____	____	____	____
A. Staff	____	____	____	____
i. Sampling staff salaries	____	____	____	____
ii. Sampling staff overhead/ social costs	____	____	____	____
B. Sampling equipment	____	____	____	____
i. Sampling equipment cost/ depreciation	____	____	____	____
ii. Sampling equipment maintenance	____	____	____	____
iii. Sampling equipment power	____	____	____	____
C. Sampling bags, forms, supplies	____	____	____	____
D. Seed cost	____	____	____	____
2. Sampling during processing/storage	____	____	____	____
A. Staff	____	____	____	____
i. Sampling staff salaries	____	____	____	____
ii. Sampling staff overhead/ social costs	____	____	____	____
B. Sampling equipment cost/ depreciation	____	____	____	____
i. Sampling equipment cost/ depreciation	____	____	____	____
ii. Sampling equipment maintenance	____	____	____	____
C. Sampling bags, forms, supplies	____	____	____	____
D. Seed cost	____	____	____	____
3. Testing samples	____	____	____	____
A. Moisture Testing	____	____	____	____
B. Purity & physical quality testing	____	____	____	____
C. Germination testing	____	____	____	____
D. Tetrazolium testing	____	____	____	____
E. Quick germination testing	____	____	____	____
F. Vigor testing	____	____	____	____
i. Equipment, furnishing cost/ depreciation	____	____	____	____
ii. Equipment/furnishings maintenance	____	____	____	____
iii. Salaries/labor	____	____	____	____
iv. Supplies	____	____	____	____
v. Working space cost/rental/ depreciation	____	____	____	____
vi. Electricity, heating, cooling, etc.	____	____	____	____
Other	____	____	____	____
Total Cost	____	____	____	____

4. IQC personnel should be highly-trained and experienced, with good understanding of all operations and how to monitor/test operations and seed. But, to save costs, have several (as required) competent and conscientious assistants.
5. Set up the IQC laboratory immediately adjacent to conditioning.
6. Ensure that IQC equipment is in good condition, before the conditioning season.
7. If there are problems with electricity supply, have an adequate generator to support IQC, so that ongoing tests are not affected.
8. Use a master seed lot record which follows the seed lot through all operations. IQC can enter its data in this record, and minimize record-keeping.
9. Use standard can or box containers, which can be re-used, to collect samples for rapid analysis during conditioning of a lot.
10. Have a safe storage for record samples of each lot after testing by IQC.
11. Do not return analyzed samples to the seed lot. This carries a risk of contaminating the lot. Collect seed from tested samples so they can be sold as grain, etc.
12. Buy supplies in larger quantities, for a quantity discount.

Minimizing Storage Costs

1. Keep storage spotlessly clean and free of loose seed. The same worker who constantly sweeps/cleans conditioning should sweep/clean the storage.
2. Mark off the storage floor into number stack rows, for positive identification of storage locations.
3. Follow storage procedures recommended in seed storage management manuals.
4. Use "stack cards" to identify the seed in each stack row.
5. Keep birds out of the storage (with transparent door "flaps" and screened windows), and keep poisoned baits in place to control rats and mice.
6. Keep storages as dry and cool as possible. Be sure the storage construction is for seed storages, not general storages.
7. Regularly check moisture content, germination and condition (free of rat damage, etc.) of all stored seed. Follow a testing schedule for all stored lots.
8. Sell stored seed according to a "first in, first out" policy, and move high-germinating but low-vigor lots quickly.
9. Construct storages with smooth walls and floors for easy cleaning.

Seed Storage Costs	Unit	Unit Cost	Units Used	Total Costs
1. Equipment	_____	_____	_____	_____
A. Equipment cost/depreciation/ rent	_____	_____	_____	_____
B. Equipment maintenance	_____	_____	_____	_____
C. Equipment fuel/power	_____	_____	_____	_____
2. Staff and laborers	_____	_____	_____	_____
A. Staff salaries	_____	_____	_____	_____
B. Staff overhead/social costs	_____	_____	_____	_____
C. Laborers	_____	_____	_____	_____
3. Storage buildings	_____	_____	_____	_____
A. Cost/depreciation/rental	_____	_____	_____	_____
B. Maintenance and repairs	_____	_____	_____	_____
C. Power, electricity, fire protection, etc.	_____	_____	_____	_____
D. Security, watchmen, etc.	_____	_____	_____	_____
4. Labeling/identifying stored seed	_____	_____	_____	_____
A. Supplies	_____	_____	_____	_____
B. Supervisor	_____	_____	_____	_____
C. Workers	_____	_____	_____	_____
5. Insect & pest control	_____	_____	_____	_____
A. Staff & supervisor salary & overhead cost	_____	_____	_____	_____
B. Chemicals & materials	_____	_____	_____	_____
C. Other supplies	_____	_____	_____	_____
D. Fumigation covers & equipment/ supplies	_____	_____	_____	_____
E. Application equipment	_____	_____	_____	_____
i. Equipment cost/depreciation	_____	_____	_____	_____
ii. Equipment maintenance	_____	_____	_____	_____
iii. Equipment power/fuel	_____	_____	_____	_____
6. Re-stacking, handling seed in/out of storage	_____	_____	_____	_____
A. Equipment	_____	_____	_____	_____
i. Equipment cost/ depreciation/ rent	_____	_____	_____	_____
ii. Equipment maintenance	_____	_____	_____	_____
iii. Equipment fuel/power	_____	_____	_____	_____
B. Staff and laborers	_____	_____	_____	_____
i. Staff salaries	_____	_____	_____	_____
ii. Staff overhead/social costs	_____	_____	_____	_____
iii. Laborers	_____	_____	_____	_____
Other	_____	_____	_____	_____
Total Cost	_____	_____	_____	_____

Seed Transport and Distribution to Dealers Costs				
	Unit	Unit Cost	Units Used	Total Costs
1. Trucks, forklifts and equipment	_____	_____	_____	_____
A. Cost/depreciation/rent	_____	_____	_____	_____
B. Maintenance	_____	_____	_____	_____
C. Fuel	_____	_____	_____	_____
2. Drivers	_____	_____	_____	_____
A. Driver salaries	_____	_____	_____	_____
B. Driver overhead/social costs	_____	_____	_____	_____
3. Pallets	_____	_____	_____	_____
4. Labor for loading/unloading	_____	_____	_____	_____
Other	_____	_____	_____	_____
Total Cost	_____	_____	_____	_____

Minimizing Seed Transport and Distribution Costs

The objective of all seed operations is to sell seed to farmers. All required seed must be delivered efficiently to each dealer or distributor in time for him to sell it to farmer customers. In doing so, transport, handling and delivery must also be cost-effective. Guidelines for least-cost timely delivery include:

1. At beginning of the distribution season, ask each retail dealer or sales outlet how much seed is needed immediately. Deliver minimum amounts; then telephone them weekly (or more often if needed), learn current seed needs, and send a truck on a planned delivery route to supply additional seed to several dealers. Load trucks with each dealer's seed in a separate pile; the first dealer's seed should be the last seed loaded into the truck, etc.
2. Use the above system of several smaller deliveries as seed are needed. Do not sell seed on consignment, and do not accept return of unsold seed.
3. Keep trucks in top mechanical condition, with engines operating efficiently. Have access to a spare, in case of failure of one truck.
4. Use trucks of most economical capacity; avoid excess truck capacity, but also avoid using small trucks requiring several trips. Use trucks with a covered van type of body, to prevent damage and loss of seed in transit.
5. Load seed onto pallets, for easy movement from storage into trucks, and for easy unloading at the dealers.

6. Build (as universally recommended) the storage with its floor at truck bed height. Install loading ramps at storages so the forklift can safely and easily go from the storage into the truck, for quick stacking/loading in the truck.
7. Help dealers build safe seed storages, not only to protect seed quality while they have it, but also with facilities for quick unloading of the truck.

Chapter 8

Conditioning Plant Area and Buildings

Location

Raw seed will be brought in by trucks and other vehicles, and clean seed will be taken out by trucks and other vehicles. Seed conditioning and storage buildings should be located so that there is easy access to and from the main road.

The plant should be located near the city, to take advantage of city fire and police protection, and city utilities (water, sewage, electricity). At the same time, it should be located on the edge of the city nearest to most seed growers and farmer customers, to minimize the traffic faced by trucks carrying seed.

Locate storages in a dry, well-drained, clean area so the storage environment is dry and pest-free. The storage should not be in a crowded, polluted, or heavily-traveled area. All-weather paved access to main roads should handle the required flow of trucks.

The plant should not be located next to housing or industrial areas where dust or fumes may affect the seed plant environment, or where rats and birds are harbored and may create problems for the seed plant. Dust in the seed plant should be controlled, so that it does not present a problem to surrounding operations. The seed plant also should not be located next to a densely-populated area where there may be problems from nearby persons.

If the plant is located in a rural area, arrangements for dependable supply of utilities must be made. It may also be necessary to provide housing for plant staff and workers, if adequate other housing is not located within reasonable distances from the plant.

Access Roads

Access roads should allow trucks to bring in raw seed, and take out cleaned seed, without obstruction, and with easy access to loading/unloading docks. Land use should be minimal, but adequate parking for waiting trucks should be provided.

If a truck scale is used, it should be located convenient to the scale office (preferably in the main office for convenience of staff) and beside the road near the compound entrance.

General Area

The area around the buildings and in the entire plant area should be as dry as possible, well-drained and with no wet or damp areas or nearby operations which release dust, fumes, heat or moisture. No trash, weedy growth, operations, neighboring buildings, etc., should be nearby where rats, insects, etc., can be harbored and move to the storage.

The general conditioning plant area should be:

1. Dry, free of surface moisture and high subsoil moisture (still must have adequate drainage!).
2. Clean, free of brush and trash (both inside the operating compound and in adjoining plots) which attract/cover insects and rats.
3. Close to the main road, with easy access by trucks, without creating a disturbance in the neighborhood or encountering traffic problems.
4. With adequate space for required buildings, roadways, operations, etc., with enough separation (between buildings inside the compound, and from buildings outside the compound) to ensure good fire protection.

The area outside the buildings should be kept well-mowed, neat and clean. As much as possible, attractive landscaping should be established.

Security

The inventory of seed stocks, buildings, equipment and vehicles need to be given reasonable protection from intentional and accidental damages or losses. The entire plant area should have an adequate fence around it. The single entrance road should have a secure gate, and a guardhouse. Adequate outside security lights should be installed, and guards and security alarms adequate for local conditions should be in place.

Surface Around Outside of Buildings

For 2–2.5 m around the outside of the building (the distance of the roof overhang, plus space for a drainage), clear off the ground and pave it with concrete. This helps keep the area clean (to minimize attraction for rats and birds), reduces rat burrowing under the building, and helps drainage.

Put a slight outward slope to this paved ramp or walkway, so water drains away from the building. It is best to have an adequate grassed (and well-maintained) drainage ditch out beyond this paved area, to drain water away from the building.

Layout

Buildings must be carefully laid out in the plant area, so that:

1. The required operating sequence can be efficiently followed.
2. There is minimum but safe distance or barriers between consecutive steps in the operating sequence.

Buildings

A building is only a shell which protects the equipment, seed, work and people in it. The building should fit the needs, not the needs fit the building.

Any type of building construction—brick, metal insulated, etc.—is suitable, so long as it protects the seed, equipment and operations, and provides the required space.

Operations Protected in Buildings

A sizeable seed enterprise with storage and conditioning operations normally has different operations, all of which require protective housing in buildings. These operations include:

1. Offices for management, administration, field men, and work-area chiefs.
2. Quality control laboratory.
3. Receiving.
4. Drying.
5. Raw seed storage.
6. Conditioning.
7. Clean seed storage.
8. Shop and maintenance/repair work.
9. Vehicle, tractor and equipment storage (usually only a shed).
10. Service building with restrooms, which should include shower and locker facilities for staff.

Placing Buildings for Efficient Work Flow

Storage should be separated from conditioning areas so as to prevent/minimize entry of dust, insects, personnel, etc., and to provide a safe distance to prevent spread of fire. However, storage should be close enough to the conditioning area so that minimum time and cost are required to move seed to/from conditioning and/or storage. Often, storage and conditioning are in the same building, separated only by a fire wall and appropriate fire door.

The cleaning/bagging area of conditioning should be located near to the storage entrance, to minimize cost and time of moving seed to/from cleaning/bagging to storage. However, they should be sufficiently separated to permit (A) fumigation in the storage area, without interrupting or endangering the cleaning/bagging work; (B) clearly separated to facilitate management; (C) separated enough so that dust from conditioning does not enter the storage, and (D) unnecessary personnel can be kept out of the storage.

The usual operating sequence at a seed conditioning/storage plant is generally as shown in the flow sequence in Figure 8.1.

The units and staff (and the buildings they are normally in) which deal with the seed are shown in the operating flow sequence shown in Figure 8.2 and the layout shown in Figure 8.3.

Combining Operations into a Building

Not all operations require a separate building. Similar operations, especially if they are in a good operating sequence, can be housed in the same building. Planning can minimize the investment required for buildings. The main requirements for housing operations are:

1. Ensure adequate fire protection between different operations, such as between conditioning and storage. This can usually be ensured by providing a firewall between the two operations, with clean housekeeping and dust control.
2. Do not put an operation where it may endanger another operation. For example, do not put restrooms or other water-involving operations into storage or conditioning; also, do not put the shop with conditioning or storage, because of the potential fire hazard.

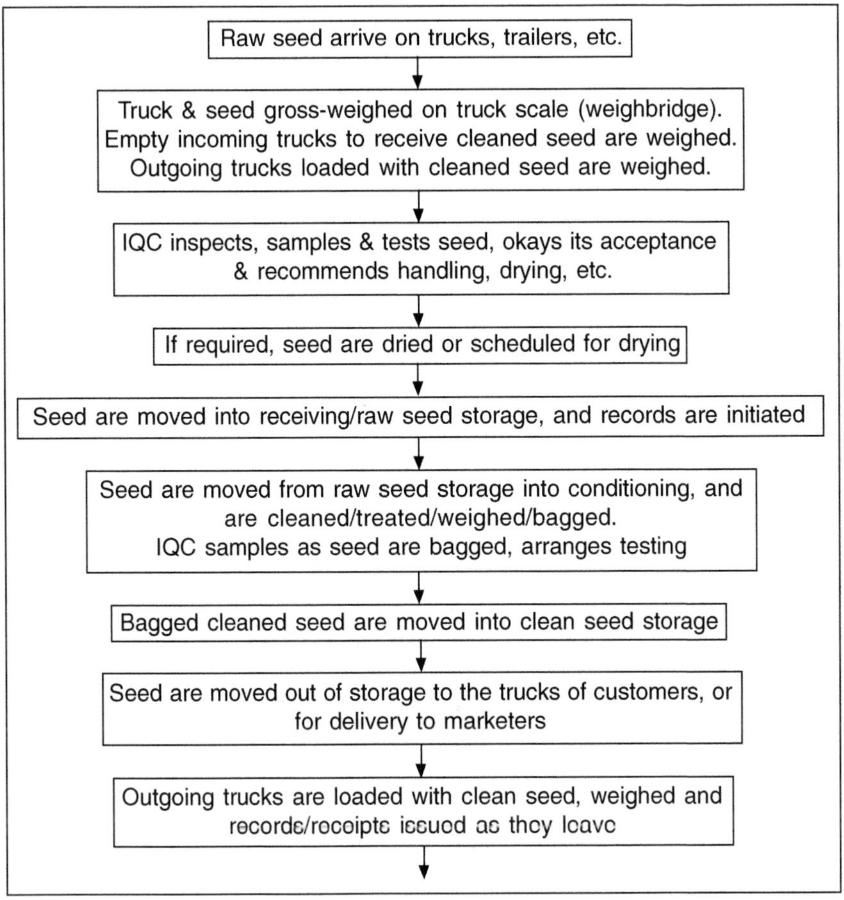

Figure 8.1 Sequence of operations from receiving raw seed until the cleaned/bagged seed are shipped out. Study of the work sequence is the first step in planning buildings needed and their placement.

Receiving, Drying, Conditioning and Storage

The receiving, drying, conditioning and storage operations are usually in adjoining buildings or sections of the same building, so that seed can be easily moved between them, with only sufficient separation to ensure fire safety, keep out dust, and permit unloading/loading ramps as required (Figures 8.4, 8.5 and 8.6).

There are many types of drying equipment and facilities. Some may be inside buildings, others under protective shed-like roofs, while some such as metal drying bins may be mounted entirely outside in the open. There should be a central control panel point, best placed in a small office.

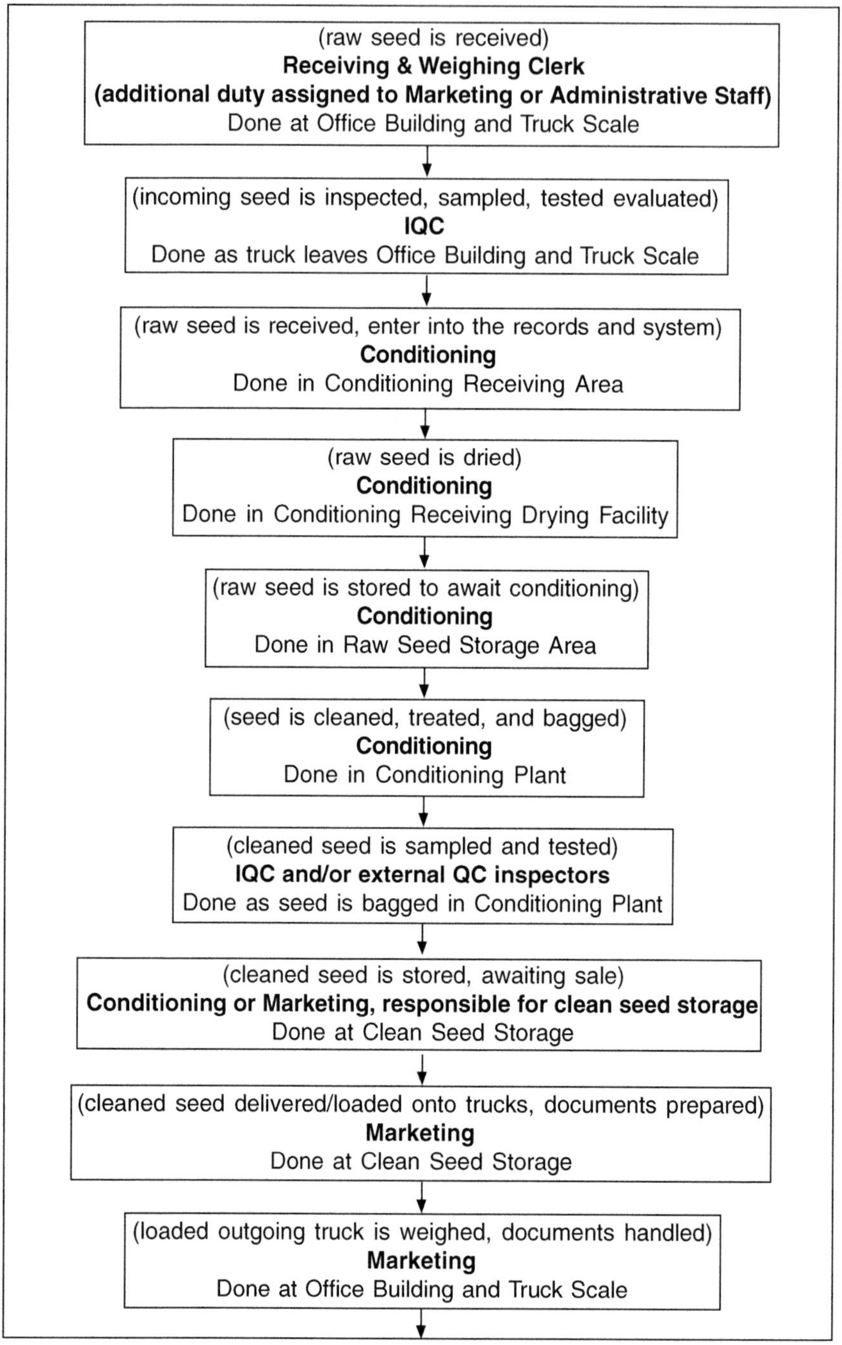

Figure 8.2 Units or departments responsible for, and work area of, operations in the seed handling sequence, and the locations or buildings where each operation is done.

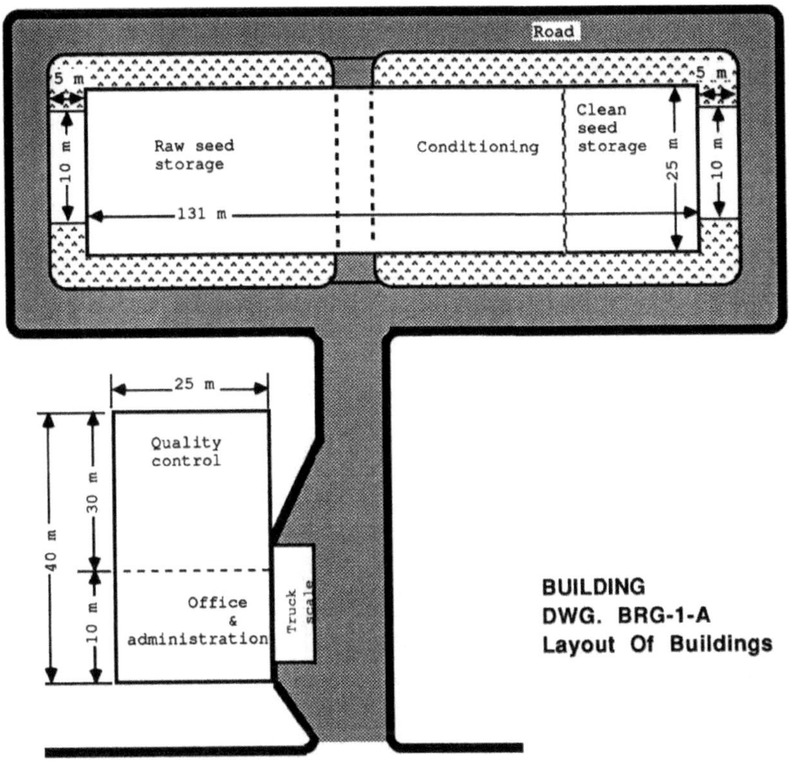

Figure 8.3 A typical layout of buildings in the seed conditioning plant area. While many different layouts are possible and efficient, depend on the size and shape of the plant area and on size of operations, the layout of buildings should permit efficient flow of the work sequence. Source: Billups and Lasheen unpublished designs.

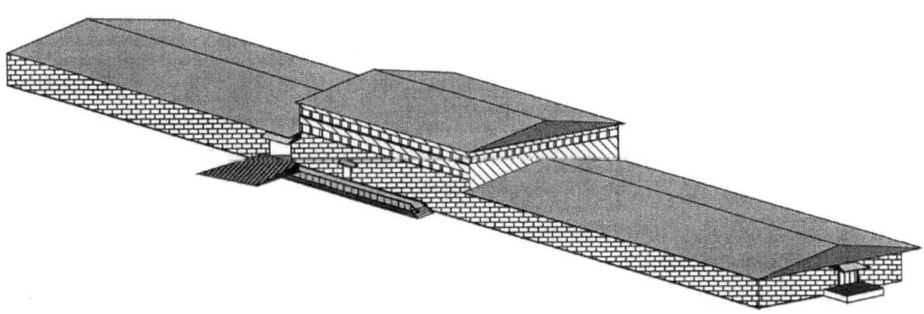

Figure 8.4 Isometric drawing of a seed conditioning building which combines receiving, raw seed storage, conditioning, and clean seed storage into one building with efficient work sequence flow. Source: Billups and Lasheen unpublished designs.

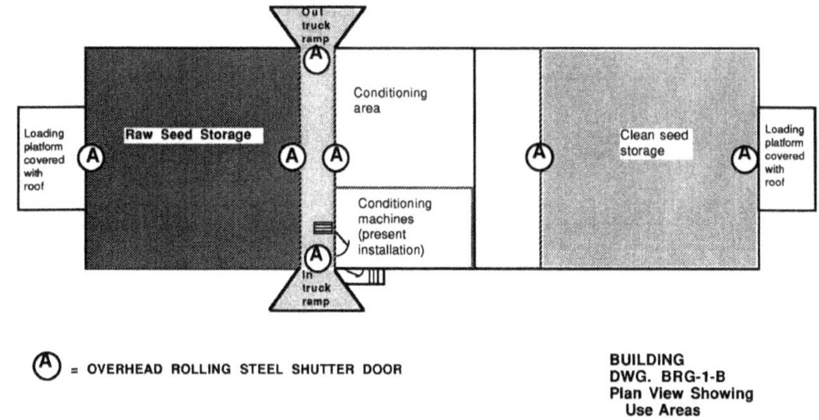

Figure 8.5 Floor plan of a conditioning building, showing adjoining but separated areas for raw seed storage, receiving, conditioning and clean seed storage. Note that trucks do not enter the storages; outsidle loading/unloading docks covered by roofs, provide truck access. Source: Billups and Lasheen unpublished designs.

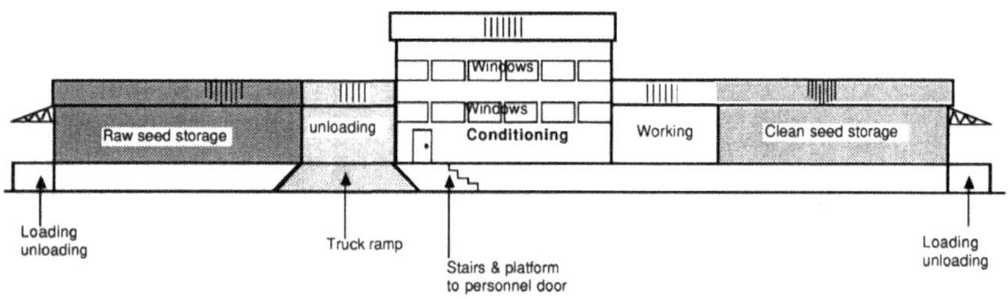

Figure 8.6 Front elevation view of seed conditioning building, showing use of different sections Source: Billups and Lasheen unpublished designs.

In the conditioning area, there should be a small office, to enclose the main control and process operating panel, and house operating records, small tools, etc. Keep in mind that the main over-ride switch must always be mounted at the machine, and that the operators must constantly monitor the machines themselves, and not remain in the office or other area.

Shop and Vehicle Shed

Shop and maintenance facilities are kept in a separate building, because of the potential for noise, water and fire. However, it should be as close as is safe to the conditioning building. A shed or building attached to the shop usually houses vehicles, tractors, etc. The restroom facilities are usually located in part of this building, to avoid having a source of water in the conditioning/storage building (Figure 8.7).

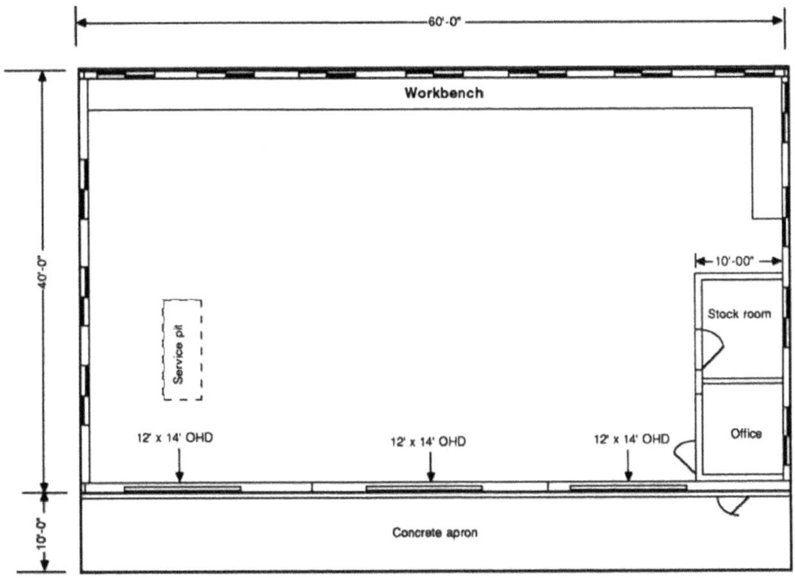

Figure 8.7 Layout plan of a workshop for a seed enterprise with major conditioning facilities. Vehicles and machines can be serviced and repaired here, and welding and similar construction can be safely done here without creating a fire hazard in seed handling or storage areas. Source: Billups and Lasheen unpublished designs.

Office and Truck Scale

The office building usually houses all offices, including management personnel, marketing personnel, field inspectors and other personnel. An important part of the office building is the reception area and sales/marketing counter where dealings with growers and buyers take place. The Internal Quality Control lab is usually in the office building.

Both incoming and outgoing seed should be weighed as part of the management control system. This usually requires a truck scale (often called a weigh bridge), where the loaded and/or empty truck is weighed. From this, the weight of seed material is determined and scale or weight tickets prepared as part of the permanent record and financial system. The truck scale has to be manned constantly during shipping and receiving seasons; it is more economic to locate the truck scale beside the office, so that a person in the office can do double duty and control the truck scale. Thus, the truck scale should be installed beside the office building, and the office building should be beside the single entry/exit road where trucks pass.

For outgoing cleaned seed, the shipping delivery door (usually the door at the opposite end from the door into the cleaning/bagging area) is located close to the road and is easily accessible to trucks, on the access road with a minimum requirement of paving and land used for roads. These are illustrated in Figures 8.8, 8.9, 8.10 and 8.11.

The storage office should not be included inside the storage building, because this is expensive and much-needed space. A small delivery/ receiving office or booth should be located in the entrance or just outside the storage, while the main storage records and manager's desk should be in another building such as the office building. Only a few authorized personnel should be allowed into the storage, where the enterprise's sale product is stored. Traffic in and out should be minimized.

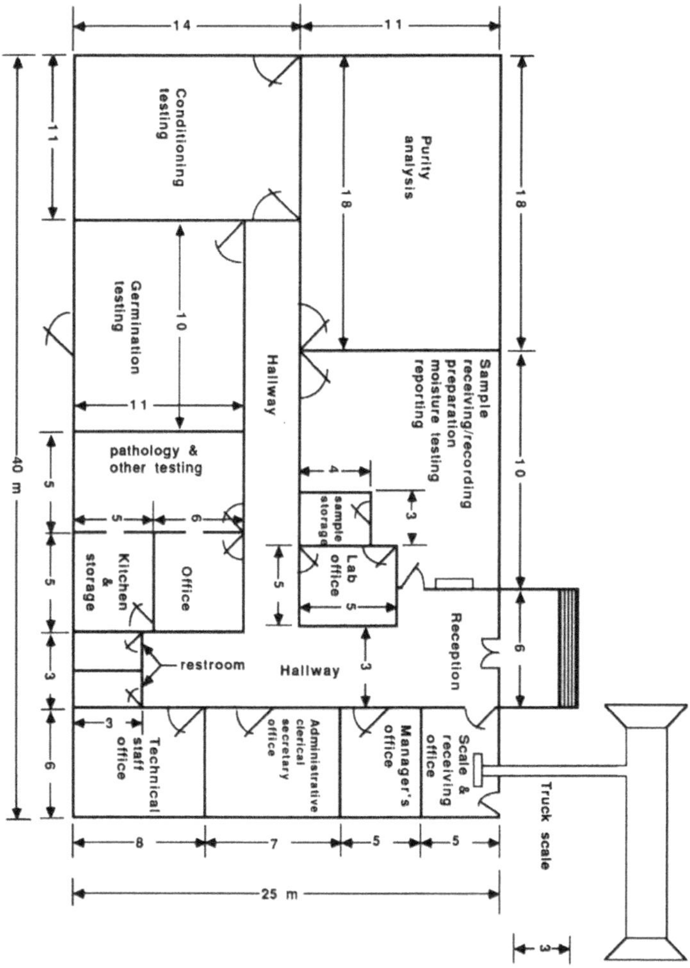

Figure 8.8 Floor plan of the office/IQC building and truck scale used in one seed conditioning enterprise. The dimensions of each room, and its use, are shown. Source: Billups and Lasheen unpublished designs.

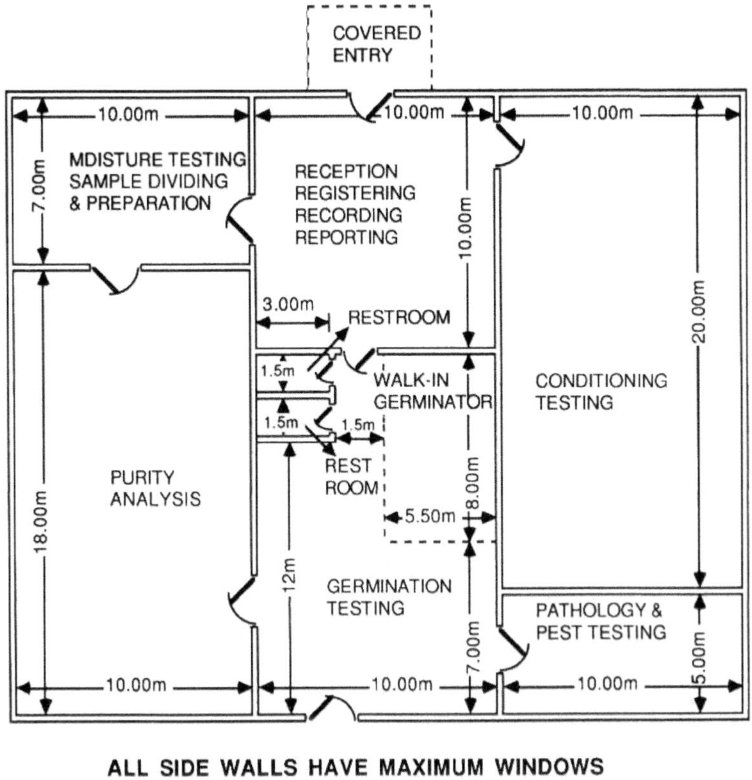

Figure 8.9 Another layout arrangement of the IQC lab in the administration building, or in a separate building. Source: Billups and Lasheen unpublished designs.

Figure 8.10 Isometric view of the same office and truck scale building of a seed enterprise. Note that for emphasis, the truck scale is shown at an elevation. It is installed at an elevation to minimize water collecting in the scale's mounting pit; however, the approach road (with a paved surface) is raised to the top level of the scale. Source: Billups and Lasheen unpublished designs.

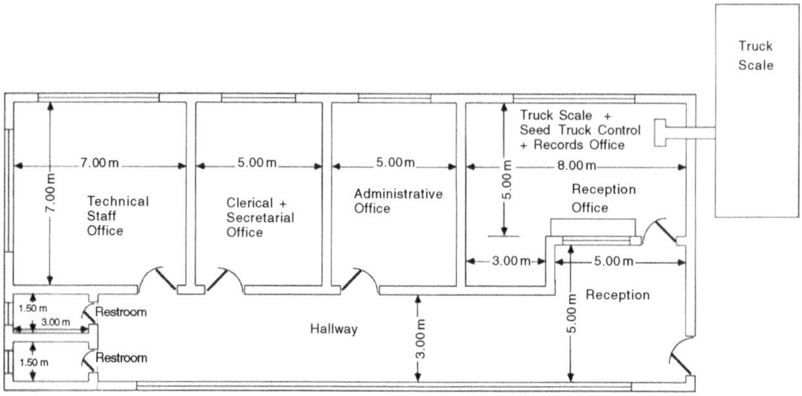

Figure 8.11 Layout of a seed enterprise office building with truck scale, but with IQC installed in a separate building. Source: Billups and Lasheen unpublished designs.

Restrooms and Water

No restrooms or other water-requiring facilities should be inside the storage. These should be located in a different building. **The only water** allowed in the storage should be a fire-control stand in each end, beside the door. These should be kept dry and in good condition, so as to minimize moisture within the storage.

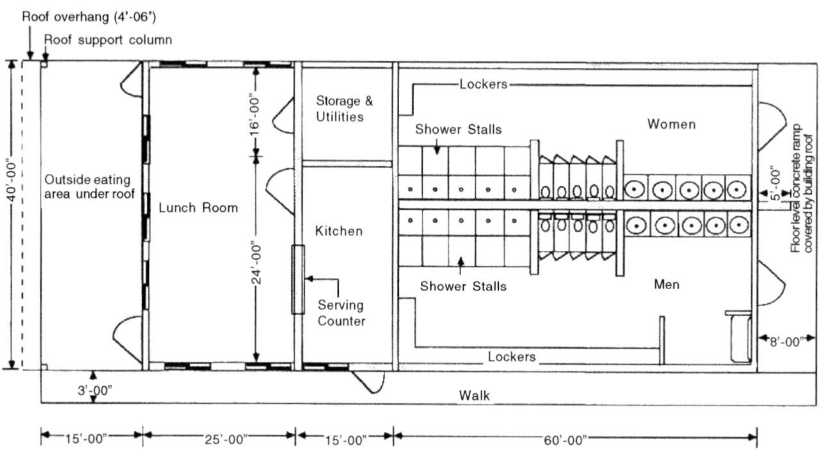

Figure 8.12 Layout of a service building at a larger seed conditioning plant. It includes showers and restroom facilities for men and women, a storage room, kitchen with food serving counter, lunch room, and an outside eating area under a roof. Source: Billups and Lasheen unpublished designs.

In the conditioning plant, water must be available near the treater. However, water inside both storage and conditioning areas must be strictly controlled to maintain a dry environment which protects seed quality and discourages rodents.

Restrooms may be located in a separate area of conditioning where water is not likely to affect conditioning or storage areas. These restrooms can serve both conditioning and storage. And, the restrooms must be constantly kept clean and dry (Figure 8.12).

Building Orientation

If the conditioning/storage building can be constructed so that its long dimension runs east-to-west (or west-to-east!), only the narrower ends will be most-exposed to solar radiation, which helps keep the interior cooler and thus safer for seed quality.

Chapter 9

Enterprise Technical Management

Profitable management includes organization and operation of both administrative and technical operations. There are specialized requirements for good management of technical operations, to ensure efficient operations, and these are covered in most sections herein.

Because of the close coordination required between seed production, conditioning, storage and marketing, their specialized management approaches all are discussed here, in terms of general guidance which can apply to a seed operation anywhere. More detailed management of conditioning is discussed in a separate chapter.

Management Requirements

Technical management of seed conditioning, or even a seed enterprise, involves:

1. Planning to decide what must be done, and how it is best done.
2. Ensuring smooth "flow" of the sequence of operations.
3. Preventing problems.
4. Minimizing costs in time, labor, and money.
5. Minimizing loss of good seed.
6. Ensuring high quality and purity of seed.

Basic technical management of seed conditioning, and of the seed enterprise, must ensure:

1. Varieties well-suited to local conditions and farmer needs.
2. Basic seedstocks which are varietally pure.
3. An efficient, quality-oriented production system.
4. Properly organized and equipped harvesting, to get seed from the field in good condition.

5. Drying, conditioning, treating and bagging to ensure high seed quality with minimum loss and operating cost and time.
6. Safe storage and transport which keep seed in good condition.
7. Adequate distribution and retail dealers to get seed to farmers.
8. And most important—trained employees who are highly motivated to turn out high seed quality and provide good service to farmers.

Components of Seed Enterprise Management

It is too easy to forget or overlook something! Management is to **organize so that nothing is overlooked!** To accomplish this, seed conditioning management must include:

1. Planning.
2. Guiding, coordinating and informing staff.
3. Records and reports.
4. Follow-up and supervision.
5. Analyzing operations to improve them continuously.

Planning

Management planning involves determining what you want to do, what you can do, how and if you can do it, and if it will be profitable. The planning stages include:

1. Market Survey and Marketing Plan
2. Financial Plan
3. Business Plan
4. Technical Plans

In the enterprise, technical plans which must be prepared and followed to achieve the business plan include:

1. Production Plan (Purchasing/Procurement Plan)
2. Conditioning Plan
3. Storage Plan

Prevent Problems

Management essentially consists primarily of preventing problems, which is most effectively done by:

1. Planning what must be done.
2. Planning what is required to do it.
3. Making arrangements in advance.

4. Guiding/directing staff carefully.
5. Monitoring the work.
6. Follow-up, follow up!

By preventing problems before they ever happen, good management achieves two savings:

1. The cost of the loss (in seed, time, operating costs, etc.) created by the problem, and
2. The cost of implementing remedial actions.

Seed Production

Use Records/Reports to Guide Management Decisions

1. Which growers or fields give high yields and are free of weeds, and can be used in the future.
2. Which seed will present problems in cleaning, so it can be discarded before money is spent on it.
3. What problems occur with each crop variety, or source of stock seed, to prevent problems in other fields, this year and in the future.
4. Problems with in each production area, and how they can be prevented. Or, determine the feasibility of moving future seed production to other areas.

Prevent Problems

1. Use the best-possible stock seed to plant seed production fields. This reduces roguing costs and lost fields and seed lots, and gives better-quality seed.
2. Select the best farmers as contract growers, and use only high-yielding fertile fields. This reduces costs for stock seed, cultivation, fertilization, weed control, and handling-management.
3. Select fields with no infestations of plant diseases or troublesome weeds. This prevents losses and problems.
4. Select growers near the conditioning plant, to minimize transport costs.
5. If climatic conditions are not a problem, select growers in the same area, to reduce transport costs. If weather (example, hail) may cause problems, spread fields over several areas so if one area is damaged you do not lose all your crop.
6. Train growers, and help them identify and rogue out undesirable plants.
7. Regularly visit all growers/fields, identify potential problems or needs, and help growers handle them properly.

8. Before harvest, help growers identify the proper stage to harvest, make sure they have the supplies (bags, etc.), equipment and labor to harvest seed to maintain high quality.
9. Check seed moisture content as it is harvested, and arrange immediate drying as required.
10. Inspect the seed fields, harvest facilities and operations, etc., to be sure they meet standards.

Conditioning

Use Records/Reports to Guide Management Decisions

1. Determine which seed, and how much, are ready for sale.
2. Determine costs of various components, and final costs of seed.
3. Identify components which need to be improved in operations and/or costs.
4. Determine if conditioning is adequate (in achieving seed quality, cost, or time required), or if changes or upgrades are needed.
5. Identify crops/varieties with conditioning problems, and means of remedying them.
6. Identify changes in production (season, location, field size, etc.) to improve cost/quality-effectiveness of conditioning and handling.

Prevent Problems

1. Analyze purity and test germination before cleaning, to see if you can clean the seed, or if it costs too much/lose too much to clean it. If so, don't clean it; dispose of it as grain.
2. Examine a small sample to see which screen sizes are best, and other cleaning needed.
3. Take seed out of storage only when you are ready to begin cleaning.
4. Thoroughly clean all machines before cleaning a new variety.
5. Frequently examine good seed and all waste products, to be sure cleaning conditions have not changed.
6. Analyze the seed after each cleaning machine, to see if it is doing its job, or is not needed?
7. Check seed for mechanical damage at all places. Install rubber "shock absorbers".
8. Label every bag with a lot number.
9. Keep good records.
10. As soon as seed are cleaned/bagged, move them into good storage.

Distribution and Marketing

Use Records/Reports to Guide Management Decisions

1. Who sold how much of what seed.
2. What seed should be produced next year.
3. Where seed was sold.
4. How seed was sold.
5. Which farmers bought seed.

Prevent Problems

1. Every sales place must safely store seed awaiting sale. You may have to loan portable insulated storage boxes.
2. Do not put too much seed at a dealer; this requires excessive stock and increases the amount of unsold seed which is lost.
3. Do not seed on consignment (*i.e.*, the subsidiary dealer only pays for the seed which is sold; you must take back the unsold seed). Get dealers to pay for seed, and give a large discount; this gives them a strong incentive to sell it.
4. During the sales season, contact dealers often, to find out if they need more seed. Or, a delivery truck can "make the rounds" between dealers once a week, with new supplies.
5. Restock each dealer before his supply is exhausted. Never leave a dealer without seed during the sales season.

Retail Dealer

Use Records/Reports to Guide Management Decisions

1. Make a detailed, realistic Business Plan, based on a realistic Market Survey.
2. Use your market survey/plan and Business Plan to guide what you buy/stock.

Prevent Problems

1. Buy small amounts, at several times, if you can.
2. Have a good storage at your store. DO NO let the seed get hot, wet, lose germination or get damaged.
3. Put seed into good storage quickly.
4. Keep seed in good storage, until the farmer buys it.
5. Let farmers see that you keep seed in good storage.

6. Do not stack seed on the floor or against walls.
7. Do not store seed with fertilizers, chemicals, fuel or pesticides.
8. Keep good records; if something goes wrong, use records to identify possible causes.
9. Have germination tested just before you buy the seed, or have it tested yourself.
10. Keep your shop CLEAN inside and outside, to repel birds and insects. Spray walls, floor, ceiling of storage with insecticides. Keep rat poison baits in place.

Stocking, Buying and Inventorying

Use Records/Reports to Guide Management Decisions

1. Help decide what to buy, how much, and when to buy it.
2. When/where to move seed, and where to concentrate sales efforts.
3. Plan partial purchases/deliveries at different times during the sales season, to minimize tying up working capital.
4. Quality and performance of each grower and supplier, and select future growers and suppliers.
5. At the end of the season, help identify sales trends to decide how much to produce the next year.

Prevent Problems

1. Based on the Marketing Plan, determine in advance how much of what can be sold.
2. Place orders for your needs as far in advance as possible.
3. Avoid orders in excess of what can be sold; this ties up your working capital, or results in a loss if seed lose germination. Note: a good safe storage is the most important asset your enterprise can have, to prevent loss of seed.
4. Be sure the seed you purchase are of high quality. Always have your own quality tests made, and do not patronize suppliers whose goods are low quality—regardless of their prices. You must consistently give farmers good value and high quality for their money.
5. If you do not have your own testing lab, arrange with the government seed testing lab to test seed for you.
6. Arrange delivery shortly before planting season begins, but be sure that you get delivery by the time you must sell seed. This shortens the time your working capital is tied up.
7. Move seed into safe storage as quickly as possible.

8. If you can get credit at no cost until the seed/inputs are sold, use it. But, if interest or a higher price is charged on credit purchases, pay cash on delivery to minimize costs.
9. If deliveries are dependable, take partial deliveries at different times during the sales season. Place a total order in advance of the sale season, so that you get the stocks you need. If you can "take an option" in advance and then purchase only what you actually can sell, this helps reduce expenses.

Chapter 10

Conditioning Organization and Staffing

Organization

The Conditioning Unit (or Section or Department) must be organized to ensure that all work is done properly and is cost-effectively coordinated. Its work includes:

1. Effective planning and management of the work.
2. Operating, adjusting and monitoring of operations.
3. Maintenance and repair.
4. Keeping records and reports and office work.

Organizational structure is usually simple, much as in Figure 10.1.

Mechanics, electricians, and carpenters are directly needed in the Conditioning Unit; however, they can also take care of all other equipment and buildings of the seed enterprise when not needed in conditioning, so the Conditioning Maintenance Unit handles all the maintenance work of the enterprise.

Storage

Receiving, drying, and raw seed storage should be the responsibility of the Conditioning Unit, so as to coordinate operations. Clean seed storage may be a responsibility of the Conditioning Unit or of the Marketing Unit. Control and operation must be coordinated so that Conditioning can move seed in, and Marketing can move seed out, while maintaining current and accurate records. In either case, the staff of the Conditioning Unit usually operate forklifts, etc., so they do the handling work of Storage.

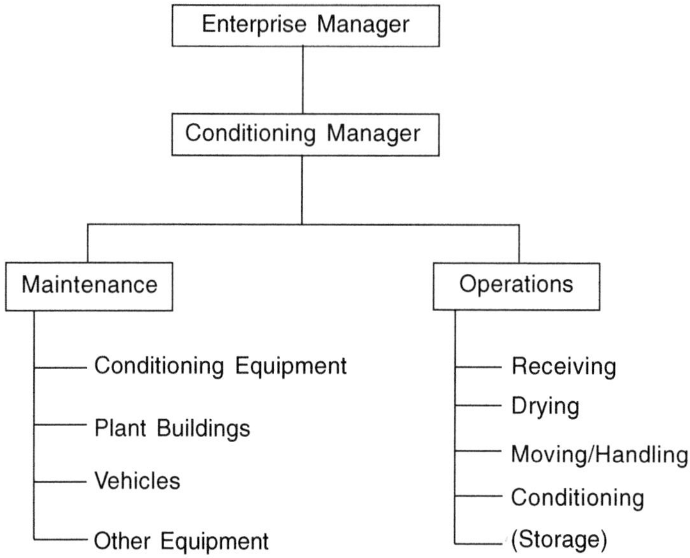

Figure 10.1 Organization of conditioning within the seed enterprise.

Personnel Requirements and Organization

The Conditioning Unit (or Section or Department) should keep personnel to a minimum number of competent and dedicated employees who can work as a team and are focused on getting out high-quality seed at minimum cost and time. Conditioning personnel usually include those shown in Figure 10.2.

Duties of Personnel

Manager: The Conditioning Manager (or Chief or Department Head) is responsible for all planning and management, keeping operations going on schedule, and supervision of all employees. Must be good at organizing, planning, motivating employees, and detailed monitoring and supervision.

Secretary-Records Clerk: Handles office work, telephone and fax, keeps records and maintains timely reports from the records. Must keep good records and be able to summarize them into reports, have drive and ability to take responsibility, and be able to work closely with all staff and get their input into keeping records complete and up-to-date.

Mechanic: Repairs and maintains machines, vehicles, etc., during work seasons, doing the heavier repair work which is not done by the operators as part of their regular maintenance. During off-seasons, plans, conducts/

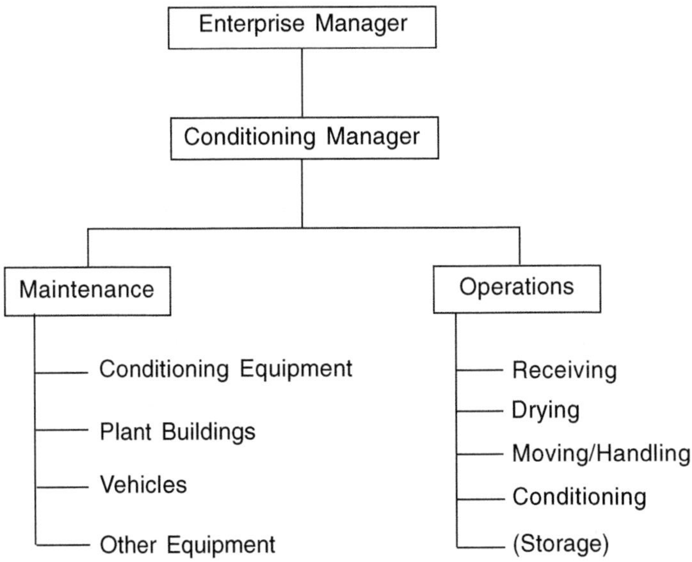

Figure 10.2 Staffing and personnel of a typical Conditioning Unit.

supervises the "deep" maintenance, ordering parts, and repair work to ensure that machines are in excellent condition for the working season. Must have good mechanical aptitude, be able to understand the workings of the machines, identify and correct problems. Welding skills are essential, and a detailed understanding of electrical repairs helps.

Electrician: Installs motors, lights, electric lines and systems, repairs problems, ensures that motors operate properly, and ensures proper power flow to all machines. Must be able to install lines, hook up machines and motors, install safety measures, ensure proper rotation of motors, etc.

The Mechanic and Electrician are often one competent person.

Shift Supervisors: When more than one daily operating shift (and/or weekend shifts) are worked, one experienced person is the Shift Supervisor, who keeps the work organized and running smoothly. The Shift Supervisor is often one of the operators who has considerable experience and the ability to organize and manage work. The Mechanic-Electrician are not shift supervisors (except in emergency situations), so that they can be free to handle emergency repairs.

Conditioning Operators: The operators are trained in adjusting, feeding, operating, monitoring, and light maintenance of the machines. Their responsibility is to constantly monitor the machines, adjust them so that seed quality is high and operations are cost-effective, to spot small mechanical problems with the machines and correct what they can and get the mechanic to correct what they can't handle. They also do routine machine maintenance and clean-up between lots, keep the treater supplied and operating, ensure that bags, bag weights and labels, etc., are correct.

Forklift Operators: Operate the forklifts to bring raw seed into conditioning and take cleaned seed into storage. In most plants, the forklift operators are also cleaner operators.

Conditioners: Move waste products as required, and handle seed and waste products, bagging, clean machines between lots, clean up conditioning and storage, do all required work, etc.

Conditioning operators, forklift operators, and conditioners are often the same persons, doing all the different kinds of work.

Permanent-Hire or Temporary Seasonal Employees

Personnel who have permanent jobs, get training each year, learn by experience and keep applying it year-after-year, and are made to feel a part of a team perform much better than temporary-hire or hourly-wage personnel. Especially in conditioning seed, it is important that all persons involved have some training and are quality-conscious and keep all details in mind to be sure the operations are properly done. Even laborers should be permanent-hire.

Although seed conditioning is a seasonal operation, the "down season" usually involves even more work, in getting the equipment and facilities ready for the next season. Also, there is always a significant amount of work of different kinds during the down season. There is enough work year-round to ensure cost-effective year-round use of employees. Keeping the same employees also generates loyalty to the enterprise, and they can continuously learn and apply their learning and experience to the detailed operations of ensuring high seed quality and cost-effective operations.

Chapter 11

Conditioning Management

Conditioning Operations

Conditioning is a stage of seed preparation. Seed conditioning—seed cleaning—is the stage in the seed supply sequence which:

1. Receives raw seed from producers, and organizes the received seed stocks.
2. Dries seed to low moisture contents safe for bulk/bag handling and storage.
3. Removes undesirable materials (trash, broken seed, weeds, other crop seed, etc.) from seed, to improve its mechanical purity.
4. Removes immature or damaged seed, to improve germination.
5. Fumigates/treats the seed to prevent damage by storage insects.
6. Applies fungicides to control seed-borne pathogens and protect seed and seedlings from soil-borne pathogens after the seed are planted. Sometimes, insecticides are also applied to prevent damage by soil insects after seed are planted.
7. The amount of seed needed by a farmer, or an amount convenient for farmers to use, is accurately measured into a protective bag or other package.
8. Seed are sampled and tested to be sure they meet or exceed the desired standards.
9. Each container of seed is labeled (tag and/or written lot number) to ensure maintaining its proper identity permanently. The labeled/written information allows tracing the bag back to a lot record, which can be used to trace the seed back to its origin.
10. Seed is then moved into a safe storage to await sale to the using farmers.

Much seed is given only a basic cleaning, often using hand methods, by growers who produce it. Also, much seed is not properly treated to safeguard it from insects and diseases. On the other hand, large (often excessively large and mechanized) conditioning plants are used in large seed programs.

An intermediate level is often best. In small enterprises, small amounts of seed can be cleaned by basic local hand methods. For somewhat larger amounts, a minimal investment in mechanical cleaners is often more efficient in cost (investment and operation) and time/labor required.

A small seed enterprise can clean seed by any combination of (1) local technologies done by contract growers, (2) custom cleaners, if for-hire cleaners operate in the local area, and (3) installing and operating its own small seed cleaning machines. This latter option often requires only a sun-drying system, a small air-screen cleaner, a simple treater, and a simple bagger-weigher system. A small minimal seed cleaning facility can often be cost/time/labor-saving, and can serve as a sales promotion tool.

Management Objectives

The critical management aspect is to get the seed cleaned up to high purity and germination in an efficient operation. The method of doing it depends on local conditions, but the cleaning process is still essential. Technical management of seed conditioning means planning and carrying out operations:

1. Cost-effectively, at the least cost.
2. Preventing cost-increasing problems.
3. Getting the best seed quality, in the required amounts.
4. Having seed ready for marketing at the right time.

This involves:

1. Plan for a smooth sequence of operations.
2. Prevent/avoid problems.
3. Have good records, and use them to guide decision-making.
4. Control costs.
5. Follow-up, monitor, check to be sure operations are done as required.

Operations Management

Management decisions in seed conditioning operations include:

1. Drying the seed to safe moisture contents before the seed heat or lose germination.
2. Fumigating insect-infested seed before serious damage occurs, or the insects spread to other seed lots or infest the storage.
3. Initiating records and lot identification numbers, to maintain identity and record control on all seed.

4. Sampling and testing seed lots to (1) determine what conditioning is required, (2) schedule conditioning so similar lots of the same crop variety can be conditioned in sequence to avoid the time and cost of equipment clean-up, (3) identify poor-quality lots so they can be discarded before conditioning, and (4) plan a conditioning schedule so the required amounts and kinds are ready when needed for sale, and (5) estimate expected amounts of cleaned seed, to determine total amounts which must be conditioned.
5. Cleaning up the equipment and facilities when changing varieties/ crops, to avoid mechanical contamination with undesirable seed.
6. Adjusting and using machines so as to remove all undesirable materials with minimum loss of good seed.
7. Applying the desired treatments at the desired rates.
8. Packaging accurate weights of seed into the desired bags, and sealing/ labeling them properly.
9. Handling the seed so as to prevent loss, deterioration, damage or losing identity.

Management Requirements

Management is based on leadership, organization, monitoring and expediting skills. Seed conditioning is managing a complex technical operation in a business-like fashion, and thus becomes more complicated. Good conditioning management requires:

1. Planning and scheduling conditioning so that all seed are ready when needed for marketing, and operating time/costs are minimal.
2. Organizing the staff, operations and work so they can proceed in an orderly, efficient and economic manner.
3. Maintain a minimal but adequate staff, and ensure that they are well-trained, efficient and dedicated.
4. Carrying out the required operations efficiently in time and cost.
5. Maintaining good records and up-to-date reports, that support operations and management decisions.

Monitoring and Follow-up

Follow-up of every operation is a critical management task. Regularly follow-up, monitor, check that operations are efficient and as required. All staff should be aware that their work will be followed-up. Be sure that each activity:

1. Is done right, properly and completely.
2. At the right time.

3. At the lowest cost.
4. And, the seed is ready for marketing when needed.

Management monitoring involves:

1. Plan all operations.
2. Keep a list of what has to be done and when it must be done.
3. Give complete orders to staff, so they fully understand.
4. Regularly, check to see how they are working. Regularly follow-up everything.
5. In a larger seed operation, be sure to use IQC for monitoring.

Planning and Scheduling

Planning conditioning operations must include all operations of the seed enterprise, as seed are delivered from production to conditioning, which must them clean and store the seed, and have it ready when marketing needs it (Tables 11.1,11.2,11.3 and 11.4).

Production (receiving) must coordinate with conditioning, get the seed ready and deliver it to conditioning in an organized, consistent manner which allows conditioning to identify and label the seed, keep it separate from other lots, initiate quality testing, dry the seed as required, and introduce it into conditioning as required.

Marketing (delivery) must advise conditioning as to when and how much of each kind of seed must be ready for marketing, and then accept delivery and handling of the cleaned seed.

Conditioning planning can then handle receiving, conditioning, delivering clean seed, keeping records and reports, and safeguarding seed quality. Written plans and schedules, prepared in advance, are the best way to ensure that nothing is overlooked. Planning worksheets, such as those shown below, can be used.

Always discuss plans with all staff, to get their input, help them understand what must be done, and to build team spirit.

Staff and Work Organization

Organization

Work should always be organized so that specific persons have specific responsibilities, are fully aware of these and trained to handle them efficiently. And, a manager should monitor all work to ensure that it is done properly and on time, and that different activities mesh together in an efficient manner.

Table 11.1 Receiving schedule and plan for raw seed coming in for conditioning.

Crop & Variety	Beginning & End Dates											
	Jan	Feb	Mar	April	May	June	July	Aug	Sep	Oct	Nov	Dec

Table 11.2 Conditioning operations schedule and plan.

Crop & Variety	Beginning & End Dates											
	Jan	Feb	Mar	April	May	June	July	Aug	Sep	Oct	Nov	Dec

Table 11.3 Shipping and marketing operations schedule and plan.

Crop & Variety	Beginning & End Dates											
	Jan	Feb	Mar	April	May	June	July	Aug	Sep	Oct	Nov	Dec

Table 11.4 Monthly conditioning operating schedule and plan for the month of ____________.

Date	Crop & Variety	Lot No. Shift	Working	Working Supervisor and Staff	Maintenance	Bags	Treatment	Remarks

More Than One Daily Shift

In busy seasons, it is usual practice to operate for more than one shift; for example, three 7–8-hour shifts or two 11–12-hour shifts. Time must always be allowed, either within the shift or after the shift, for maintenance. A mechanic should be on duty at all times, in case of breakdowns. When more than one daily shift is worked, the assistant mechanic may serve as the mechanic for the extra shift.

Shift Supervisors

Each shift must have a competent supervisor to monitor, guide and control the work to ensure operating efficiency and seed quality. An experienced and competent operator may be promoted to shift manager for additional shift(s).

The Conditioning Manager still has responsibility for overall operations, and must be on-call 24 hours a day during round-the-clock operations.

Workers/Conditioners

"Daily hire" labor almost never knows the operations and equipment, and has little or no loyalty to the enterprise or the drive to do the best work. It is always, in the long run, more profitable to have good facilities which require minimum labor, and then maintain permanent-hire workers who are trained and have the enterprise's interests at heart.

Plan labor needs in advance so that adequate workers will be in place to carry out all work efficiently. Planning includes number of workers, number of daily shifts, vacation time, allowing for sickness, etc.

Equipment, Spare Parts, and Supplies

During the off-season, all equipment should be overhauled and the entire facility brought up to excellent condition before the start of the cleaning season.

Preparing operating plans also helps identify the operations to be done, and helps:

1. Prepare an estimate of the spare parts which may be needed, and
2. Supplies (treatment, bags, tags, labels, sewing machine thread, etc.) to be needed.

This helps procure the needed materials in advance, so there are no delays during the busy season.

Operations

In advance, the conditioning operations should be planned, so as to handle raw seed when they are delivered, and then to ensure having them ready when they are needed for marketing to farmers.

Staff and operators should be well-trained so they understand at all times what must be done, and their individual responsibilities. Adequate monitoring and supervision must be provided.

Machines

Although operators are trained and trustworthy, the conditioning shift manager must constantly move about in the plant, to ensure that operators are truly closely observing the machines, and to check the general status of operations. It must be emphasized that seed lots sometimes vary in raw seed composition, and operating conditions (for example, strings from opened bags may plug a feed hopper) change. The operators must constantly monitor the machines, and the manager must monitor the entire operations.

Problems

As much as possible, prevent problems by good planning, management, prior preparation, and monitoring. Problems cost money in lost seed, time, profits, sales, and cost money in correcting the problem. If a problem happens, you lose 2 times: the loss itself, and the cost of remedying it.

To minimize potential problems, during the off-season all equipment should be thoroughly upgraded and brought to good condition. Advance preparations, to the maximum extent possible, should be done before the conditioning season starts. If a machine indicates that it may be less than 100% reliable, it should be replaced in the off-season.

During operations, operators should be trained to watch for unusual sounds or occurrences, so that problems can be immediately identified and rectified before they become serious.

When problems occur, the minimal part of the operations should be immediately shut down and the problem resolved. This requires having spare parts in stock, and a competent mechanic on hand.

Management Supervision

Seed Quality

The internal quality control program (IQC) is an essential management tool. IQC should, at regular and frequent intervals (such as every 3–4 hours) evaluate conditioning operations and sample/test seed to determine the efficacy of operations and seed quality. If an operation is below standard or unnecessary, IQC can order it to be changed or eliminated.

IQC should also test and evaluate every seed lot before it is conditioned. If the lot is of poor quality, it may be more cost-effective to dispose of it before spending money to clean it. IQC can also estimate what conditioning each lot requires, how much weight will be lost, and advise conditioning on what to handle the lot.

Initial Machine Settings

Previous records of similar lots, combined with tests and trials by IQC should be used to indicate what initial machines, components such as screens, and adjustments should be used when conditioning begins on a specific lot or variety and crop. This will usually save considerable time and labor.

Determining Proper Separation— Sampling During Conditioning

During conditioning, the clean seed and waste materials from each machine should be sampled every few hours, and examined to be sure that all undesirable materials are removed and excessive losses of good seed do not occur.

Stopping and Changing Conditioning

If less-than-desirable work is done at any time by any machine, it should be re-adjusted, or stopped and components (screens, etc.) changed. If an adequate holding bin is installed over each machine, the machines ahead of the stopped machine can usually continue running, with their seed output held in the stopped machine's holding bin.

Moving Cleaned Seed Into Storage

As soon as seed is properly cleaned, treated, bagged and labeled, it should be moved from the conditioning area into safe storage. Large amounts of cleaned seed should not accumulate in the conditioning area, as the bags may be damaged by operations, dust may collect on the bags, and other problems may occur.

Records

A Lot Record is normally established and maintained for each lot. This record continues the coded identification of seed from its production field, and includes detailed information on all operations during conditioning. It usually includes storage and sales information, so that this one record (without requiring additional or duplicate records) can follow the lot through storage and marketing. It becomes a permanent record on the lot, and is the first record consulted when information on the lot is required.

Complete records, showing in detail all operations, must be maintained for each seed lot conditioned. These records should show everything done to the lot, in a manner which enables later determination of the cost of each operation. And, records should be identified by lot numbers which allow tracing any bag back to its source, so that the cause of any problem can be identified and prevented in future work.

Technical records need more: they should show:

Everything that was done to each lot/variety of seed.
How much it cost to do each operation.

Reports

Reports are prepared by summarizing the data in records, and providing firsthand information on operations. Regular reports are prepared, to show how much seed of which crops/varieties were conditioned during the report period, and advise management and marketing on what seed is now ready for marketing.

Frequent reports should be prepared and provided to management, so they know:

1. What seed is conditioned, ready for marketing, and how much.
2. What seed is not ready—why not, and when it will be ready.
3. How operations are going.
4. Are we "on schedule" with the Conditioning, Business, etc., Plans?

Management should regularly hold meetings with all involved staff, to brief them on the current situation (as shown by the reports), and what must be done.

Analyzing and Using Records as Management Tools

Management can use the records and reports of conditioning to:

1. Determine which seed, and how much, are ready for sale.
2. Determine the costs of various components of conditioning, to determine final costs of seed.
3. Identify components of conditioning which need to be improved in operations and/or costs.

4. Determine if conditioning is adequate for the needs (in achieving seed quality, cost factors, or in time required to get seed ready), or if operations need to be changed or facilities upgraded.
5. Identify crops/varieties which present conditioning problems, and possible means of remedying them.
6. Identify possible changes in production (season, location, field size, etc.) which can improve cost/quality-effectiveness of conditioning and subsequent handling.

Chapter 12

Planning and Preparing for Conditioning

Conditioning is a complex operation of many activities which must be completed properly and in a specific, timely sequence. It is complicated by the fact that conditioning can begin only after seed is harvested, but seed must be ready in time to distribute to farmers before the planting season. Careful planning is essential, to ensure that nothing is overlooked or poorly executed, that seed quality is high, costs are kept as low as possible, and seed is ready when needed for marketing.

Planning Requirements

A detailed in-advance conditioning plan is essential for each working season. It must include all requirements for operating and managing conditioning work:

1. Getting equipment and all components such as screens into good condition before the conditioning season starts, to avoid breakdowns and reduced output.
2. Stocking spare parts which may be needed during the conditioning season, plus a reserve for unexpected occurrences.
3. Stocking adequate amounts of supplies, such as treatment, bags, lubricants, sewing machine thread, dryer fuel, forklift fuel, etc.
4. Planning the sequence of conditioning, to minimize down-time due to clean-up between different kinds and varieties.
5. Ensuring adequate workers and operators.
6. Ensuring adequate supervisors or shift managers, to have a responsible person in place at all times.

7. Ensuring that IQC is staffed, equipped and supplied so that it is ready to monitor operations and test seed at all times.
8. Ensuring that all staff and workers are well-trained and informed of what their duties are.
9. Ensuring that a conditioning plan or schedule is prepared and realistic, so that the required work is accomplished by the required dates.
10. Arranging delivery of seed from growers in an orderly fashion, which meets the plant's receiving, drying and handling capabilities. This includes ensuring that adequate trucks are available and in good condition.
11. Ensuring safe and adequate on-farm handling, labeling and drying of seed before it is delivered to the conditioning plant.
12. Arranging for the required tags and labels to be on hand when cleaned seed is bagged.
13. Arranging to handle required records and reports.
14. Coordinating with Marketing to move seed from storage to the distribution system, if required.
15. Arranging to handle dust and waste products.
16. Arranging for insect and pest control—supplies, fumigation, working space, etc.

Coordination With Production and Marketing

The **Production** Department or staff must ensure that seed is of high quality, properly dried and handled, kept uncontaminated and is properly identified, and is then delivered to conditioning in a manner which allows Conditioning and IQC to handle it safely. This requires scheduling on-farm handling and delivery, and sometimes the harvesting, plus ensuring safe on-farm arrangements.

A schedule of when each grower's seed should, or must, be delivered to the conditioning plant should be worked out in collaboration with the Production Department and growers.

The sole purpose of handling seed is to be able to sell it to using farmers. The **Marketing** Department or staff must prepare a sales plan, and advise the Conditioning Department or staff of how much of what kind of seed must be ready by what dates and in which places. This requires advance planning, which forms the basis of preparing the conditioning schedule.

Planning the Conditioning Schedule

As information from Production (on crop, variety, estimated amount, date of harvest, condition of seed, etc.) and Marketing (crop, variety of seed needed to distribute by what dates) is received, a conditioning schedule should be prepared. This is a major plan, and can determine whether or not conditioning is done on time.

The conditioning schedule should show what crop varieties/lots/ Certified classes will be conditioned in what sequence, and requirements for handling it. This is best-prepared in the form of a table, as shown in the chapter on Records.

To ensure timely and complete cleaning of the required seed, the conditioning schedule should show:

1. Crop.
2. Variety.
3. Estimated date to be received.
4. Estimated amount of seed.
5. Drying required.
6. Estimated time required for drying.
7. Raw seed storage.
8. Sequence and date of conditioning.
9. Requirements for conditioning (bags, thread, tags, treatment, labor, operators, etc.)
10. Estimated time required for conditioning.
11. Estimated final amount of clean seed, in bags and weight.
12. Estimated date ready for marketing.

As the conditioning schedule is put together, there should be constant interaction with Production and Marketing, to be sure that the conditioning schedule is realistic and fits the needs.

Clean-Out

The conditioning plant can be a major cause of admixture and contamination of otherwise pure seed. Every time that seed of a new variety, crop, or higher Certified class is conditioned, the entire plant must be cleaned up thoroughly **before** the new seed are conditioned.

The conditioning plan or schedule must allow adequate time for each clean-up of the plant, and an inspection by the Certification agency or the IQC.

Clean-up must include floors, bins, machines, elevators, conveyors, pits, and the entire system. Clean-up work and time can be reduced by proper construction of the building and floors, selection of equipment, construction of bins, clean-up equipment and facilities, and installation arrangements. However, in the typical medium-larger conditioning plant, at least 2–3 days (or operating shifts) are required, depending on number and experience of staff involved.

A major benefit of planning the conditioning schedule is that it helps schedule cleaning lots of the same seed kind in sequence without intervening clean-up, to minimize the time lost in clean-up between lots.

Planning Staff and Labor Requirements

Lack of required personnel can seriously reduce seed quality, or cleaned seed output. Arrangements, in advance, must be made to ensure that adequate supervisory staff, operators, mechanic (and electrician and carpenter, if required), and workers are available for each of the daily shifts to be operated. In the busy season, the plant usually operates for 18–20 hours per day, seven days a week.

The staff and labor schedule must allow for reasonable off time for all staff, depending on their requirements. And, in case of emergencies, replacements must be available. Extreme efforts to ensure workers safety pays off in this, and in other, areas.

Having the entire staff and all workers as fulltime employees has been most effective and in the long range, more cost-effective. By hiring them permanently, they can continuously receive experience and on-the-job training which makes them more efficient in cost and time. And, their loyalty to the enterprise also is a benefit.

Cross-Training Staff

An extremely effective management system has been to "cross-train" staff and workers; *i.e.*, each person is trained to do the jobs of other persons in closely-related activities. This provides a built-in reservoir of trained staff in case of unexpected needs.

Anecdotal experiences report that it is "better to hire mechanics and then teach them the required seed maintenance, repair and operations". This ensures better maintenance and having full staff who can identify potential machine problems and prevent them before they become serious.

Planning for Maintenance

During Conditioning

During conditioning, all machines must be kept in top operating condition to avoid breakdowns and ensure good separations at the best capacity. Preventing problems is the best and most cost/time-effective maintenance.

Maintenance during operations must be planned in advance. It consists of:

1. All operators should be trained and have good mechanical skills, and must constantly monitor all equipment. At any time when they hear a different sound or notice a very small variation in any part of any machine, they should immediately examine it, identify the cause, and correct the condition before it becomes serious.
2. Operators should be able to make small preventive repairs. An advanced-skills mechanic should be on the staff, and on call at all times. The operators should immediately notify the mechanic when anything "different" is noticed, and the mechanic and/or the operator should take care of it immediately.
3. Even when three daily shifts are operated, 2–4 hours should be allotted for lubrication and other preventive maintenance. The mechanic, with the assistance of operators, should spend the daily maintenance period in checking for loose bolts, loose belts, rattling loose parts, etc., and keep them in good condition. A daily complete check of all machines while they operate is needed.
4. Commonly-needed spare parts should be ordered in advance of the cleaning season, and kept in stock for immediate replacement as needed.
5. A maintenance schedule should be prepared and followed. The work done and spare parts used should be recorded in the maintenance schedule, and replacement spare parts immediately ordered. The schedule and summary of maintenance, following the models shown in the chapter on Maintenance, should be used in planning to replace machines as they age.

Off-Season

Management planning should ensure that during the off-season, major maintenance is done to bring all equipment into "nearly new" condition. This should include replacing all worn parts, repainting as needed, replacing old or outmoded or inadequate machines, changing installations, etc.

Stocking Supplies

Planning must identify all supplies needed, and order them far enough in advance to ensure that they will be available when needed. It can be extremely costly to have to shut down because there are no bags or seed treatment or other supplies.

Table 12.1 gives an example of advance planning to order and stock operating supplies in advance of the conditioning season.

Table 12.1 Conditioning supplies needed and to be ordered.

Crop & Lot No.	Item	Date Needed	No. Needed	No. in Stock	No. to be Ordered

Stocking Spare Parts

Every manufacturer provides lists of spare parts which are often replaced and should be kept in stock. When a machine is purchased, the recommended spare parts should also be purchased. Then, each time a spare part is used, a replacement part should be immediately ordered.

Table 12.2 shows a planning summary which can be helpful in planning spare parts provision.

Table 12.2 Estimated needs for spare parts.

Machine & Item	Source	Date Needed	No. Needed	No. in Stock	No. to be Ordered	Delivery Time

Chapter 13

Conditioning Records

Management Tool—Not Just a Historical Record!

Records are often laborious, done late, and too tedious to understand quickly. They become nothing more than a historical record, and have little usefulness in day-to-day management and operations. The manager often makes decisions based on little factual knowledge, and his management involves much guesswork.

Records are a historical record of what the enterprise does. But, this is a secondary and economically less valuable use. The real benefit of operating records is to provide up-to-date information which can be summarized and given to the manager and operating staff, to guide them before problems or losses occur, so operations can be managed and conducted so as to avoid unfavorable occurrences.

Seed Condition and Identity

To protect the enterprise and its customers, managers and responsible staff must know details of everything which happened to the seed, in the field and after harvest. Complete information helps ensure that all the proper things were done, and helps identify—and prevent recurrence—what happened and who was responsible when something goes wrong. The only way you can know what happened is to keep complete records.

Exact identity of each bag or container of seed allows you to know exactly what is in that bag, so you can be sure of giving customers what they want and need, and to protect the enterprise's reputation. To do this, you must be able to identify each bag and trace it back to its lot and production field. The only way you can do this is to keep complete records on each field, each seed lot from the field, and each bag in the lot.

Management and Operations

You are in business to earn a profit. The only way and enterprise can be profitable is to (1) sell what customers want at a price they can afford, and (2) operate efficiently and cost-effectively so expenses and losses are minimal, and profit is highest.

This requires complete information on all operations and seed. This information must be complete, easily understood, timely, up-to-date, readily available, and useful.

The only way to get such management information is to have a complete record system, with regular summary management reports with current information.

Operating Communication

In a larger operation which involves several or many people, the enterprise operates more profitably and efficiently when all concerned persons have complete knowledge of objectives, what is being done, how it affects their responsibilities, and how their performance affects the enterprise's operations.

Records (and reports based on them, to all concerned) are a critical part of the internal communications required to keep everyone operating at maximum efficiency—and to let the manager know they are doing this! Records and reports do not stay with the person or section doing them—they must go to all concerned.

Continuity of the Enterprise

Many managers of small enterprises "keep their records in their head". But—what happens if the manager has an accident and is unable to continue managing the enterprise? Without knowledge of details of the enterprise's operations, his family or associates cannot keep the business going—and thus may lose the income which has made their standard of living possible.

A good record system prevents this loss of continuity, and helps keep the enterprise going and providing the income needed by the associates and family of the manager.

Records Required

Every manager, and every staff member with responsibilities, requires current, up-to-date, complete but brief, easily and quickly understood, accurate information on what was done, what is occurring, how it is being done, and what its results are. The information can then be used to make decisions which help avoid problems and ensure maximum profitability.

This requires complete records, but not unnecessary information and tedious record systems which do not provide information quickly and easily. A record system should be easily kept, provide the required information, and enable quick reports (often, reports can be in the same form as the record).

What to Record

Keep a record of anything that affects:

1. The cost of seed or operations;
2. The identity of seed;
3. What happened to the seed;
4. The current status of the seed; and
5. The quality of the seed.

Who Keeps Records and Makes Reports

In most seed enterprises and operations, the manager himself will have a great input into maintaining records, and thus can compile reports which help him gain a more realistic appraisal of the progress and efficiency of the operations.

Financial records and reports are maintained and submitted by a trained accountant, or at least a bookkeeper who works under supervision of an accountant.

Technical records and reports should be maintained by the person responsible for the operations, if this does not create an impossible workload. In larger enterprises, a clerk can handle the records and reports (sometimes of more than one section or area of technical activities), under the close supervision of the person responsible for the activities.

Monitor Records and Operations

A primary responsibility of a manager is to monitor all operations under him, and be sure they are correct. Monitoring records/reports operations/preparation/submission helps get the desired information and avoid mistakes.

The manager—of the enterprise, or of conditioning operations—must monitor record-keeping and report-preparation as well as the actual operations, to ensure that everything runs on schedule, is correct, and provides good information.

Operating Logs/Registers

An excellent management tool is an operating log for each operation. This should record what was done, when/how it was done, who did it, and other useful information, in brief tabular form. Management can use these (and summary reports from them) to determine how efficient the operations are, to plan operations, to identify causes of problems, etc.

Suggested formats for operating logs/registers are shown below. Only a single page format is shown; the log/register is built up by adding more pages as operations continue. Each page is numbered, and each entry is given a line number. This allows each operation to be located in the log/register, helps tie the record system together, show what was done in the recorded operation, and the operations before/after the recorded operation.

Files

For each grower and supplier, a permanent file should be established, with each season/year's records kept as a single group in this file. This shows his long-term performance, quality and cost of his seed, etc. This helps enterprise management make reliable decisions on which growers and suppliers to use.

For each seed production field, a separate file should be established and kept separately until the harvested seed is delivered. At this point, this field file is placed in the file of the grower.

For each seed lot, a separate lot card and all other documents relating to the lot (test reports, sales, etc.) should be kept in this file. Maintain this file as an "active" file until all the lot is sold and should have been planted. In some countries, the seed law requires the seller to maintain records for three (3) years after all seed in the lot has been sold. At this time, this lot file can be placed in the grower/supplier file; if necessary to conserve space, all unnecessary documents can be discarded at this time.

Select the Records Needed

Don't overload yourself with paper work! Shown below are records which have been used to keep the kind of information that affects seed and/or operations, and thus which should be recorded and reported to the manager. Not every seed enterprise needs all these records. Select the basic records needed on operations. Then, as needed, add new records and reports.

Use Good Recording Staff and Computers

Having a record and report system does not make it useful in management. Time and effort are required to keep records current and prepare reports which help make decisions. The manager is usually too involved in operating the enterprise to keep all the required records. So, the question arises—Who keeps the records? In general, the person responsible for each activity keeps records, so they are complete and up-to-date, and remind the responsible person of what was done and helps him keep up with what must still be done.

However, the load of record-keeping may be so great—especially in larger operations—that keeping them would interfere with the responsible person's work. In this case, a clerk or other person may be assigned to keep records and make reports, by working closely with the responsible person. Often, in a smaller enterprise, one such clerk can keep all the required records. The manager's wife (or son or daughter) is often the record-keeper.

A good computer can enable the record-keeper to keep much more complete records in much less time. Some commercially-available software (such as financial management) can keep several records with minimal entries.

No matter how small the enterprise, a computer is a good investment, especially if it includes a printer and e-mail connections. In selecting a computer, never go for the minimum; be sure that your computer has enough features to give you full support and enable you to grow.

Records Must be Inter-Related

Records are inter-related, because they apply to different stages in handling the same seed. Further, the operating records are the basis which provides information for the financial records. The lot numbers and other record identification, are essential to permit accurate inter-relation identification of records, which in turn permits good financial and operating management of the enterprise.

Model Record Formats

Shown below are model formats for records, which have been useful to seed operations. Note that:

1. Most of the examples include only a few (usually 3–4) blank lines for information to be entered. In real-time use, these blank lines would fill the entire page of the record/report.
2. Modify the heading of columns or information blanks as needed for specific operations.

Planning Records

Conditioning work plan: Plans cleaning operations by crop and variety, to ensure having seed ready when needed for marketing, minimize down time for clean-out, ensure having bags and supplies ready when needed, arrange for labor, and other management responsibilities (Table 13.1). This record shows information for all planned seed to be conditioned: entry number, crop, variety, field/lot number, amount of seed, date to start conditioning, date to finish conditioning, estimated time required for conditioning, clean-up before and after conditioning, inputs/supplies/labor needed, and remarks

Operating Records

Receiving receipt: Provides the initial document/receipt of each lot of seed delivered to the conditioning plant for cleaning; initiates seed conditioning records, and provides the grower/deliverer a receipt (Table 13.2). This record shows, for the lot it concerns: crop, variety, field number, lot number assigned, crop year of production, identifying number put on containers, date received, number of bags or containers received, net weight, seed moisture content, signature of person receiving the seed, disposition of the seed, remarks, date receiving completed, total net weight received, and signature of person receiving the seed.

Sampling and pre-cleaning analysis: Shows quality components found in analysis of an individual sample of the raw, un-cleaned seed lot. Provides a basis for determining cleaning requirements and estimating cleaning loss; also used to determine if the seed is acceptable or should be rejected (Table 13.3). This record for each sample shows (for the lot it represents): crop, variety, field/lot number, date sampled, sample number, total amount of the seed lot, test date, purity, germination, test results (% found and kind of pure seed, other crop seed, weed seed, inert, damaged seed, insects, immature seed, other), cleaning recommendations to remove each kind of undesirable material, estimated cleaning loss in each step, estimated total cleaning loss, estimated clean seed out-turn, original and estimated final germination, and the involved persons.

Raw seed receiving log/register: Provides a log of all seed lots received, with basic information required to manage handling of each lot of the seed received (Table 13.4). This record shows a summary of all lots, including separately for each lot: page number in register, time period covered; and for each separate lot entry number, date and time, receiving receipt number, crop, variety, field number, assigned lot number, net weight, moisture content, disposition/storage, remarks, and identifies the person receiving each lot.

Drying operations log/register: Logs the operation of the dryer(s), and of the drying done to each lot of seed (Table 13.5). This record shows: register page number, time period covered; and for each lot separately, entry item number, date(s), hours of dryer operation, crop, variety, lot number, initial moisture content, final moisture content, any additional drying needed, and remarks.

Insect control log/register: Records details of the use of insect control (fumigation, treatment, spray, etc.) on different seed lots (Table 13.6). This record shows: register page number and the time period it covers; separately for each seed lot entry item number, crop, variety, insecticide treatment and rate of application, reason (insect pests) for treatment, beginning and end date of treatment, and results.

Cleaning operations log/register: Logs the work of the seed cleaning plant, with information on cleaning of each seed lot (Table 13.7). This record shows: register page number and time period covered; separately for each seed lot, it shows item entry number, start and finish dates, number of hours of operation, crop, variety, starting weight, finished clean seed weight, cleaning machines used, adjustments of each machine, treating and rate, clean seed bag used, bag net weight, total number of bags, and disposition or storage location.

Lot blending record: Records, with appropriate detailed information, the small lots which were blended into a larger lot (Table 13.8). This record shows: (and becomes part of the permanent lot record of the lots involved) crop, variety, date(s) of blending, final lot number, final net weight of the blended lot, number of bags blended, blending method, reason for blending; for each lot going into the blend shows lot number, amount, the % this sub-lot is of the final lot; % each of germination, pure seed, other crop seed, weed seed, and inert; and remarks.

Storage and Shipping Records

In some enterprises, storage and shipping are part of the duties of the Conditioning Department. In others, storage and shipping are handled by the Marketing Department. In either case, the following model records are useful.

Planning

Storage needs for planned operations: Summarizes expected seed to be stored, storage period, and amount/kind of storage required. Helps plan use of storage, and obtaining additional storage (Table 13.9). This record shows register page number and period covered; for each lot it shows separately the crop/input, variety/kind, amount, expected date received, storage needed, expected date taken out of storage, storage period, kind of storage needed, storage available, and remarks.

Planned shipping/sales delivery schedule: Gives details of planned shipping or sales of stored commodities; helps plan storage use, storage location of different lots, and arranging transport and labor/handling (Table 13.10). This record shows register page number and period covered; for each lot

Table 13.1 Conditioning work plan. Source: Gregg, Records and Reports. Unpublished.

Year/Season ____________ Operating Period From ____________ to ____________

No.	Crop	Variety	Field/ Lot No.	Amount	Date to Start	Date to Finish	Est. Time Rq'd	Clean-up Before/ After	Inputs/ Supplies/ Labor Needed	Remarks

Table 13.2 Receiving receipt. Source: Gregg, Records and Reports. Unpublished.

Receipt No. ______ Crop ________ Variety ________ Field No. ____________
Lot No. Assigned ______ Crop year ________ Ident. Put on Containers ________

Date	No. Bags/ Containers	Net Weight	Moisture %	Receiver's Signature	Disposition	Remarks

Date Delivery Completed ____________ Total Net Weight Received ____________
Signature of Receiver ________________________________

Table 13.3 Sampling and pre-cleaning analysis. Source: Gregg, Records and Reports. Unpublished.

Crop ____________ Variety ____________ Field/Lot No. ____________
Date Sampled ____________ Sample No.____________ Amount of Raw Seed ____________
Test Date: ____________ Purity ____________ Germination ____________

Cleaning Recommendations and Test Results:

Material	Name/ Description	% Found	Cleaning Recommendations	Estimated Cleaning Loss
Pure Seed				
Other Crop Seed				
Weed Seed				
Inert				
Damaged Seed				
Insects				
Immature Seed				
Other				

Estimated Cleaning Loss: % ____________ Estimated Clean Seed Out-turn ____________
Germination ____________ % Estimated Germination of Final Lot ____________ %
Sampler ____________ Analyst ____________

separately, it shows crop/input, variety/kind, amount, expected date shipping/selling starts, expected date shipping/selling ends, peak shipping/selling period, shipping method, amount expected to be picked up by buyers, and remarks.

Operating

Receiving receipt: Acknowledges receipt of seed into storage, and provides basis for following documentation and management of storage (Table 13.11). This record shows receipt number and date, who it was received from, type of containers; separately, item number, crop and variety or type of goods, field/lot number, bag size, number of bags, total weight received, disposition or storage, and remarks. It is signed by the person delivering the goods, and the person receiving them.

Shipping/delivery receipt: Documents removal of seed from storage; used to manage and verify disposal of stocks and inventory of stored materials (Table 13.12). This record shows date, delivery receipt number, sale/ship order number, to whom sold, or delivered; for each item it shows item number, crop/variety or other goods, lot number, number of bags or containers, total net weight, and remarks. It is signed by the person receiving the goods, and by the person who turns them over to the receiver.

Table 13.4 Raw seed receiving log/register. Source: Gregg, Records and Reports. Unpublished.

Page No. ________ For Period From ________ to ________

No.	Date/Time	Receiv. Receipt No.	Crop/ Variety	Field No.	Assigned Lot No.	Net Weight	Moisture Content	Disposition, Storage Location	Remarks	Receiver Initials

Table 13.5 Drying operations log/register. Source: Gregg, Records and Reports. Unpublished.

Page No. ________ For Period From ________ to ________

No.	Date(s) (-)	Hours Opern.	Crop/Variety	Lot No.	Drying Temp.	Initial % Moisture	Final % Moisture	Additional Drying Needed	Remarks

Table 13.6 Insect control log/register. Source: Gregg, Records and Reports. Unpublished.

Page No. ________ For Period From ________ to ________

No.	Crop	Variety	Treatment & Rate	Reason For Treatment	Date		Results/Remarks
					From	To	

Table 13.7 Cleaning operations log/register. Source: Gregg, Records and Reports. Unpublished.

Page No. ________ For Period From __________ to ____________

No.	Dates (-)	Hours Opern.	Crop/ Variety	Start Weight	Cleaned Weight	Cleaning Machines Used	Screens etc., Used	Treating Chemical/Rate		Bagging Net No.		Disposition or Storage
										Wt.	Bags	

Table 13.8 Lot blending record. Source: Gregg, Records and Reports. Unpublished. (becomes part of lot permanent record)

Crop ____________ Variety __________ Date(s) of Blending __________ Final Lot No. __________ Final Lot Weight __________

No. of Bags _____ Blended as: Raw Seed Lot _____ Cleaned Seed Lot _____ Blending Method ______________________________

Reason for Blending ____

Blending Components (Original Small Lots Going Into the Blended Lot):

Lot No.	Amount	% of Final Lot	Germ. %	Pure Crop %	Other Crop %	Weed Seed %	Inert %	Remarks

Table 13.9 Storage needs for planned operations. Source: Gregg, Records and Reports. Unpublished.

Page No. ____________ For the period ______________________

Crop/ Input	Variety/ Kind	Amount	Expected Date Received	Storage Needed	Expected Date Out	Storage Period	Kind of Storage Needed	Storage Available	Remarks

Table 13.10 Planned shipping/sales delivery schedule. Source: Gregg, Records and Reports. Unpublished.

Page No.__________ For the Period______________

Crop/Input	Variety/ Kind	Amount	Expected Date Start Ship/Sell	Expected End Ship/ Sell	Peak Ship/ Sell Period	Ship Method	Buyer Pick-Up	Remarks

Table 13.11 Receiving receipt. Source: Gregg, Records and Reports. Unpublished.

Receipt No.________ Date:________
Received From ______________________________
Type of Containers ______________________________

No.	Crop/Variety or Other Goods	Field/Lot No.	Bag Size	No. Bags	Total Net Wt.	Disposition	Remarks

Receiver ______________ Supplier/Deliverer______________

Table 13.12 Shipping/delivery receipt. Source: Gregg, Records and Reports. Unpublished.(sales invoice may be used instead of this, to avoid a second form)

Date__________ Delivery Receipt No. __________ Sale/Ship Order No.__________

Sold/Delivered to __________

No.	Crop/Variety or Other Goods	Lot No.	Bag Size	No. Bags/ Units	Total Net Wt.	Remarks

Receiver __________ Storage Operator __________

Table 13.13 Stack/bin card. Source: Gregg, Records and Reports. Unpublished.

Storage Location __________ Crop __________ Variety __________

Date	Amount In	Amount Out	Balance on Hand	Remarks

Table 13.14 Example of a different format for a storage stack card. Source: Gregg, Records and Reports. Unpublished.

STORAGE STACK CARD NO. __________

Storage location __________

Crop __________ Variety __________ Lot No. __________

Seed condition __________

Bag type __________ Bag net weight __________ Bags/pallet __________

Date	No. Bags In	No. Bags Out	Total Bags Remaining	Invoice, Order No.

Stack/bin card: Maintains current status of kind and amount of seed/ other sales products in a stack or bin (Tables 13.13 and 13.14). This record shows storage location identification, crop, variety, and for each date (of inventory or seed movement) the amount of seed moved in, or amount of seed moved out, balance on hand, and remarks.

Receiving log/register: Records details of commodities received into storage; helps manage storage and receiving operations, and provides data on what/when commodities come into storage Table 13.15). This record shows register page number and period covered; for each received lot, the crop (or input), variety (or kind), number of bags or units, total weight, date received, agency it was received from, expecting shipping/selling dates and period, location in storage, and remarks.

Table 13.15 Storage receiving log/register. Source: Gregg, Records and Reports. Unpublished.

Register Page No. ____________For the Period ____________________________

Crop/ Input	Variety/ Kind	No. Bags or Units	Total Wt/ Details	Date Received	Received From	Expected Ship Date/Period	Storage Location	Remarks

Table 13.16 Storage inventory. Source: Gregg, Records and Reports. Unpublished.
(use same form for daily, weekly, monthly inventory report)

Date ______________________ Inventory Record Page No. ______________________

No.	Lot No.	Crop/Variety or Input/Kind	No. Bags/ Units	Bag Net Wt.	Total Wt.	Germination %	Test Date	Storage Location	Condition/ Remarks

Table 13.17 Delivery/shipping log/register. Source: Gregg, Records and Reports. Unpublished.

Register Page No. ___________________________ For the Period ___________________________

No.	Lot No.	Crop & Variety	Shipped/ Sold to	Date	Amount			Ship. Receipt No.	No. Bags Remaining
					No. Bags	Bag Wt.	Total Wt.		

Table 13.18 Internal Quality Control testing schedule for stored seed. Source: Gregg, Records and Reports. Unpublished.

Schedule Page No. ____________ For the Period ______________________

No.	Lot No.	Crop & Variety	Year Grown	No. of Bags	Scheduled Re-Testing (date; X when tested; plus 0 if germination below standard)											
					J	F	M	A	M	J	J	A	S	O	N	D

Table 13.19 Quality of seed in storage and recommended disposition. Source: Gregg, Records and Reports. Unpublished.

Date ______________________

Date Tested	Crop	Variety	Lot No.	Amount	Quality	Recommendations (Discard, Keep for Sale, etc.)	Remarks

Storage inventory: Summarizes amount of sales commodities in storage on a given date, their location and condition/readiness for sale (Table 13.16). This record shows date of inventory and inventory page number; fore each item, as of the inventory date, it shows item number, lot number, crop, variety, number of bags or units, weight per bag/unit, total weight, germination percent and test date, storage location, and remarks.

Delivery/shipping log/register: Records date, kind and amounts of sales commodities removed from storage (Table 13.17). This record shows register page number and time period covered; for each item, it shows item number, lot number, crop, variety, shipped/sold to whom, date sold, number of bags and bag net weight and total weight sold, shipping receipt number, and number of bags remaining.

Internal quality control testing schedule for stored seed: Provides a reminder of dates each lot must be re-tested, and a log/register of quality of each lot on the required testing date (Table 13.18). This record shows schedule page number and period covered; for each lot it shows entry item number, lot number, crop, variety, year grown, number of bags, and date of scheduled retesting, with space for entry when tested.

Seed lot disposition recommendation: Prepared by IQC, based on seed quality (Table 13.19). This record shows date of report; for each lot, it shows date the lot was tested, crop, variety, lot number, total amount of seed presently in the lot, quality of the seed lot, recommended disposition of the lot, and remarks.

Seed Lot Permanent Record

One of the most important records is shown below, the permanent and all-inclusive record maintained separately for each seed lot. This record (Table 13.20) includes everything done to that particular lot, from planting the production field to selling the seed included in this lot.

This record is important because it:

1. Provides substantiated back-up data in case of customer complaints about the seed.
2. Provides information on the quality of seed from the grower.
3. Gives details of how certain contaminants were removed in conditioning, and thus provides information for cleaning future similar lots.
4. Gives estimates of the cleaning loss associated with cleaning the lot, and thus gives practical information for accepting or setting a price on similar lots in the future.
5. Gives information on the operation, and hours of operation, of each conditioning machine.
6. Provides a basis for planning future conditioning requirements.

7. Gives information on lot quality and cleaning requirements, to help select similar farmer/grower seed lots which can be conditioned in sequence without delays to clean-up the conditioning line.

Table 13.20 Seed lot permanent record. Source: Gregg, Records and Reports. Unpublished.

Crop ________ Variety ________ Lot no. ________ Record no. ________

1. *Production*

Grower ________

Crop Year Grown ________ Field No. ________ Field Record/Certif. No. ________

Stock Seed Planted ________ Class, Lot No. ________

Ha ________ Yield ________ Inspection Reports No. ________

(Potential) Problems Noted ________

2. *Receiving*

Date(s) ________ Receipt No. ________

Receiving Log/Register Page/Line Nos ________

No./Type of Containers ________ Net Weight ________ Moisture % ________

Sampling ________

3. *Pre-Processing Analysis of Raw Seed Lot*

Date ________ Moisture ________% Pure Crop Seed ________% Other Crop Seed ________%

Weed Seed ________% Inert ________% Shriveled/Broken/Damaged Crop Seed ________%

Estimated: Cleaning Loss ________%Germination ________% Undesirable Materials

Which Cannot be Separated ________

4. *Drying*

Pass No.	Date	Drying Time (hrs)	Initial % Moisture	Final % Moisture	Remarks
1					
2					
3					

5. *Insect Control*

Log Page, Line No.	Treatment & Rate	Reason For Treatment	Date		Remarks
			From	To	

6. *Cleaning*

Date(s) ____________________Opern. Hours ____________________
Proc. Log/Register Page & Line No.s. ____________________________

Machines Used	Settings, etc., Used	Weight of Waste	Remarks

7. *Blending*

Blended ___(Yes/No) Reason for Blending ____________________________________
Blending Components (Original Small Lots Going Into the Blended Lot):

Lot No.	Amount	% of Final Lot	Germ. %	Pure Crop %	Other Crop %	Weed Seed %	Inert %	Remarks

8. *Treating*

Treatment(s) Used ____
Rate ______________ Opern. Hours ______________ Label Applied ______________

9. *Bagging*

Date(s) __________ Opern. Hours ____________ Bag Type ______________________
Bag Net Weight ________ No. of Bags ________ Total Lot Bagged Net Weight ________
Bag Marking/Label ____

10. *Official Test*

Date __________ Lab ____________________________________ Test No. __________
Pure Crop Seed _____% Inert _____% Other Crop Seed _____% Weed Seed _____%
Other Varieties _____% Germination _____% Name of Other Seed ______________

11. *Storage*

Date Into Storage ____________ No. Bags ____________ Total Weight ____________
Storage Location/Description ____
____

12. *Sale & Shipping*

No.	Date	Sold/Shipped to	Sales Log /Reg No.	No. Bags Sold/Shipped	No. Bags Remaining
1					
2					
3					

13. *Carryover*

No. Bags __________ Net Weight __________ Carryover Storage ______________

Re-testing:

Date __________ Lab __________ Test No. __________ Germination __________%

Date __________ Lab __________ Test No. __________ Germination __________%

Date __________ Lab __________ Test No. __________ Germination __________%

14. *Losses in Storage and Handling*

Date	No. Bags Lost	Weight Lost	Cause of Loss

Chapter 14

Conditioning Reports

Management

A seed enterprise must be able to (1) sell what customers want at a price they can afford, and (2) operate efficiently and cost-effectively so expenses and losses are minimal, and profit is highest.

This requires complete information on all operations, whose form is complete but brief, easily understood, useful, timely and up-to-date so it can be used (1) to manage ongoing operations, and (2) know what has been done and what must still be done.

The only way to get such management information is to prepare, and give to staff, complete current reports which are based on information in a current and complete record system.

Management Tool and Operating Guide

Every person in the enterprise, down to the operators of the conditioning equipment, must have full knowledge of what is being done, how it should be done, and what the status is. Records are often historical records (which no one uses regularly!), and have little usefulness in day-to-day management of an enterprise despite the information they contain.

The manager, or every person and operator who does any kind of work must constantly make a decision on how to do his particular tasks. To be efficient and avoid costly mistakes, regular, timely and complete reports must be prepared and given to all who have responsibilities in the areas covered by the reports.

The real benefit of operating records is to provide up-to-date information which can be summarized into reports which are given to the manager and all persons involved and/or with responsibilities, to guide their operations and decisions before problems or losses occur, so that they can avoid unfavorable occurrences.

Post Current Data on a Chalkboard

On a prominent and constantly-accessible wall in the conditioning hall or area, install a large chalkboard. On this board, post pertinent information about the work which is currently going on—crop, variety, lot number, bag size, conditioning required, treating, required seed quality, time allowed for cleaning, etc.

Seed Condition and Identity

Each report must include exact and complete identity of the seed it covers, and give its condition of being cleaned, quality test results, treatment, etc.

Reports, and the records they are based on, make it possible to identify each bag and trace it back to its lot and production field.

Profitability

With up-to-date summarized and condensed but complete management reports based on real-time information, the enterprise's manager and staff can make realistic decisions based on real-life facts of each aspect of the operations. This enables making effective management decisions in time to maximize profits and prevent problems—thus helping keep the enterprise alive, serving its customers well, and expanding its sales and operations.

Operating Communication With Staff

In a larger operation which involves several or many people, the enterprise operates more profitably and efficiently when all concerned persons have complete knowledge of the objectives, what is being done, how it affects their responsibilities, and how their performance affects the enterprise's operations.

Reports are a critical part of the internal communications required to keep everyone operating at maximum efficiency—and to let the manager know they are doing this! Reports must go to all concerned.

Report Requirements

Each report should include current, up-to-date, complete but brief, easily and quickly understood, accurate information on what was done, what is occurring, how it is being done, and what its results are. The information should be adequate to support informed decisions which help avoid problems and ensure maximum profitability.

The record system should be of such a format that reports can be quickly and easily prepared. Often, reports can be in the same form as the record.

Reports Needed

Reports should always be in a summarized form to quickly give an overview of all aspects covered. However, each report should have enough detail to help make the management decisions required to keep work moving efficiently.

A regular, timely reports should be prepared on each aspects of conditioning.

Who Makes Reports

In most seed enterprises, whether small, medium or large, the manager(s) is(are) often too involved in the actual work of the enterprise to handle details of keeping records and summarizing them into management-oriented reports. Few seed enterprise managers are trained in financial matters and accounting sufficiently to enable them to maintain financial records as required by most tax laws.

Technical and operating reports, such as those covering aspects of seed conditioning, should be maintained by the person responsible for the operations, if this does not create an impossible workload. In larger enterprises, a clerk can handle both the records and reports (sometimes of more than one section or area of technical activities), under close supervision by the person responsible for the activities and using information from all responsible persons.

In small enterprises and operations, the manager himself will have a great input into maintaining records, and can compile reports which help him gain a more realistic appraisal of the progress and efficiency of operations.

Monitoring Operations With Reports

A primary responsibility of a manager is to monitor all operations, and be sure they are correct. Monitoring records/reports operations/preparation/submission helps ensure getting the desired information and avoiding accidental (and deliberate!) mistakes.

The manager—of the entire enterprise, or of only a section of its operations—must monitor record-keeping and report-preparation as well as the actual operations, to ensure that everything runs on schedule, is correct, and provides good information.

Operating Logs/Registers

An excellent management tool is keeping operating logs for each operation. This should record what was done, when/how it was done, who did it, and other useful information. Managers and responsible staff can use summary reports from the logs to determine how efficient operations are, to plan operations, identify causes of problems, etc.

Suggested formats for operating logs/registers are shown for operations where these are useful. Only a single page format is shown; the actual operating log/register is built up by adding more pages as operations continue. Each page is numbered, and each entry is given a line number. This allows each operation to be located in the log/register; this helps tie the record system together, show what was done in the recorded operation, and the operations before/after the recorded operation.

Report Formats

In using this guideline for good records and reports, please note the following:

1. Most of example reports shown here include only a few (usually 4) blank lines intended for information to be entered. In real-time use, these blank lines would fill the entire page of the record/report.
2. Sometimes, it may be useful to modify the heading of columns or information blanks shown in these records and reports. Headings given here are those found necessary in many seed programs and enterprises, and may not fit exact needs of one enterprise. Modify them in the way which makes them most useful to specific needs.
3. Many of the reports shown here in the sections for Planning, Operating or Reporting are the same in more than one of these 3 categories. In actual operations, there is no need to duplicate them; use one record/report for all.

What Records and Reports Do

Planning

Conditioning Work Plan: Plans cleaning operations by crop and variety, to ensure having seed ready when needed for marketing, minimize down time for clean-out, ensure having bags and supplies ready when needed, arrange for labor, and other management responsibilities (see Table 11.2, in chapter 11).

Operating Logs Used as Reports

These operating logs can be used directly as reports of operating status, and thus save unnecessary time spent in preparing separate reports.

Raw seed receiving log/register: Provides a log of all seed received, with basic information required to manage handling of the seed received (Table 14.1). This report shows register page number and period covered; separately for each lot, it shows item number, date and time of receiving, receiving receipt number, crop, variety, field number, lot number assigned to this lot, net weight, moisture content, disposition and/or storage location, remarks, and initials or name of the person receiving it.

Drying operations log/register: Logs the operation of the dryer(s), and of the drying done to each lot of seed (Table 14.2). This report shows register page number and period covered; separately for each lot, it shows item number, dates of drying, number of hours of drying operations, crop, variety, lot number, drying temperature, initial moisture content, final moisture content, any additional drying needed, and remarks.

Insect control log/register: Records details the use of insect control (fumigation, treatment, spray, etc.) on different seed lots Table 14.3). This report shows register page number and period covered; separately for each lot, it shows item number, crop, variety, lot number, insecticide treatment and rate, reason for treatment, start and end date of treatment, results, and remarks.

Cleaning operations log/register: Logs the work of the seed cleaning plant, with information on cleaning of each seed lot (Table 14.4). This report shows register page number and period covered; separately for each lot, it shows item number, begin and end dates of cleaning, total hours of operation, crop, variety, lot number, starting weight, cleaned seed weight, cleaning machines used, screens/adjustments used, treating chemical and application rate, bag net weight, number of bags, and disposition or storage location.

Reports

Weekly (or other needed period) seed cleaning report: Summary report of all seed cleaning operations, providing information on the machines used, basis for costing of seed cleaning, and information on the quality and amount of seed cleaned and made ready for sale (Table 14.5). This report shows begin and end dates covered by this report, person reporting; separately for each lot it shows crop, variety, date cleaning finished, lot number, number of bags, bag net weight, total net weight, total operating time, disposition or storage location, cleaned seed sample number, and remarks.

Table 14.1 Raw seed receiving log/register report of seed received. Source: Gregg, Records and Reports. Unpublished.

Page No. ________ For Period From ________ to ________

No.	Date/Time	Receiv. Receipt No.	Crop/Variety	Field No.	Assigned Lot No.	Net Weight	Moisture Content	Disposition, Storage Location	Remarks	Receiver Initials

Table 14.2 Drying operations log/register report of drying. Source: Gregg, Records and Reports. Unpublished.

Page No. ________ For Period From ________ to ________

No.	Date(s) (-)	Hours Opern.	Crop/ Variety	Lot No.	Drying Temp.	Initial Moisture %	Final Moisture %	Additional Drying Needed	Remarks

Table 14.3 Insect control log/register report of insect control operations. Source: Gregg, Records and Reports. Unpublished.

Page No. ________ For Period From________ to ________

No.	Crop	Variety & Lot No.	Treatment & Rate	Reason For Treatment	Date		Results/Remarks
					From	To	

Table 14.4 Cleaning operations log/register report of seed conditioning operations. Source: Gregg, Records and Reports. Unpublished.

Page No. ____________For Period From ____________ to ____________

No.	Dates (-)	Hours Opern.	Crop/Variety, Lot No.	Start Weight	Cleaned Weight	Cleaning Machines Used	Screens, etc., Used	Treating Chemical/ Rate	Bagging		Disposition or Storage
									Net Wt.	No. Bags	

Table 14.5 Weekly (or other needed period) seed cleaning report. Source: Gregg, Records and Reports. Unpublished.

Report for the Period From ____________ to ____________ Reported by ____________

Crop	Variety	Date Finished	Lot No.	No. Bags	Bag Net Wt.	Total Wt.	Operating Time	Disposition or Storage	Clean Seed Sample No.	Remarks

Storage and Shipping

Storage and shipping are often included in the responsibilities of the Conditioning Department, although sometimes these are activities covered by the Marketing Department. These reports are included in case the Conditioning Department handles these activities.

Receiving log/register: Records details of commodities received into storage; helps manage storage and receiving operations, and provides data on what/when commodities come into storage (Table 14.6). This report shows register page number and time period covered; separately for each lot, it shows crop, variety, lot number, number of bags, total net weight, date received, received from, expected date it will be sold or shipped out, storage location, and remarks.

Storage inventory: Summarizes amount of sales commodities in storage on a given date, their location and condition/readiness for sale (Table 14.7). This report shows date of inventory and inventory record page number; separately for each lot it shows item number, crop, variety, lot number, number of bags, bag net weight, total net weight, germination and test date, storage location, remarks and any comments on condition of the lot.

Delivery/shipping log/register: Records date, kind and amounts of sales commodities removed from storage (Table 14.8). This report shows registr page number and period covered; for each lot separately it shows item number, lot number, crop, variety, shipped or sold to, date of shipping/sale, number of bags, bag net weight, total net weight, shipping/sale receipt number, and total number of bags remaining in stock.

Quality of seed in storage and recommended disposition: Helps monitor performance of the existing storage facilities in terms of seed deterioration in storage, and helps prevent accidental sale of below-standard seed, or using storage space for unsalable lots (Table 14.9). This report shows date of the recommendation and separately for each lot it shows date tested, crop, variety, lot number, amount of seed, quality factors, recommendations for use/disposal, and remarks.

Table 14.6 Storage receiving log/register report of seed received into storage. Source: Gregg, Records and Reports. Unpublished.

Register Page No. ______________ For the Period ______________

Crop/Input	Variety/Kind Lot No.	No. Bags Or Units	Total Wt/ Details	Date Received	Received From	Expected Ship Date/Period	Storage Location	Remarks

Table 14.7 Storage inventory report as of______________. Source: Gregg, Records and Reports. Unpublished. (use same form for daily, weekly, monthly inventory report)

Date ______________ Inventory Record Page No. ______________

No.	Lot No.	Crop/Variety or Input/Kind	No. Bags/ Units	Bag Net Wt.	Total Wt.	Germination		Storage Location	Condition/ Remarks
						%	Test Date		

Table 14.8 Delivery/shipping log/register report of seed disposition. Source: Gregg, Records and Reports. Unpublished.

Register Page No. ______________ For the Period ______________

No.	Lot No.	Crop & Variety	Shipped/Sold to	Date	Amount			Ship. Receipt No.	No. Bags Remaining
					No. Bags	Bag Wt.	Total Wt.		

Table 14.9 Report of quality of seed in storage and recommended disposition. Source: Gregg, Records and Reports. Unpublished.

Date ____________

Date Tested	Crop	Variety	Lot No.	Amount	Quality	Recommendations (Discard, Keep for Sale, etc.)	Remarks

Chapter 15

Maintenance

Modern conditioning plants contain motors, machines, elevators, instruments, and many other mechanical and electrical devices that will give longer trouble-free service if they are properly maintained. Equipment manufacturers provide manuals that give detailed instructions for proper lubrication and maintenance for each equipment item. These manuals should be readily available in the plant for each machine, and persons responsible for operation should be familiar with essential maintenance procedures.

Other equipment such as vehicles and forklifts also require regular maintenance. Buildings and other infrastructure require maintenance and repairs.

All facilities—conditioning equipment, vehicles, buildings, etc.—wear out with use and must ultimately be replaced. For this purpose, the accounting system includes a depreciation allowance.

The seed enterprise and seed conditioning facilities require repair and maintenance of motors, machines, electrical installations, wood structures such as building frameworks, steel structures such as bins and supports, and vehicles.

Maintenance Organization

A basic internal repair and maintenance staff is essential; an enterprise cannot afford the cost and delay of bringing in repairmen from outside agencies. Because conditioning deals with machinery and buildings, it is usually more efficient to include the maintenance program in the Conditioning Unit, so that the maintenance staff has constant involvement with the part of the seed enterprise which has the greatest need for maintenance work.

A special maintenance section may be organized, but is usually not needed; the mechanic and other maintenance/repair staff are simply staff within the Conditioning Unit.

Maintenance Staff

The Senior Mechanic is the head of the maintenance operations, and is responsible for organizing and carrying out maintenance and repairs, and supervising other maintenance workers. He usually does the majority of the work, especially the more technical aspects. The mechanic(s) should be able to do mechanical work on seed conditioning machines, welding, vehicle repair, and general metal work. For cost-effectiveness, the internal maintenance staff should be able to handle all work except for major overhauls of vehicles, which can be done by outside workshops.

There are also usually an additional mechanic (who may be responsible for specific work such as vehicle maintenance), a carpenter for wood work, and an electrician capable of handling motors, wiring installation, switch and control installation and maintenance, etc. These staff can also do double duty as conditioning operators during the peak seasons.

The ideal situation would be to have a Senior Mechanic supervisor and one or more Mechanics, all of whom can handle mechanical repairs, electrical work, carpentry, welding, and other work, as well as double as conditioning operators during peak seasons.

Personnel Responsibilities for Maintenance

Responsibilities for different maintenance activities must be specifically assigned.

1. *Daily maintenance*: The simple aspects of daily maintenance (cleaning, lubricating, adjusting, tightening, etc.) should be the responsibility of the equipment operators. They should have more experience and "feel for the equipment", and be able to do this level of maintenance work effectively. The daily lubrication and other maintenance activities should be done at the end of the operating shift by the operator, under the supervision of the mechanic.
2. *General machine monitoring for repair needs*: The operators who have regular contact with the machines should monitor their machines, listen for different sounds, watch for problems, etc., on a constant basis. The operators do not sit in a control room or break room; they should constantly move around their machines, checking their work and observing their operation. This allows them to catch any problem immediately, and correct it before it becomes serious.

3. *Weekly and monthly maintenance*: This should be done by the mechanic, assisted by the operators.
4. *Annual or off-season maintenance, repairs, installation changes, etc.*: Major maintenance, changes in installation, new machines, repainting and renewing, etc., are usually done in the off-season. This work is done by the Mechanic(s), with assistance from the operators.

Operator Maintenance

Operators are key to good maintenance of all seed cleaning, conveying and handling equipment. They must ensure that machines work properly, downtime is minimal, and no machine is allowed to run with something which can become a serious problem. They have 3 major responsibilities:

1. Constantly observe all machines, listen for unusual sounds and watch for anything out of the ordinary.
2. Immediately stop machines and make minor adjustments and corrections before the problem becomes serious.
3. As an ongoing activity, contact the mechanic for immediate repairs which something outside their capabilities occurs or more serious maintenance repairs are required.

Maintenance Facilities

The enterprise **must** have adequate tools and facilities to carry out the required work. All required power tools and hand tools must be immediately available when maintenance work must be done.

An adequate workshop must be maintained. To minimize fire and other hazards and work interference, the workshop should be in a separate building, outside the conditioning and storage areas. . However, equipment such as saws, welders, etc., should be movable so they can be moved to the conditioning or other areas for repairs, especially during the off-season when most major repairs are made.

Maintenance Scheduling

Maintenance is an ongoing need. To ensure complete and timely maintenance, a regular maintenance, inspection and repair schedule must be established for all equipment. The schedule must be formalized to ensure proper implementation, and responsibilities must be specifically assigned to personnel, and follow-up records and reports required. For each machine, a schedule of frequency of each maintenance task should be established and followed (Table 15.1).

Table 15.1 Equipment maintenance schedule. Source: Gregg, Records and Reports, unpublished.

Machine:
Person Responsible:
Note: In column showing frequency, check for daily; for others, put day of week/month/year date it should be done.

No	Action	Daily or Each Use	Weekly	Month-ly	Annual
1		X			
2		X			
3		X			
4					
5					
6					
7					
8					

Timeliness of Maintenance Work

Immediately upon identifying a potential problem or maintenance need, the machine should be stopped and repaired as required. This will minimize the down-time of the conditioning operation, and usually results in less cost.

Spare Parts

Ordering and receiving spare parts always takes time. To minimize down-time, an adequate stock of spare parts should be maintained. Most equipment manufacturers provide lists of most-frequently-needed parts, and under local working conditions, experience may show that other parts are often needed. These should be kept in stock in the seed enterprise's workshop.

As parts are used, new parts should be ordered from the equipment manufacturer or other established sources.

A record of all spare parts needed and maintained for each machine should be kept up-to-date, to facilitate re-order of parts and to provide an indication of the cost of operating the machine (Table 15.2).

Obtaining Spare Parts

To ensure having ready access to spare parts, contacts should be made with suppliers who service each machine used. Information on sources of spare parts should be fully maintained and kept up-to-date, to avoid confusion when parts must be ordered. Some parts which are commonly used, such as bearings, may be found in local markets. An example of a Spare Parts Source Register is shown in Table 15.3.

Table 15.2 Spare parts record. Source: Gregg, Records and Reports, unpublished.

Machine, Make and Model No.:
Serial No. & Date Purchased:
Electricity:
Source of Parts:

No.	Part Name & Number	No. used	No. in Stock	No. Ordered
01				
02				
03				
04				
05				
06				

Maintenance Records

Maintenance and repair records are an essential part of management. They not only give management a detailed account of the cost and time required to keep the equipment operating; such records also are a good evaluation of the staff's ability to maintain and operate their equipment safely.

Equipment and building maintenance records include:

1. *Vehicle annual maintenance and repair log*: Maintains a complete yearly log of all maintenance and repairs, and costs, on a single vehicle. Shows item and total annual costs, and helps determine need for replacement (Table 15.4).
2. *Building annual maintenance and repair log:* Maintains a complete yearly log of all maintenance and repairs on a single building. Shows item and total annual costs, and helps determine need for replacement (Table 15.5).
3. *Equipment annual maintenance and repair log:* Maintains a complete yearly log of all maintenance and repairs, and costs, on a single piece of equipment. Shows item and total annual costs, and helps determine need for replacement (Table 15.6).
4. *Equipment annual maintenance schedule:* Maintains a yearly schedule of the kind of maintenance which should be done on a single piece of equipment, by month of the year (Table 15.7).

Table 15.3 Spare parts source register. Source: Gregg, Records and Reports,unpublished.

No.	Machine	Serial No.	Name & Address of Parts Supplier & Contact Person	Fax No.	Tel. No.	E-mail

Table 15.4 ______________ vehicle annual maintenance and repair log. Source: Gregg, Records and Reports, unpublished.

Year of Log ______________ Vehicle & Kind. ______________ Serial No. ______________
Use ______________ Year Manufactured ______________ Year Purchased ______________ Cost ______________

No	Date	Maintenance/Repair	Parts Used	Total Cost	Down Time	Remarks
		TOTAL				

Table 15.5 ______________ building annual maintenance and repair log. Source: Gregg, Records and Reports, unpublished.

Year of Log ________ Building ________ Use ________ Serial No. ________ Year Obtained ________ Cost ________

No	Date	Maintenance/Repair	Materials Used	Total Cost	Down Time	Remarks
		TOTAL				

Table 15.6 ____________ equipment annual maintenance and repair log. Source: Gregg, Records and Reports, unpublished. (use a separate form for each equipment item. Maintain all in a Maintenance and Repair Log Book)

Year of Log ____________ Equipment & Model No. ____________ Serial No. ____________
Use ____________ Year Manufactured ____________ Year Purchased ____________ Cost ____________

No	Date	Maintenance/Repair	Parts Used	Total Cost	Down Time	Remarks
		TOTAL				

Table 15.7 Maintenance schedule. Source: Gregg, Records and Reports, unpublished.

Machine:
Person Responsible:
Note: When maintenance is done (other than daily), put date done and initials of person doing it.

Maintenance	J	F	M	A	M	J	J	A	S	O	N	D

Maintenance and Repair Reports

Management should get regular periodic maintenance and repair reports on all equipment and buildings, as a basis for management decisions on planning replacements. Reports can use the same form as the individual item maintenance/repair log. In a large enterprise, a special form can be used to request replacement of an item, but in small/medium enterprises with good communication between workers and managers, this is not necessary.

Planning Replacements

The depreciation record for each machine, plus an evaluation of the maintenance cost, down-time experienced, and present condition of each equipment item, must be considered. Ultimate replacement of each equipment item must be considered, and must be planned well in advance. When a machine is replaced in the conditioning line, it should be replaced during the off-season, and while the machine is still operating, so that conditioning operations are not upset or delayed.

When an entire conditioning line or plant reaches the point where it should be replaced, this should be done in a manner which permits the old plant to operate cost-effectively until the new plant is fully installed, "shake-down" operations are conducted, and it is brought into on-line, continuous-operation condition.

Depreciation

As equipment and buildings age, their value declines and the cost of maintenance and repair increases. Accounting systems have established a system, called depreciation, for calculating the life and cost of buildings and equipment items, as a financial system of determining when the economic life of an item has been fulfilled.

A depreciation and replacement system must be based on accurate records, which include:

1. *Asset register*: Provides an accounting/management summary of all assets and their depreciation schedules (Table 15.8).
2. *Annual summary of depreciation of fixed assets*: Provides a summary of all depreciation charges (Table 15.9).
3. *Buildings depreciation summary schedule*: Summarizes depreciation schedule of all buildings (Table 15.10).
4. *Equipment depreciation summary schedule*: Summarizes depreciation schedule of all equipment (Table 15.11).
5. *Vehicle depreciation summary schedule*: Summarizes depreciation schedule of all vehicles (Table 15.12).

Table 15.8 Asset register. Source: Gregg, Records and Reports, unpublished.

Date	A/C Code	Description	Cost Additions	Cumulative Cost	Location	Ref.	Depreciation Amount Per Year	Depreciation To Date	Current Book Value

Table 15.9 Annual summary of depreciation of fixed assets, 200_. Source: Gregg, Records and Reports, unpublished.

COST/VALUATION	Land	Buildings	Plant and Machinery	Furniture and Fittings	Computer Equipment	Motor Vehicles	Total
Balance as of last year New additions this year							
Balance as of date (this year)							
DEPRECIATION Balance as of date last year Charge for this year Period							
Balance as of date (this year)							
NET BOOK VALUE As of date (last year)							
As of date (this year)							

Table 15.10 Buildings depreciation summary schedule. Source: Gregg, Records and Reports, unpublished.

No.	Date	Building	Year Constructed /Bought	Cost	Years Depr. Schedule	Salvage Value	Annual Depr. Cost	Year to be Replaced	Current Book Value	Remarks
		TOTAL								

Table 15.11 Equipment depreciation summary schedule. Source: Gregg, Records and Reports, unpublished.

No.	Date	Machine	Year Bought	Cost	Years Depr. Schedule	Salvage Value	Annual Depr. Cost	Year to be Replaced	Current Book Value	Remarks
		TOTAL								

Table 15.12 Vehicle depreciation summary schedule. Source: Gregg, Records and Reports, unpublished.

No.	Date	Vehile	Year Bought	Cost	Years Depr. Schedule	Salvage Value	Annual Depr. Cost	Year to be Replaced	Current Book Value	Remarks
		TOTAL								

Maintenance Tips

The following are general guidelines to maintenance which is important to the seed conditioner but are not included elsewhere herein.

Fuel-Operated Motors

Fuel-operated motors operate equipment such as forklifts, trucks, emergency generators, etc. These should be lubricated and maintained according to the manufacturer's recommendations in the equipment operator manual In addition, the air filter system should be serviced regularly to ensure a ready supply of clean air to the combustion chamber. Plugged air filters make starting difficult and reduce efficiency and life of the motor.

Always follow the manufacturer's recommended maintenance procedures! General maintenance precautions include:

1. Check and tighten all nuts and bolts regularly.
2. Check oil level in crankcase frequently.
3. Clean air filters regularly.
4. Clean sparkplugs and adjust spark gap according to the manufacturer's instructions.
5. Be sure fuel tank, fuel lines and carburetor are clean.
6. Use only clean fresh fuel. Avoid using old fuel, as water may form in it.
7. Adjust the idler and main fuel jets so that the motor runs smoothly without excessive exhaust smoke.
8. Clean trash and dirt from finned areas around cylinders and heads, and from the entire motor, to ensure clean efficient air cooling.

Electric Motors

Electric motors operate on single-phase or three-phase current, at different voltages according to the system used in the country. In a conditioning plant, all motors and electrical fixtures should be dust-tight explosion-proof, to reduce the danger of dust-caused explosions. Small machines usually use fractional horsepower single-phase motors, while large machines use three-phase motors of differing horsepower output.

Always follow the manufacturer's recommended maintenance procedures! General maintenance precautions include:

1. Check line voltage, or have a dial volt meter installed to show the line voltage. Where line voltage fluctuates frequently, install a voltage stabilizer between the line and the motor.
2. Be sure all connections are correct and motors are turning in the proper direction.

3. When electric power goes off in the main line, it may come back on with reversed polarity. After power goes off and then comes back on, check each and every motor to be sure it is turning in the proper direction. Paint on each motor a prominent arrow to show the proper direction of rotation.
4. Check and follow the manufacturer's instructions for proper lubrication of motor main shaft bearings.
5. Be sure all wiring is installed by licensed electricians, can carry the required power load, has no potential shorts, and is dustproof.
6. Use the safe recommended types of motor starters for dusty locations and the type of motor.
7. Check motors regularly to ensure that all connections are tight.
8. Remove accumulations of dust and trash regularly.
9. Safely shield all belts and pulleys.
10. Switch off the main power supply before making any electrical repairs.
11. Have a safety over-ride shutoff switch at each machine, to ensure worker safety during cleaning or maintenance of the machine. Do not have the over-ride switch at the main control panel.
12. Be sure all motors are properly fused and properly grounded.

Conditioning Machinery

Basic maintenance procedures are listed under the technical description of most machines. Always follow the manufacturer's recommended maintenance procedures!

Some items to look for in purchasing a new machine are the design facilities for maintenance. These include pre-lubricated sealed bearings, central greasing points, etc. (Figures 15.1, 15.2).

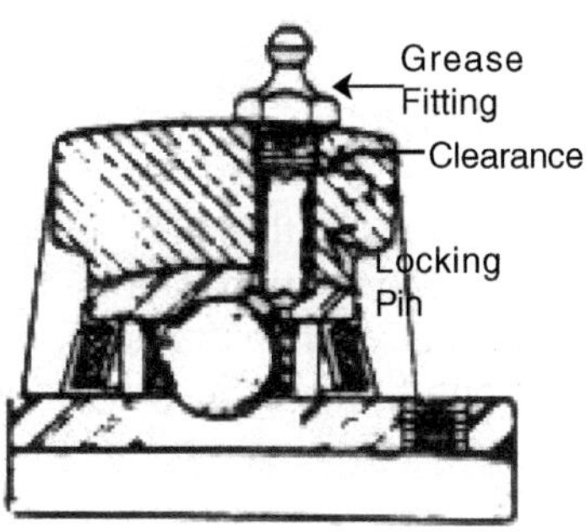

Figure 15.1 Cross-section of a sealed bearing. Source: Sealmaster, Morse Chain Division, Borg-Warner Corp. promotional materials.

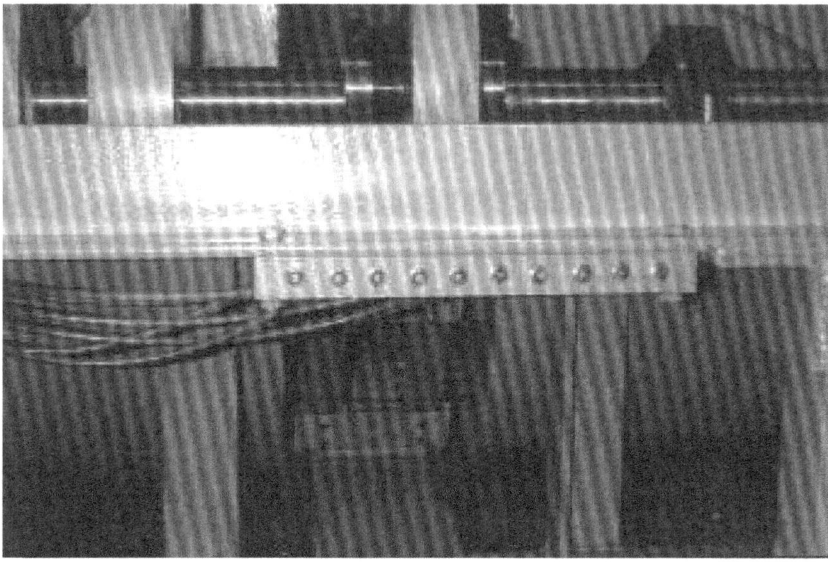

Figure 15.2 Central "grease console" on a Crippen gravity separator. All fittings are located here; tubes carry grease from the fittings to each bearing which requires grease. Source: Crippen International, Inc. descriptive literature.

Bearings

Most bearings come from the factory lubricated with a good industrial grease. Some bearings are sealed and require no lubrication. Lubrication intervals for other bearings depend on the use, speed, environmental conditions, etc. In general, one recommendation is to re-lubricate bearings every 6 weeks with a medium-weight industrial grease. Do not over-grease, so as to avoid breaking the grease seals.

Gear Boxes and Gear Reducers

In general, follow instructions on the name plate. Lubricate with a high-grade petroleum-based rust/oxidation inhibited 90-weight gear oil. Under normal operating conditions, change the lubricant every 2,500 hours or 6 months, whichever is soonest. Drain the gear reducer, flush with kerosene, clean the magnetic drain plug, and refill to the proper level with new lubricant.

Do not overfill, as this causes over-heating. Do not underfill, as this causes gear failure.

Check the oil level vent plug regularly, and be sure they are open and clean.

Belts

Regularly, check belts for wear and proper tension. Before checking them, be sure the over-ride switch is off and that no one will turn the machine on during maintenance. When checking belts, also check sheaves, and replace worn sheaves to ensure that belts fit properly and give the required operating speed.

Remove the belt drive guard to inspect the belts and sheaves. Adjust to a "reasonable" belt tension; if it is too tight, it may damage bearings. Slack belts wear excessively, cause slippage, and deliver less power. If required, adjust sheave position so that the belts run straight.

Dust, rust and oil damage belts; be sure belts are kept clean and dry.

Replace belts which show wear. When replacing multiple belts (two or more belts on the same sheaves), used matched belt sets, as they will run smoother and last longer.

Always replace belt guards after completing the maintenance.

Vehicles

All vehicles—cars, trucks, forklifts, tractors, etc.—should be maintained on a regular schedule. The schedule should include daily, weekly, monthly, mileage, and annual check-up lists for preventive maintenance, and immediate identification and repair of all problems identified in the check-ups or during use.

Always follow the manufacturer's recommended maintenance procedures! As with other equipment, a maintenance checklist should be established and followed, such as the widely-used maintenance checklist shown in Table 15.13.

IQC Testing Equipment

Always follow the manufacturer's recommended maintenance procedures! General maintenance procedures are not listed in the IQC chapter, so available basic requirements of this delicate equipment are listed here. Generally:

1. (A) handle equipment gently and carefully, (B) use as per instruction manual, and (C) keep clean, dust-free and dry.
2. Most equipment is factory calibrated. For repairs/maintenance beyond minor maintenance for small calibrated equipment, carefully pack the equipment and return it to the factory for repair/maintenance and recalibration.
3. Develop, and add to, Maintenance Guidelines as possible.

Table 15.13 Vehicle maintenance checklist.

Maintenance Component	To check	Frequency of check	Date checked, condition
LUBRICATION:			
Engine	Engine oil filter		
	Engine oil change		
Transmission	Transmission oil change		
	Transmission oil filter		
	Differential/transaxle lubricant change		
	Chassis lubrication		
	Fluid levels check		
	Wheel bearings clean, repack, adjust		
TIRES (include spare)	Condition		
	Air pressure/tread depth		
	Rotate		
	Wheel balance		
	Alignment		
ENGINE	Tune-up, performance		
	Spark plugs		
	Air filter		
	Fuel filter		
	PCV valve		
	Crankcase filter		
	Canister filter		
	Emission control		
	Exhaust analysis		
BRAKES	Inspect brake system		
	Brake fluid level, condition		
	Flush/bleed/adjust brakes		
	Check, adjust parking brake		
COOLING SYSTEM	Level/condition		
	Antifreeze		
	Pressure test		
	Radiator cap		
	Hoses, clamps, thermostat		
	Power flush		
	Heater operation		
DRIVE BELTS	Fan and accessory belts		
	Camshaft/timing belt		
	Belt tension/adjustment		
BATTERY	Electrolyte level		
	Condition		
	Connections, cables, posts		
	Battery protection treatment		

contd. . . Table 15.13

Maintenance Component	To check	Frequency of check	Date checked, condition
AIR CONDITIONER	Performance test		
	Discharge, evacuate, recharge		
	A/C filter, drier		
	Leak test		
STEERING,	Inspect, adjust, repair		
SUSPENSION	Inspect, adjust, repair		
EXHAUST SYSTEM	Inspect, repair		
LIGHTS, HORN	Lamps, bulbs		
	Aim, check, adjust headlight low, high beams		
	Horn operation		
WINDSHIELD	Washer level, operation		
	Wiper blades		
	Glass		
STATE INSPECTION	Safety		
	Exhaust emission		
	Vehicle registration		
OTHER			

4. One competent person should be assigned responsibility for maintenance of each item.
5. Lab supervisors should regularly check that responsible persons are carrying out maintenance.

Personnel are usually less familiar with maintenance of IQC lab equipment, so the following recommendations of manufacturers are listed as examples in Table 15.14.

Table 15.14 Examples of recommended maintenance schedules for some common lab equipment. Sources: From manufacturers' information and instructions, from Seedburo.

No	Action	Daily or Each Use	Weekly	Monthly	Annual
15.14.1 Westrup Lab Air & Screen Cleaner LA-LS					
1	Keep clean, free of dust, trash, moisture, etc.	X			
2	Clean screens after use, being careful not to bend the perforations.	X			
3	Check/empty all containers, air liftings, air pipes, etc.	X			
15.14.2 Westrup Lab Brushing Machine LA-H					
1	Clean/replace filter in dust aspirator as needed, according to usage.				
2	Keep clean and free of trash, dust, moisture, etc.	X			
3	Check/empty/clean all containers, pipes, etc., after each use.	X			
4	Clean mantle (cylinder) and brushes after each use.	X			
15.14.3 Westrup Lab Indented Cylinder LA-T					
1	Keep machine and spare liners clean and free of trash, dust, seed, and moisture.	X			
2	Clean/empty all containers after each use.	X			
3	Carefully clean out all cylinders/mantles after each use. Keep them free of rust.	X			
4	As needed, oil the thread spindles.			X	
5	Check tension of V-belts, adjust as needed.			X	
6	Check cylinders/mantles for wear, replace as needed.				X

15.14.4 *Torsion Balance Model DWM2-1SEP*					
No	Action	Daily or Each Use	Weekly	Monthly	Annual
1	Keep clean and free of dust, seed, trash and moisture.	X			
2	Check, re-adjust level and balance.	X			
3	To carefully move balance by hand, put lower front knob to arrest, loosen 2 arrest knobs under balance, put shipping arrests in position, tighten arrest lock knobs, put weight loader knob to 9 or 90	X			
15.14.5 *Ohaus Portable Electronic Balance Model CT-600L*					
1	Keep housing and platform clean dry, and free of foreign material. Clean with a cloth dampened with a mild detergent.	X			
2	Keep calibration masses in a safe dry place.	X			
3	Unplug the AC adapter when not in use.	X			
4	For long-term storage, remove the batteries.				
5	As required, follow recommendations in "Trouble shooting" chart.				
15.14.6 *Delmhorst Cotton Moisture Tester Model CM-2*					
1	When not in use, store in its carrying case in a clean dry area.	X			
2	Carefully avoid leakage btween electrode poles. Keep dirt, moisture, perspiration, etc., off the insulation between poles. Keep electrode insulation dry and clean.	X			
3	If service is required, do not attempt it. Return the tester to Delmhorst Instrument Co.				

No	Action	Daily or Each Use	Weekly	Monthly	Annual
15.14.7 Seedburo Count-A-Pak Seed Counter Model 801-10/C					
1	Keep clean, free of seed, dust, dirt and moisture	X			
2	If feeder does not work, heck springs on vibratory feeder, for breaks especially around the bolt hole. Replace springs following procedure shown in operating manual.				
3	Carefully avoid bumping chutes or using them as handles; this may damage diverter assembly.	X			
4	If a solenoid fails to energize, follow operating manual instructions.				
5	If power fails, check fuse.				
15.14.8 Seedburo Compact Sling Psychrometer					
1	Use cold water only. DO NOT use hot water to wet wicks.	X			
2	Replace worn or dirty wicks, as needed.				
3	Replace broken thermometers, as needed.				
15.14.9 Seedburo L-D Germinator Model 548 and 549					
1	Keep cabinet clean inside and out, by wiping as needed with a cloth or sponge dampened with a mild soapy solution. Do not use abrasives, acids or strong cleaners.	X			
2	Do not flood electrical parts or allow them to become wet.	X			
3	Do not use wax or strong cleaners on door gaskets or vinyl trim.	X			

No	Action	Daily or Each Use	Weekly	Monthly	Annual
15.14.10 South Dakota Seed Blower Model 757					
1	Keep slip joints in the lucite air column sections clean, to assure good seating of the joints.	X			
2	Protect air screens at all times, and keep them clean, unplugged and undamaged.	X			
15.14.11 Ohaus Triple Beam Balance					
1	Do not lubricate the knives or bearings; keep them free of dust, grease, trash.	X			
2	Check zeroing of the balance.	X			
3	Keep balance clean, free of dust/dirt and moisture.	X			
15.14.12 Ohaus Explorer Balance					
1	Keep clean, free of seed, dust, dirt and moisture. If necessary, clean with a cloth dampened with a mild detergent.	X			
2	When balance will not be used for an extended period, unplug from electrical outlet.	X			
3	Keep calibration masses in a safe dry place.	X			

15.14.13 Machine: Motomco Moisture Meter Model 919

No	Action	Daily or Each Use	Weekly	Monthly	Annual
1	Keep entire machine clean and dry, free of dust, dirt, seed, moisture.	X			
2	Check calibration before each use.	X			
3	A film deposit will gradually build up on the inside surface of the test cell. Periodically, remove this by cleaning inside of test cell with a soft cloth dampened with water. Do not use alcohol or other solvent.			X	
4	Do not attempt repairs on moisture meters. Return the meter to the manufacturer's factory service unit for service.				
5	Every 3 years, return moisture meter to the manufacturer's service unit for adjustment and technical maintenance.				
6	Check test cell hanger monthly, and adjust as required, according to instructions in manual.			X	
7	Examine dump cell to determine if both wings drop simultaneously. If not, correct as described in manual. Keep dump cell clean.			X	
8	If knobs slip, tighten set screws to hold them securely to their shafts.				

15.14.14 Hoffman Controlled Environment Chamber Model SG 222 SSV SC

No	Action	Daily or Each Use	Weekly	Monthly	Annual
1	When few samples are in chamber, clean inside cabinet, following recommended procedure.			X	
2	Clean outside surfaces, following recommended procedure.		X		
3	Inspect and clean condenser, with brush and vacuum, to ensure adequate air flow.		X		
4	Keep ample space for air circulation around the unit.	X			
5	When vaporizer is filled, use wood toothpick to clean out water inlet hole in bottom of shroud. Rinse out and clean the water container.				
6	If not enough steam is produced, check and clean out (with wood toothpick) water inlet hole.				
7	If vaporizer spits, makes loud boiling noises, jets of steam, hot power cord, etc.: reduce mineral content in water (use distilled or demineralized water, melted snow, etc.)				
8	Check electrodes in vaporizer, and scrape off deposits of minerals. Can soak overnight in vinegar.			X	
9	Extend vaporizer cord full length; do not coil or bunch it, as it will heat.	X			
10	If electrodes in vaporizer are worn and shorter than 4 inches, replace the entire vaporizing head.				
11	Recording unit: periodically lift the pressure writing stylus and wipe away any accumulation of chart surface material.			X	
12	Clean contaminants off the temperature sensing bulb, when they build up.			X	

No	Action	Daily or Each Use	Weekly	Monthly	Annual
15.14-.15 Westrup Lab Gravity Separator LA-K					
1	Keep machine clean and dry, free of dust/trash and moisture.	X			
2	After each 40 hours operation, grease the bearing of the pull rod with a good-quality machine oil.				
3	Keep the deck clean. Do not allow perforations to plug up.	X			
4	Operate the machine in dust-free air. DUSTY AIR WILL CLOG THE DECK OPENINGS.	X			
5	Clean the fan if vibrations occur.				
6	Check the v-belt. Replace if needed.		X		
7	Check deck cleanliness; if needed, use compressed air, from top side, to clean deck openings.		X		
15.14.16 Grainger Speedaire Single State Air Compressor, Cast Iron Series, Model 3JR85C					
1	Check air system & electrical components for damage, deterioration, weakness, or leakage. Tighten unions, repair as required.	X			
2	Check all fasteners for proper tightness.	X			
3	Check safety valve, keep free from paint, dirt, etc. Pull ring on safety valve, allow to snap back into normal position.		X		
4	Check compressor oil level, refill as needed.	X			
5	Drain water from air tank.	X			
6	Check, clean/replace air filter.	X			
7	Blow dirt from inside motor.	X			
8	Check belt tightness, adjust as per instruction book.			X	
9	Every 3 months, change oil.			XXX	
10	DO NOT attach air tools to open end of hose until start-up is completed and unit checks OK.		X		

No	Action	Daily or Each Use	Weekly	Monthly	Annual
15.14.17 Tornado Taskforce External Filter Commercial Vacuum					
1	Keep machine clean and filter bag dry.	X			
2	Empty filter bag and tank.	X			
3	Inspect carbon brushes in motor unit and replace before they are worn down to less than ¼ inch.			X	
15.14.18 Dole Portable Moisture Tester Model 8400 & Model 8400-221					
1	Change battery.				X
2	Keep clean, free of dust, trash, seed and moisture.	X			
15.14.19 Whirlpool 16 cu. ft. refrigerator					
1	Clean inside and out with wet cloth and mild detergent.	X			

Table 15.15 Examples of spare parts for some lab equipment. Source: Manufacturers

15.15.1 Westrup Lab Air & Screen Cleaner LA-LS 5 years recommended parts in parentheses after part name

No.	Part Name & Number	No. Used/in Stock/Ordered	No. Used/in Stock/Ordered	No. Used/in Stock/ Ordered
1	54585, C-1, Vibrator 24516-06-A (1)			
2	54575, C-3, Vibrator triac control ES15 (1)			
3	53321, D-1, Eccentric ball bearing PFT 47-YAR2042F (2)			
4	53323, D-3, Eccentric bearing housing FBB 204, lock ring (1)			
5	53203, D-5, Eccentric ball bearing 6007-2RS, (1)			
6	53323, D-7, Eccentric bearing housing FBB 204 (1)			
7	53323, D-8, eccentric bearing lock ring (1)			
8	8106020, E, Back spring (2)			
9	8106010, F, Front spring (2)			
10	51401, G-3, V-belt 8x5x710, (2)			
11	54041, G-4, Coil set for motor (2)			
12	55260,I, Filter bag (2)			
13	55263, K, Canvas bag (2)			
14	54836, L, rubber balls, 112 pc set) (1)			
15.15.2 Westrup Lab Brush Machine LA-H Spare parts for 5 years shown in parentheses after part name				
1	B-3, Tube for vacuum cleaner compl. (3)			
2	B-4, Filter bag AEG type 15 (4)			
3	F-2, V-belt pulley A 160/1 (2)			
4	F-8, Brushes, set (5)			
5	F-8 Beaters, set (5)			
6	H-1, Cage (3)			

15.15.3 Westrup Lab Indented Cylinder LA-T Spare parts for 5 years in parentheses after part name and number.

No.	Part Name & Number	No. Used/in Stock/ Ordered	No. Used/in Stock/Ordered	No. Used/in Stock/ Ordered
1	B-1, Vibrator 24516-06A (1)			
2	B-3, Vibrator triac control ES 15 (1)			
3	C-1, Cylinder closing hook (2)			
4	C-2, Cylinder closing device (2)			
5	D-2, Var. pulley Piccolo VPS10 (1)			
6	D-3, V-belt SPZ 8x5 L=1500 mm (1)			
7	D-4, V-belt SPZ 8x5 L-630 mm (1)			
8	D-6, Ball bearing 6201-2RS (1)			
9	D-7, Lock ring (1)			
10	D-8, Lock ring (1)			
11	E-6, Scale rubber disc (4)			
12	E-7, Scale split-pin (4)			
13	E-8, Scale tightening disc (4)			
14	G-1, Plexiglas guard (1)			
15	H-2, Trough rubber disc & bearing 6205-2RS (1)			

15.15.4 Ohaus Portable Electronic Balance Model CT-600L Recommended spare parts

1	90524-77, AC adapter 220 v. (1)			
2	77577-01, Platform 4.75 inch diameter (1)			
3	4590-30, Scoop, 4" x 3.5", black (1)			

15.15.5 Seedburo Count-A-Pak Seed Counter Model 801-Spare parts recommended for 5 years

No.	Part Name & Number	No. Used/in Stock/Ordered	No. Used/in Stock/Ordered	No. Used/in Stock/ Ordered
1	Springs for 7" SHB vibratory feeder , 99-MHDW0067 1/16 inch leaf spring (3)			
2	Springs for 10" SHB vibratory feeder, 99-MHDW0067, 1/16 inch leaf spring (3)			
3	Springs for 10" SHB vibratory feeder, 99-MHDW0049, 1/32 inch leaf spring (3)			
4	Glass lenses, clear glass, 99-MLN0001 (2)			
5	Light source assembly 99-MSUB0017 (1)			

15.15.6. Seedburo Compact Sling Psychrometer Recommended spare parts

1	Spares Kit (incl. 2 thermometers, 1 spare container full of wicks)			

15.15.7 Seedburo L-D Germinator Model 548 and 549 Spare parts listed

No.	Part Name & Number	No. Used/in Stock/Ordered	No. Used/in Stock/Ordered	No. Used/in Stock/ Ordered
1	54H/A, heating element, 150 W, 220 V, 0.68 amps.			
2	54-TR, Lighted on/off switch, 220 V			
3	513-LTK, knob, thermostat			
4	514-LTK, dial plate, thermostat			
5	523-LTK, cord with plug, 220 V			
6	532-LTK, thermometer, 30-120 F			
7	534-LTK, thermometer, brass retaining nut			
8	611-LTK (model 548) or 613-LTK (model 549), perforated tray			
9	616-LTK (model 548) or 618-LTK (model 549) ice/water pan			
10	644-LTK, rubber feet (set of 4)			
11	652-LTK, door handle, with screws			
12	654-LTK, door latch, with strike			
13	660-LTK, plexiglass window			

15.15.8. Ohaus Triple Beam Balance replacement parts

No.	Part Name & Number	No. Used/in Stock/Ordered	No. Used/in Stock/Ordered	No. Used/in Stock/Ordered
1	3045-00, Scale Plate, SS, 6" dia. (1)			
2	1039-02, Scale Pan, 6" dia. X ¾" (1)			
3	1064-30, Scale pan cross (1)			
4	107703, Scoop, SST, 12"x 6"x2.75" deep (1)			
5	1101-20, Scoop, polypropylene, 12"x6"x2.75" deep (1)			
6	3043-20, Specimen pan (1)			
7	3052-00Specimen pan cover (1)			
8	2570-21, Specimen pan cross (1)			
9	1023-00 Balance cup (1)			
11	1034-00, Friction plate (3)			
12	1247-00, Friction plate (1)			
13	1008-23, Bearing (4)			
14	1035-03, Bearing cover (4)			
15	3134-02, 100g poise assembly (1)			
16	9704-00, Tare bar assembly (1)			

15.15.9 Ohaus Explorer Balance replacement parts

1	Power pack, cord set required for UK, Europe, Australia, 490203-01 (1)			
2	Cord set, 230 V ac, European plug, 76212-00 (1)			

15.15.10 Westrup Lab Gravity Separator LA-K Parts recommended for 5 years				
No.	Part Name & Number	No. Used/in Stock/Ordered	No. Used/in Stock/Ordered	No. Used/in Stock/ Ordered
01	545855 (Feed) vibrator (1)			
02	545755 (feed) Triac feed control (1)			
03	54522 (Feed) sensor (1)			
04	54041 (Motor & Drive) Coal for motor (1 set)			
05	51405 (Motor & drive) V-belt 8x1000 (1)			
06	53321 (Motor and drive) shaft bearing (2)			
07	53205 (Motor & drive) ball bearing (2)			
08	53395 (Motor & drive) ball link (1)			
09	6197011 (Motor & drive) spring 5x30x199 (4)			
10	55263 (Air supply) Canvas transition (1)			
11	55252 (Air supply) Canvas bag (1)			
12	54817 (Air supply) Vibration damper (4)			
15.15.11 Tornado Taskforce External Filter Commercial Vacuum				
01	11780, 230 volt carbon brush (2)			
15.15.12 Dole Portable Moisture Tester Model 8400 & 8400-221				
01	Battery, 9 volt, transistor, NEDA type 1604			
15.15.13 Dazor Lamp with Magnifier, Desk Model, Adjustable Arm				
01	Bulb			

Chapter 16

Safety in Conditioning

With many machines with moving parts, dust, flow of bulk seed, operating adjustments on machines in different places, dust, combustible wastes and seed, pests, elevated bin access, and many other activities, there are many potential dangers for personnel, equipment, and buildings.

An organized, carefully-implemented safety program is essential, as is constant supervision and detailed personnel training. Following are some general safety requirements designed to prevent—not treat after they happen!—personnel safety problems.

Managing Safety

An organized approach to safety must be established and implemented. Safety management must cover all aspects, with regular checks and constant attention to keeping safety factors in place and all staff fully advised and careful.

Prepare for Emergencies

1. Be prepared and know what to do if a fire starts, someone is injured, a machine breaks down, etc.
2. Keep a first aid kit and fire extinguishers of the proper types handy (Figure 16.1).
3. Sponsor, arrange, and have all staff take Red Cross emergency first aid training courses.
4. Be sure all staff know what to do if anything happens.
5. Keep emergency numbers for doctors, ambulance service, hospital, and fire department near your telephone.

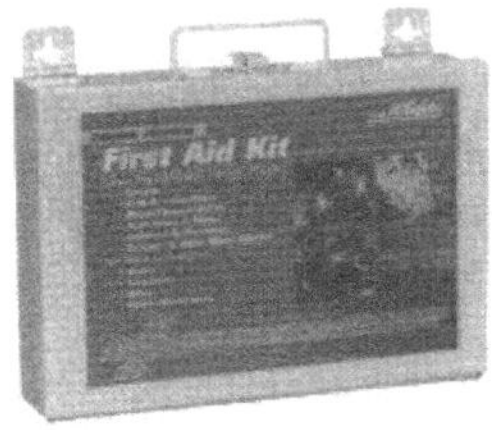

Figure 16.1 Adequate first-aid kits should be placed in all appropriate areas of the plant, and all staff should be trained in first aid. Source: Kathy Reading, Seedburo catalogs.

Official Safety Inspections

Many municipalities or other political divisions have safety requirements for different kinds of facilities, and inspect facilities to certify them or require improvements. These inspections and requirements are not just a problem; they should be considered as positive efforts to eliminate risks and dangers. Ensure that regular inspections are made, recommendations are implemented, and official certification/approval is obtained.

Staff Safety Training

Every employee must know exactly what he/she must do, without fail and on a regular basis. All staff and all workers must be trained in operations, and must be trained in safety matters. This includes: general safety precautions; how to do their assigned work quickly and properly, while still focusing on safety for personnel, seed and equipment; what to do if something—even any small thing—out of the proper and safe conditions occurs, or is noted; and, very importantly, how to report it to management.

Regular updating and renewing training is essential. . Regular safety reviews and drills should be conducted to renew their awareness of dangers, and their familiarity with what they should do.

Arrangements With Medical and Fire Agencies

Management should make arrangements with the local fire department, hospital, clinic, and doctors to handle emergency situations. This should be done by a retainer contract arrangement, or other agreement which ensures immediate attention. And, these persons should be asked to make periodic inspections of the enterprise's facilities, to make recommendations for improving conditions.

Clean-up Policy and Program

The seed enterprise must have a strict policy and practice to eliminate immediately any trash, spilled seed, or other materials from all areas of the plant, inside and outside. Any employee who leaves "a mess" must be dealt with severely.

Constant cleaning: Within the plant, at least one laborer should be assigned to clean the conditioning area floors, walls, walkways, etc., constantly. The same or other worker should also constantly keep the storages clean. The conditioning plant and all storages, access ways, and outside areas must be kept spotlessly clean.

Clean-up between lots/varieties/crops: An essential practice, dealt with in detail elsewhere, is the intensive clean-up of conditioning and handling equipment before conditioning a new seed kind. This is essential to prevent contaminating the new seed with leftover seed from the last kind handled.

All seed and other materials removed in this clean-up should be considered as waste and disposed of accordingly.

Clean-Out and Clean-up Equipment

Cleaning equipment must be adequate and in good condition. This includes push brooms, counter brushes, other brushes and brooms to reach all corners and areas, scoop shovels, dust pans, trash cans, industrial vacuum cleaners, and compressed air with hand-operated nozzles. Compressed air with "quick-snap" hose connectors should be piped throughout the conditioning plant, including on top of bins and near all machines, to facilitate clean-out.

Trash Receptacles and Disposal

Small trash receptacles should be placed at useful locations throughout the plant and storages, to receive wastes as they are cleaned up. These should have ratproof covers. Larger trash containers be designed for easy dumping into trucks, or transportable bins should be easy to haul trash away from the plant site.

Many plants convey and elevate waste materials into one or more trash bins which can dump into trucks or bags.

When wastes are dumped into trucks for transport to safe outside dump areas, the load must be well-covered with a tarpaulin or other cover to prevent materials from blowing out.

Suitable waste products such as shriveled or immature untreated seed may be kept separate and used in livestock feed manufacture. Weed seed should be burned, properly crushed, or otherwise carefully destroyed so they cannot germinate. Chaffy trash can be burned or used as compost. Waste material of any kind should be carried away from the plant site for disposal. Waste material incinerators are sometimes installed on the plant site, but this must be carefully located and operated to avoid creating fire hazards.

Dust Control

An adequate dust control system should be installed at the time the plant is built. If this has not been done, the necessary components should be added at the earliest opportunity.

The dust control system should move all dust/chaff-laden air to collectors outside the building. Dust piping should take all exhaust air from machines such as the air-screen cleaner, aspirator, or pneumatic separator, and pipe it out without allowing dust to escape from leaks in the pipes.

If the crop seed are quite dusty such as groundnuts, or have disagreeable dust such as sugar beets, dust exhaust fans and pipes should remove dust from all machines. Exhaust systems are often installed to remove dust from elevator heads and bin tops.

In some cases, the machine is enclosed with a cover, and a dust exhaust system removes dust. The vacuum gravity separator has a hood enclosure as an essential component, with dusty air carried to an outside collector. A dust exhaust hood is often installed above the positive-pressure gravity, to remove dust and very light chaff.

Dust and light chaff is removed from the exhausted air by systems of dust collectors or filters. These must be carefully selected so they can handle the volume of air and kinds of dust/waste exhausted.

Common problems with dust/trash collection systems which must be avoided are:

1. Wastes are not properly discharged from the collector, and build up inside the collector and ruin its action. Trash/dust must be allowed to discharge from the collector, into a bin or container placed beneath the collector. The bin must be regularly checked and emptied as required.
2. Trash is spilled around the collection system, and is not kept cleaned up. This attracts insects, birds and rats, and creates a fire hazard.

Personnel Safety

Potential Hazards

International safety standards require that equipment be built to allow safe operation. Operators be trained initially and periodically brought up-to-date, to minimize the possibility of injury or illness from equipment operation. Precautions established for operations should guard against:

1. Physical injury from rotating or vibrating parts of equipment.
2. Electrical shock.
3. Excessive noise.
4. Heavy lifting.
5. Inhaling dust.
6. Inhaling chemical fumes.

Preventing Injuries

Physical injury should be prevented by adequate guarding and hazard labeling of the equipment. The installed guards should always remain on the equipment during operation. Safety electrical lock-outs need to be provided for each machine so the operator can prevent someone from accidentally starting the machine when it is being serviced.

Electrical shock can be prevented by following internationally accepted wiring standards. Electrical connection boxes should remain sealed, unless during necessary service by a qualified electrician, so the operators are protected and to ensure that dust and insects do not invade the wiring connections.

Excessive noise is first a machinery selection issue where the decibel level should be kept under 85 in the conditioning area. Some equipment outside the conditioning area may exceed that level and special design precautions need to be taken so enforcement of the use of proper ear protection for the operators and guests is strictly followed. Some items that may violate the excessive noise standards are aeration fans, dryers, and dust collection fans.

Heavy lifting can be prevented by training the workers on proper load lifting limits, lifting procedures and the use of safety belts when necessary. Proper lift equipment needs to be on hand at all times to help with moving pallets, seed bags, chemical containers, etc.

Inhaling dust can be minimized by proper selection of the conditioning equipment and the use of air extraction equipment to remove the dust from the separating machines and keep the machine under negative air pressure to lessen emissions. Once collected dust needs to be properly separated from the exhaust air and disposed of correctly.

Chemical fumes may be a serious hazard so only authorized personnel, who have been trained in the dangers and proper use of the chemical compounds, should be allowed to manage and use this area. The chemical application equipment must be selected correctly and operated according the manufacturers standards so the operators do not come in direct skin contact with any chemicals or fumes.

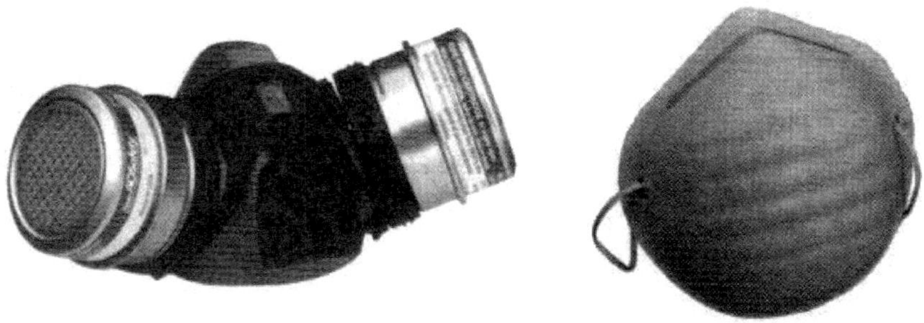

Figure 16.2 A dust respirator and a dust filter mask often used to protect personnel in dusty areas of the seed plant. Source: Kathy Reading, Seedburo catalogs.

Safety Equipment

Adequate safety equipment should always be readily available, and staff should know how to use it. This includes:

1. Keeping guards in place over all belts, pulleys, and moving parts.
2. Having first-aid kits complete, up-to-date, and located in several readily-accessible places.
3. Requiring use of adequate dust respirators for dusty operations, and appropriate gas masks for fumigation and chemical work (Figure 16.2).
4. Maintaining an effective dust removal system throughout the plant.
5. Placing adequate warning labels in all places of potential danger.

Building Safety

Door and Wall Protection

Protective door guard posts and lower wall rails, as described in the chapter on building construction, should be used to prevent damage to the doors and walls.

Operating Forklifts and Vehicles

All forklifts and vehicles should be kept in good maintenance condition. Only safe forklifts should operate inside buildings, except for covered receiving areas. Potential sources of sparks, leaks, etc., must be eliminated. Operators must be well-trained, work in a safe manner even if it requires more time for an operation, and should receive regular upgrading instructions in safe operations, maintenance, and handling.

Truck Access

Trucks should enter only the covered area of the bulk receiving dump pit. Trucks should never enter storages or conditioning, as they bring in moisture and engine fumes, and hazards of hitting someone or something. Conditioning and storage buildings should be constructed with floor level at truck bed height. Trucks should remain at covered loading platforms outside the building, and all loading/unloading done by forklift.

Physical Security

The plant area should be surrounded by a fence adequate to keep out most unauthorized persons. Entry should be at specific gates which have guards posted.

Electrical Safety

Although a good seed conditioning plant has little uncontrolled dust, it is a dusty operation. All electrical installations should be dust-tight explosion-proof industrial level. All portable electrical equipment (vacuum cleaner, blower, hand tools, etc.) must be kept in good condition, with no bare or frayed wires. When wall plugs are not in use, their dust-tight covers should be in place. Water and moisture are normally eliminated from a seed facility; any dampness found should be eliminated, not only for seed quality, but also for personnel safety.

Dust Control and Ventilation

Breathing dusty air for prolonged periods can cause health problems. This is especially true for dust generated from certain crop seed such as sugar beet. Air within the conditioning plant must be kept as clean and dust-free as possible. Clean, new dust filters should be provided for operators who are involved in short-term activities where dust is generated, and there should be a rule requiring the use of filter masks.

Dusty air should be exhausted out of the building, in accordance with the specifications provided by each manufacturer and the design engineer who planned and supervised establishment of the plant. If the crop seed handled creates dust from machines without fan/pipe dusty air exhaust systems, special dust removal systems should be installed; examples include an exhaust hood over the gravity separator or stoner deck; exhaust hood over elevator feed hoppers; suction exhaust pipes from elevator heads, bins, other machines, etc.

There should be definite exhaust and ventilation systems at the seed treater and bagger, to ensure that chemical dust or fumes are exhausted out of the building.

In addition, ventilation fans should maintain a gentle movement of air from inside the plant to the outside, and replacing it with clean fresh air from outside. Even in cold seasons, inside air should be healthy for workers, even if it requires additional air heating. Ventilation fans should be screened to keep out insects and birds.

Equipment Safety

General

Before servicing, cleaning, or inspecting a machine, lock out the start/stop station so that only the maintenance operator has control of the machine. The lock-out switch should be located on the machine.

Personnel Safety Around Machines

The following guidelines for personnel safety around machines should be carefully observed.

1. Follow the machine manufacturer's safety instructions:
   - A. Read and follow all safety messages in the operator's manual and follow all safety signs on the machine (Figure 16.3).
   - B. Maintain safety signs in readable condition; if they are damaged, replace them, and be sure all equipment components include proper safety signs (Figure 16.4).
   - C. Learn how to operate and adjust each machine and properly and safely. Do not let anyone operate without instruction.
   - D. Keep each machine in proper working condition.
   - E. Do not make unauthorized modifications which may impair the function, safety or life of a machine (note however, that some highly-experienced seed conditioning operator/mechanics sometimes make good modifications. When this is done, it is best to notify the manufacturer and get his approval, possibly even adoption, of the modification).
   - F. Safety guards are provided with all machines to protect operators and maintenance personnel. Enclosures over belts, pulleys, and other moving parts should remain in place, except when service personnel are performing maintenance.

2. Check machines before operating them:
   A. Be familiar with all controls.
   B. Check condition and tightness of drive belts and chains.
   C. Be sure there are no foreign objects in/on/leaning against the machine, its bin and spouts.
   D. Be sure feed to the machine is set up properly and is not obstructed.

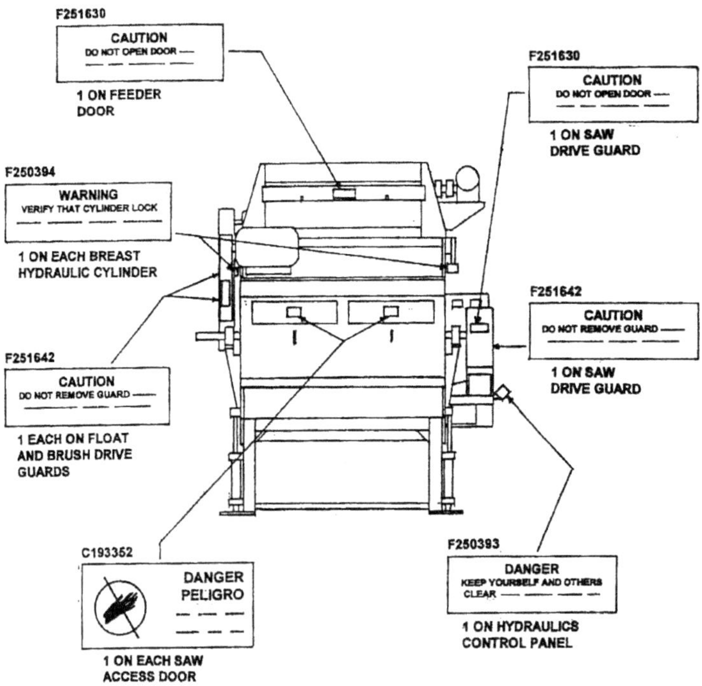

Figure 16.3 Location of danger warning decals on the Continental Eagle LE-176 cottonseed delinter. Source: Don Clark, Continental Eagle Corp. Inc. descriptive literature.

Figure 16.4 Typical safety warning signs on machines. Source: Crippen Internatiional descriptive literature.

E. Be sure all discharge spouting from the machine is properly set up.
F. Be sure all shields are in place.
G. Inspect driveline shield for free rotation.
H. Be sure all bolts and nuts are tight.
I. Be sure that all components (for example, cylinders in a cylinder separator, screens in an air-screen cleaner, are properly secured and tightened.
J. If in doubt (and always as a good preventive measure), turn the drive pulley a few rounds by hand to check for obstructions.

3. Wear protective clothing:
   A. Wear close-fitting clothing which cannot be caught by moving parts, and safety equipment appropriate to the job.
   B. Prolonged exposure to loud noise can cause impairment or loss of hearing. Wear a suitable hearing protective device such as earmuffs or earplugs to protect against loud noises.
   C. Operating equipment safely requires full attention of the operator. Do not wear radio or music headphones while operating machine.
4. Avoid injury from moving parts:
   A. Never wear loose fitting clothing.
   B. Keep hands, feet, hair, and clothing away from drive belts, pulleys, and other moving parts.
   C. Never attempt to clear obstructions or objects while machine is in operation.
5. Understand warning signs and words (Figure 16.5):
   A. A warning word—DANGER, WARNING, or CAUTION—is used with the safety-alert symbol. DANGER identifies the most serious hazards.
   B. DANGER or WARNING safety signs are located near specific hazards. General precautions are listed on CAUTION safety signs. CAUTION also calls attention to safety messages

Workspace Around Machines

Workers must have safe access to all parts of each machine, for maintenance, repair, adjustment, and clean-up between varieties. All discharge spouts must be easily accessible for sampling the discharged products.

Many separators have discharge spouts which are too low to allow for adequate spouting or sampling if the separator is installed on the floor of a single-level plant. In such cases, the separator must be installed on a stand. The stand must have adequate safe floor space around the machine for operator access.

Figure 16.5 Widely-used cartoons illustrating the 4 main concepts of personnel safety: 1–avoid moving parts; 2–read and understand instructions; 3–wear protective clothing; 4–heed warning signs. Source: Crippen International.

If loose seed or trash are on floors, they may cause workers to slip and fall. Floors of worker platforms and access steps should be of expanded metal grating, so that loose seed will fall through instead of creating a hazard. And, one worker should be responsible for constant clean-up to remove loose seed hazards.

Access to Cleaners and Separators

Safe access must be installed for workers to reach all parts of each separator and cleaner for clean-out, repair, maintenance, and adjustment. Temporary measures such as movable steps or ladders should not be used. Permanent, safe work platforms at levels which permit quick easy access should be built around each machine. These platforms should have sturdy floors of expanded metal grating, which allows loose seed to fall through to the main building floor, without creating hazards for workers on the machine platforms. Safe access stairs should let workers reach the machine platforms quickly (Figures 16.6 and 16.7).

In some plants, all machines are placed on a single level of platform, with adequate space around each machine. A few stairs at strategic points allow operators to get up to the platform. In other plants, each machine is on a separate platform, with its own stairs.

All work platforms and stairs should have guardrails adequate to prevent workers from falling off.

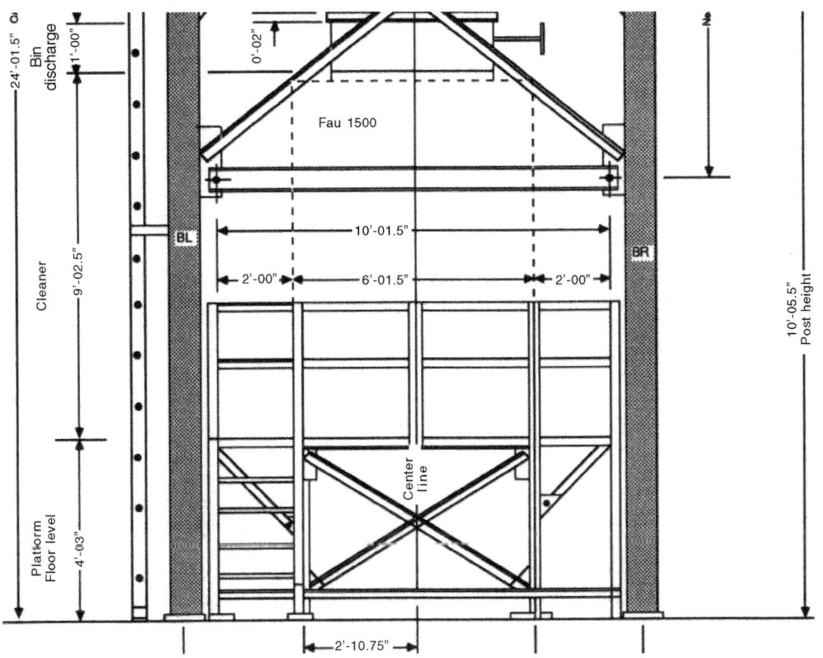

Figure 16.6 Front view of the installation of an air-screen cleaner, showing position of access stair, guard rails, and adequate side work space. Source: Billups and Lasheen unpublished designs.

Access to Bins

Safe access must be installed for workers to reach tops of bins for clean-out (Figure 16.8). Temporary measures such as movable ladders or steps are dangerous and cause work delays.

In some plants, a permanent, safe ladder is installed beside each bin, so workers can safely get to the bin top to clean it out. At the top of the ladder, the side supports should extend well above bin tops, to provide safety handholds for workers.

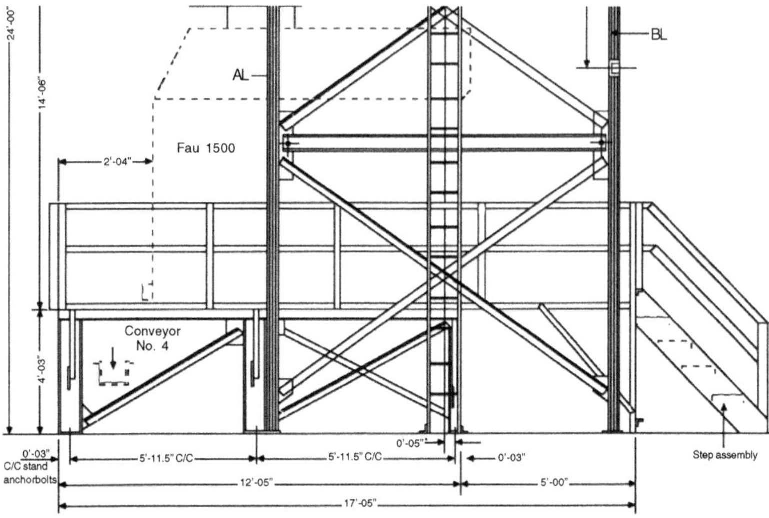

Figure 16.7 Side view drawing of access to, work space around, and installation of an air-screen cleaner. Note (1) stair to the work space, (2) guard rails on the stair and around the work space, (3) support system for the stand and the cleaner combined into one support structure, (4) location of ladder to the bin top, (5) height of the cleaner is adequate for handling (bagging or conveying) waste products, and (6) mounting of a vibrating conveyor beneath the cleaner to take the cleaned seed to the next elevator. Source: Billups and Lasheen unpublished designs.

There must be safe work space at the top of bins, for workers cleaning out the bins. Some plants have covers on bin tops to provide worker access; this also helps contain dust generated when seed flow into the bin. However, bin covers must have hatches for workers to enter the bin, and an air outlet to relieve air pressure built up when the bin is filled.

In some plants, the tops of all bins are at almost the same level, and a single work platform with a floor of expanded metal grating allows operators to reach all bins.

There should be ladder steps inside the bin, for workers to enter the bin. The steps should be of round steel rods, and installed in such a way that all surfaces are smooth and will not hold seed.

Access to Elevators

Maintenance and clean-out is required at elevator heads. Since these are usually the highest work points in the conditioning plant, safe access must be provided by ladders or catwalks. A safe work platform, with adequate

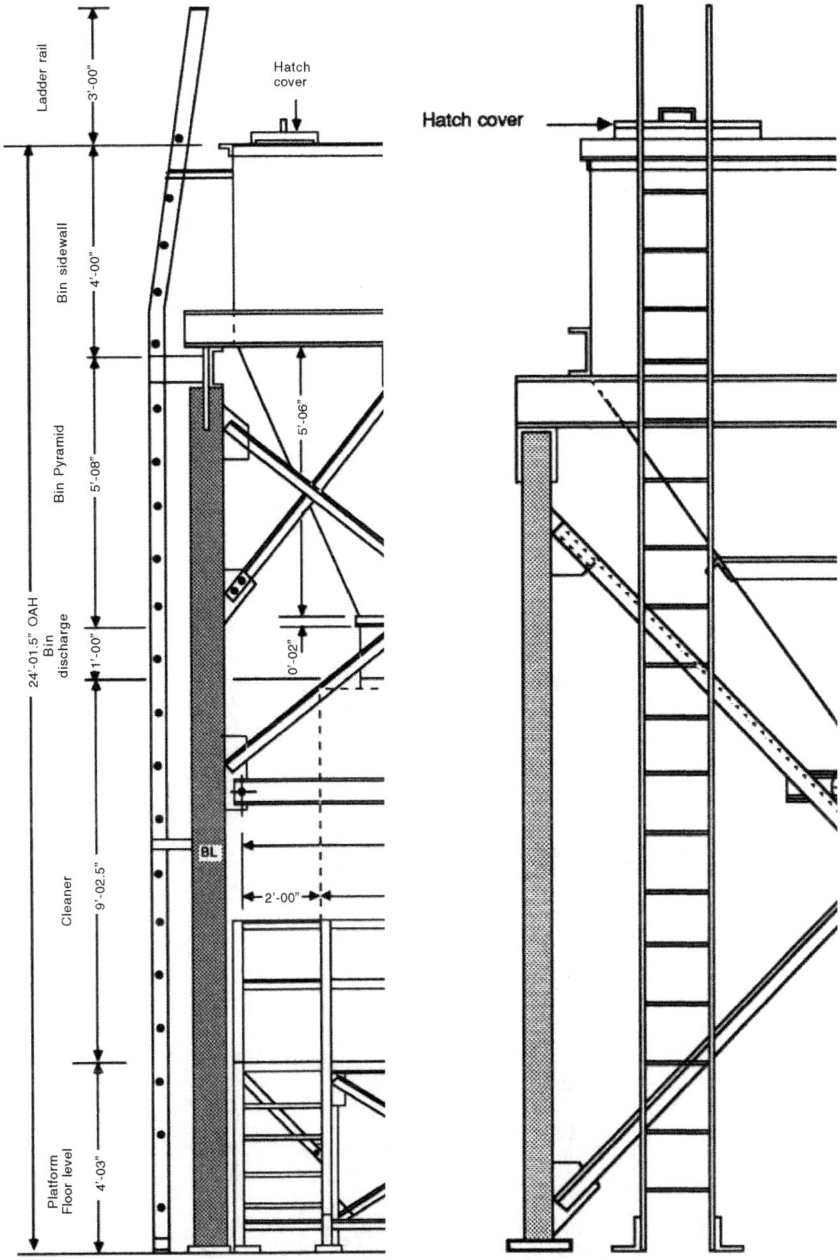

Figure 16.8 Side view (on left) and front view (on right) of the access ladder for operators to reach to top of the bin over a separator. Note that, for safety, the frame of the ladder extends well above the top of the bin and provides a safe support for workers. Source: Billups and Lasheen unpublished designs.

guardrails, must be provided at the elevator head for the safety of workers (Figures 16.9 and16.10).

Guardrails

Every working platform should have guardrails sufficient to prevent workers from falling off. At the same time, guardrails should not interfere with operations such as changing screens in the air-screen cleaner. Access steps should also have guardrails. They should be strong enough to support workers in any case of contact.

Figure 16.9 Access ladder and work platform with guardrails at the head of an elevator. Source: Lyle Hollander, Sudenga Industries technical materials.

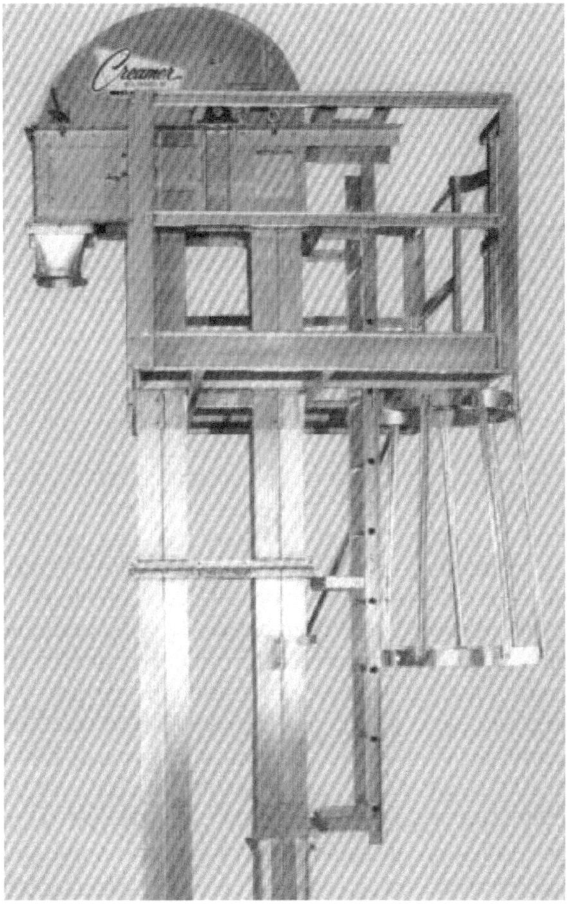

Figure 16.10 Access to the head of an elevator, showing safety features of ladder to the head, guard around the top of the ladder, and work platform with guard rails. Source: Chris Ansel, Creamer Metal Products.

Catwalks

Sometimes, it is more time-consuming and less risky to have fewer ladders to climb, by using catwalks to allow workers to move safely among several raised points such as elevator heads and bin tops. Catwalks are walkways with guardrails, providing safe access between raised points. Electrical outlets and "quick-snap" compressed air outlets are often provided at convenient points on the walkways, to facilitate work and clean-out (Figures 16.11, 16.12 and 16.13).

Long Pipes

Long pipes and spouts should be avoided whenever possible, as seed can receive damaging impacts from long falls. Whenever a long pipe for spouting seed cannot be avoided:

1. Be sure to install a seed letdown ladder at the end of the pipe, to absorb seed impact without damage to seed.
2. Brace the pipe sufficiently (Figure 16.14) to prevent falling or damage.
3. Do not allow workers to scale the pipe or its support structure.

Seed Safety

Safety of seed and seed quality is discussed in detail in other chapters. Primary aspects are as follows.

Moisture and Water: To protect seed, all operating areas should be kept completely dry. Free moisture also may cause workers to slip and be injured by falls on the smooth-finished floors. All seed working areas should be kept dry. The buildings should have raised floors; outside drainage should prevent moisture from entering through doors. Windows should be protected by roof overhangs or other means to keep out rain water. Trucks and outside vehicles should not enter seed working and storage areas; the only truck access area should be to dump pits to receive bulk seed. Restrooms and showers for workers should be in a separate but nearby building.

Fumigation: Always use only fumigants which are safe for seed, and apply them with the procedures and times of treatment as specified by the manufacturer. Location and handling of fumigation should be safe for workers.

Pest Control: Should be adequate to prevent damage by insects, rats, birds, etc.

Chemical Damage: Should be prevented at all times. Do not store or hold seed in areas with, or near to, fertilizers, herbicides, chemicals, salt, etc.

Fire and Explosion Safety

Grain seed dust is flammable and will flame up so quickly that it sometimes results in an explosion. In moving and handling both seed and grain, fine dust is abraded from the surface of seed and waste materials in the seed lot. Waste materials—hulls, glumes, stems, etc.—are also flammable. Both dust and trash present a fire and explosion hazard. In addition, dust is a health hazard to personnel, and trash/waste attracts insects and rats, and provides cover for them. Dust, trash and waste materials must be constantly and

Figure 16.11 A catwalk installed around an auger conveyor in an outside storage bin system. This allows workers to service all bin tops without climbing a ladder to each bin. Source: Lyle Hollander, Sudenga Industries technical materials.

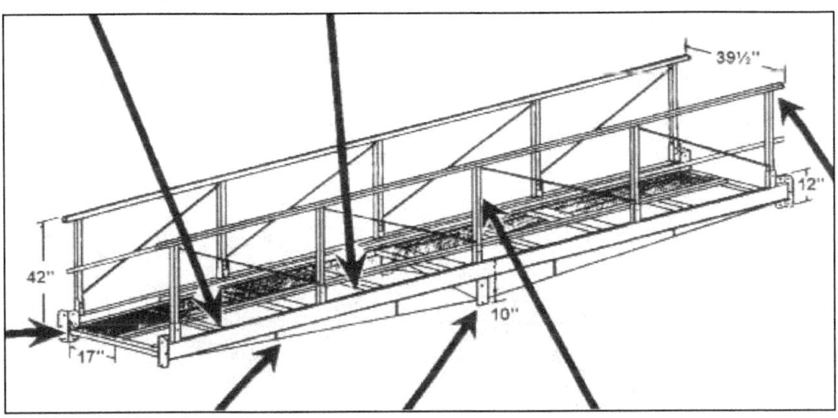

Figure 16.12 Structure of a safe catwalk system. Source: Lyle Hollander, Sudenga Industries technical materials.

Figure 16.13 Catwalk extending from bin tops to the receiving elevator and elevator head. Source: Randolph Block, Screw Conveyor Corporation.

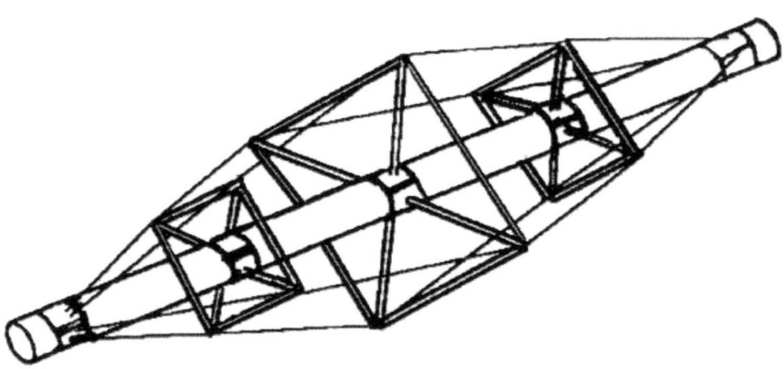

Figure 16.14 Support framework for a long pipe. Angle-iron frames hold the cables which support the center sections of the pipe. Source: Lyle Hollander, Sudenga Industries technical materials.

completely controlled, to eliminate fire hazards, personnel danger, and minimize damaging pests.

Building Construction and Design

Buildings should be constructed of fireproof materials such as reinforced concrete slabs, concrete blocks, or bricks. Local recommendations on distance between buildings should be observed. Fire doors, metal fireproof doors which close in case of fire, should be installed in doors between areas, such as between storage and conditioning.

Floors and walls should be smooth-finished, to make them easier to keep clean and to eliminate cracks or depressions where trash can accumulate.

Diesel is a common fuel for dryers. The diesel tank should be installed outside with ready access to fuel delivery trucks but at a safe distance from buildings and the dryer burner, on a concrete pad which prevents growth of weeds around the tank. The area around the tank should be kept clean and closely mowed.

Electrical System

All electrical installations should be dustproof and explosion-proof, approved for dusty installations. Switches and all boxes should be dust-tight, and low-voltage outlets should have dustproof covers.

Outside Areas

The plant should be constructed in an area which is not close to houses or other buildings which may cause a fire hazard, or cause problems with rats, dust, birds, trespassing, or other potential problems.

The outside plant area should be well-drained to minimize moisture, and seeded to a lawn-type grass which can be kept closely mowed. Landscaping should be low-growing plants or flowers; avoid large shrubs.

Keep all outside areas free of trash, weeds, loose seed, old equipment or vehicles, and anything else which may attract pests or create fire hazards.

Fire-Fighting Equipment

Adequate firefighting equipment should be readily available throughout the plant. Chemical fire extinguishers, kept charged and ready, should be placed throughout the plant next to conditioning machines, dryers, etc. An extinguisher should be mounted properly on all forklifts. Water-type and chemical extinguishers should be placed frequently in storage and handling areas.

Fire hose box installations and fire alarm activators (which alert the nearest fire station) should be strategically located throughout the plant and storages. Sprinkler systems are not commonly used in seed plants, because of the damage to seed if the sprinklers operate.

Waste Products From Separators

All waste products should immediately be removed from the conditioning area, and stored in a manner which prevents fire hazards or attracting insects and pests.

First Aid Kit

A complete first aid kit should be conveniently located in each work area, and kept fresh and fully stocked. Different kinds of first-aid kits can be purchased locally or assembled with help from local medical specialists. Contents of a large industrial first aid kit available from zeemedical.com are listed in Table 16.1.

Table 16.1 Contents of "Large Plastic First Aid Cabinet" first aid kit for industrial use. Source: zeemedical.com. (This and other kits available on internet at zeemedical.com)

Item	Name	Original Amount	Refill Amount
0203	Clean Wipes 50/Bx (Zee)	1 bx	1 bx
0204	Antiseptic Swabs 50/Bx (Zee)	1 bx	1 bx
0219	Antiseptic Spray, 4 oz.	1 ea	1 ea
0305	Tape, 2" X 5 Yd. 3 Cut Spool (Zee)	1 ea	1 ea
0370	Tape, Elastic 1" X 5 Yd. Spool	1 ea	1 ea
0501	Cotton Tip Applicator 3", Ns, 100/Vl	1 ea	1 ea
0606	Eye Wash, Sterile 4 Oz. (Zee)	1 ea	1 ea
0608	Eye & Skin Buf. Flushing Sol. 8 Oz	1 ea	1 ea
0618	Thera Tears Eye Drops	1 pk	1 pk
0700	Butterfly Bandages, Medium, 20/Box	1 bx	1 bx
0713	Fingertip Bandages XLG, 25/Box	1 bx	1 bx
0714	Fingertip Bandages, 40/Box	1 bx	1 bx
0716	Fabric Knuckle Bandages, 40/Box	1 bx	1 bx
0731	Sheer Strip 1", 100/Box	1 bx	1 bx
0740	Elastic Strip Bandages, 50/Box	1 bx	1 bx
0743	Large Patch Bandages, 25/Bx	1 bx	1 bx
0910	Sterile Dressing 5" x 9"	1 ea	1 ea
0920	Gauze Pads 3" X 3", 10/Bx (Zee)	1 bx	1 bx
0944	Elastic Roller Gauze N/S 3" X 4.5 Yd	1 ea	1 ea
0996	Zee Flex 3" X 5 Yds	1 ea	1 ea
1417	Pain-Aid 100/Bx (Zee)	1 bx	1 bx
1420	Ibutab 100/Bx (Zee)	1 bx	1 bx
1428	ZEE Anti-Diarrheal Caplets, 2mg, 12/bx	1 bx	1 bx
1435	E.S. Un-Aspirin 100/Bx (Zee)	1 bx	1 bx
1446	Antacid, Trial 100/Bx (Zee)	1 bx	1 bx
1451	Pept-Eez 42/Bx (Zee)	1 bx	1 bx
1464	Soothe-Aid Lozenges 25/Box	1 bx	1 bx
1486	Dilotab II, 100/Bx	1 bx	1 bx
1801	3-in-One Antibiotic Ointment, 25/Box	1 bx	1 bx
1810	Blistex Ointment, 25/box	1 bx	1 bx
1817	Hydrocortisone Cream, 25/Box	1 bx	1 bx
2219	Dermafleur Packets, 25/Bx	1 bx	1 bx
2331	Emergency First Aid Pocket Guide	1 ea	1 ea
2353	Small Economy Instant Ice Pack	2 ea	2 ea
2605	Bandage, Triangular 40" N/S 1/Un	1 ea	1 ea
2651	Water-Jel Burn Jel 6/Bx	1 ea	1 ea
3044	Nitrile Gloves, Large, 2 pairs	1 ea	1 ea
3521	Kit, Scissor, 4-1/2" Wire Loop Handle	1 ea	1 ea
3537	Splinter Out 10/Pk	1 ea	1 ea

Chapter 17

Preventing Seed Loss

Profitable Operations Management

In terms of seed and seed quality, management of a seed conditioning plant has four primary objectives:

1. Achieve the highest possible seed quality.
2. Prevent admixture or contamination of seed.
3. Operate cost-effectively and time-efficiently.
4. Prevent loss of seed or seed quality.

If seed quality is lost, the seed is lost. If seed are lost, the entire investment made in producing them is lost, and the potential income from marketing that seed is lost. A constant and major effort must be made to prevent losses.

Causes of Seed Losses

Seed—and thus profit—losses occur from:

1. Insect damage.
2. Damage or spillage by birds and rodents.
3. Loss of identity of seed.
4. Moisture damage, resulting in loss of germination, mold, or heating.
5. Admixture or contamination during handling or conditioning.
6. Damage (accidental or deliberate) or theft by unauthorized or excessive personnel in the operating and storage areas.

Pests Which Damage Seed

Losses from insect infestation or bird or rodent damage can be costly unless effective control and prevention measures are maintained constantly as an ongoing program. Effective prevention of damage includes (1) an effective killing program, (2) sanitation within the plant, and (3) effective measures to keep them out.

Effective Killing Program

Insect damage can be prevented by treating seed with insecticides prior to storage, fumigating seed when they are received, are in sacks or bins, and by treating bins, storage areas and conditioning areas with insecticides.

Rodent and bird populations can be reduced by keeping the outside area clean and free of trash, by making buildings rodent/bird-proof, keeping poison baits or water in place, trapping, and by fumigation. Pre-baiting with non-poisonous foods makes poison baits more acceptable. The use of anti-coagulants poison baits is safe and effective for ridding the plant of rodents.

Plant Sanitation

The first and most important prevention is good sanitation inside and outside the plant. This not only reduces fire and contamination hazards, but also reduces food and cover for insects, birds and rodents. A sanitation program should include means of protecting seed from high moisture, immediate removal of all screenings and refuse, storing wastes in covered metal containers; immediately cleaning up all spilled seed; storing sacked seed on pallets; avoiding stacking in corners and against walls; and keeping the outside area free of trash, weeds, wet places, and wastes.

Effective Exclusion Measures

Keeping insects out of the plant is difficult, but can be greatly improved by: removing screenings and litter from the outside area surrounding the plant, screening all outside openings, and fumigating possibly-infested seed as they enter the plant.

Rodent exclusion measures include use of concrete floors, masonry walls, metal flashing on doors and sills, and covering all openings and potential rodent entryways with 0.25-inch mesh steel hardware cloth or sheet metal. When a new warehouse is being constructed, include a 12-inch or 35-cm smooth-finished ratproof ledge approximately 1 meter or 36 inches about ground level, at the top of the foundation, all around the building.

Constantly be alert for signs of insect or rodent populations, and have a good control program in effect at all times.

Preventing Insect Damage

Protecting seed from insect attack is constantly necessary, to preserve seed quality unit it can be planted, and especially during storage from harvest until planting. Stored seed makes an ideal diet for a large group of insects known popularly as stored-product insects. These pests are adapted to live on foods low in moisture content. The germ (embryo) of seed, with its high food value, is particularly attractive to these insects and is often the first part of the seed which is damaged.

Kinds of Insects Attacking Seed

The Indian meal moth *Plodia interpunctella* (Hbn), the cadell moth (*Tenerbroides mauritanicus* L.), and the flat grain beetle (*Laemophaloeus pusillus* (Schonh)), feed most on the germ or embryo of seed.

The rice weevil (*Sitophilus oryza* L.) and the granary weevil (*Sitophilus granaria* L.) cannot breed in seed with moisture content of 8% or less, and they soon die if restricted to such seed for food. Mites, which are not true insects, require food with a high moisture content; they are not troublesome in seed unless the moisture content is above 12%. Moths such as the India meal moth, the Mediterranean flour moth (*Ephesitia kuehniella* Zell), the almond moth (*Ephestia cautella* Walker), and insects of the bran beetle type such as the sawtoothed beetle (*Oryzaephulus suprinamensis* L.), and the flour beetles (*Tribolium* spp) are capable of breeding in seed below 8% moisture content. Flour beetles are not troublesome in dry clean seed that is free of broken kernels, since the young larvae find it difficult to attack undamaged seed.

Maintain Good Storage Conditions

Seed should be kept in a dormant condition to preserve its viability. High moisture content and high temperature speed up the life processes of seed and hasten loss of germination in storage. For this reason, the fundamental requirement for preserving seed is to store them at low moisture content. Low temperature is also helpful, and is second only to moisture content in importance.

Factors which speed up the life processes of seed also speed up the life processes and activities of seed-infesting insects. The higher the moisture content of seed and the higher the temperature, the faster the insects reproduce and attack seed. Seed that are cool, dry and free of trash provide not only a favorable environment for seed storage, but also an environment which is unfavorable for insect development.

Eliminate Sources of Infestation

Most insect pests of seed are good fliers, and infestation may occur in the field before harvest, especially in warm humid climates, where insects may survive year round in the fields. Usually, however, infestation occurs after seed is harvested and is placed in storage. Insect pests of stored seed are rather general feeders and will breed in nearly every type of dried food and every storage place. Infestation results for storage in bins or storages which are not thoroughly clean and free of insects, or by storage near infested materials.

Packaged seed carried over from one year to another is particularly susceptible to infestation by the Indian meal moth. This moth, one of the most troublesome insect pests of stored seed, lays eggs on or near seed packages. The caterpillars or worms that hatch from the eggs are exceedingly small and easily enter seed packages through even the smallest openings. The caterpillars "web up" the contents of the seed bag, and when fully grown, cut their way out through the sides of the package, to infest other stored seed.

Preventive Control

Prompt harvesting of seed will reduce the opportunity for field infestation by insects. Seed infested in the field should be treated or fumigated immediately after harvest t prevent further damage, and should have as low a moisture content as possible. Storage at temperatures below 70°F also helps control insects.

Sanitation

After harvest, seed should be dried to safe moisture contents and the stored in clean, insect-free, weatherproof storages. Any nearby sources of insect infestation should be eliminated. Steel bins that are easy to clean and can be tightly caulked are suitable for controlling insect infestation. Wood bins should be thoroughly cleaned; walls and floors of all storages should be treated with a residual spray before the bins are refilled. This will kill most insects that emerge from burrows in the wood. Spray around door frames of steel bins where insects may be concealed.

Use an effective insecticide recommended in your area. Apply enough spray to be able to contact and kill any insects present. Use appropriate masks or other operator protection as recommending in your area.

Bagged seed storages should be of good construction, easily cleaned and suitable for fumigation. Eliminate spaces—holes, rough spots, etc.—in walls and floors which may provide cover for insects. Fill all cracks around posts, in walls and floors. Paint walls, and seal floors.

Old wood floors and badly-worn concrete floors are difficult and expensive to keep clean and to control insects. They should be replaced with smooth-finish concrete floors.

Eliminate corners, trash, holes, etc., which insects may use as cover. And, strict sanitation in and around storages is essential. Bagged seed should be stacked on pallets, far enough away from walls to allow inspection and cleaning. Spray floors, walls and ceilings periodically with residual insecticide sprays.

Fumigation

Fumigation is a rapid and effective method of destroying insects in stored seed. Some fumigants, however, may affect the viability of seed, particularly if fumigant dosage is high, seed moisture or temperature is high, and the exposure period is long.

To avoid damage and loss of seed viability due to fumigation, seed should be fumigated only if moisture content is 12% or less; seed should be in bulk or in porous (jute, etc.) bags to facilitate penetration of the fumigant and subsequent aeration; exposure should not exceed the period recommended by the fumigant manufacturer; and fumigation temperatures should not be above 85°F.

Hydrogen phosphide is a generally effective fumigant and normally does not affect seed germination. It can also be left in place for longer periods without damaging the seed.

Caution: All fumigants are toxic to humans if breathed. Fumigants should be applied only by trained persons who are familiar with precautions necessary to protect themselves and others from harmful exposure.

Insecticide Dusts

Mixing seed with an insecticide dust provides economical and lasting protection against insect damage. Treatment is simple, and when done properly, is not hazardous to workers or injurious to seed.

Several insecticides dusts have been used to protect seed. Learn and use those recommended by the competent authority in your area. Some dusts are active insect poisons, while others affect insects physically rather than chemically.

Active insecticide or poisonous dusts are effective regardless of seed moisture content.

Chemically-inert dusts are thought to kill insects by causing breaks in their waterproof fatty coverings; their body moisture then evaporates through these breaks, causing them to die. Effectiveness of inert insecticide dusts declines as seed moisture content increases above 12 percent.

Dusts may be applied by any method which will ensure a uniform coverage of the seed. Commercial seed treaters give good coverage for bulk seed. All workers involved in applying dust insecticides should be equipped with adequate respirators.

Slurries and Sprays

Seed can be treated with slurries or sprays of recommended insecticides. Any treated seed should be handled in accordance with the manufacturer's recommendations, and should not be used for food or feed.

Cold Storage of Seed

Most insects are inactive at temperatures below about 50°F, so seed stored at lower temperatures usually escape insect damage while they are held in the cold storage conditions. Low moisture content in seed is also important for insect control. Fungi also injure seed, even at low temperatures, if the relative humidity of the air around the seed approaches 80 percent.

When seed are taken out of cold storage into a higher-temperature ambient environment, moisture from the air's humidity will condense on the cold seed and cause a rapid rise in seed moisture to unsafely high levels. Such seed will deteriorate so rapidly that benefits from cold storage may be lost.

The safest way to remove seed from cold storage is to take them out of the cold storage and immediately place them in a dehumidified storage (RH of 25% or less) and hold them here until the seed reach ambient temperature. Then, they will not absorb excessive moisture from the air's RH (but will slowly absorb enough moisture to reach equilibrium with the ambient air), and thus will not deteriorate so rapidly.

Insect-Resistant Containers

Some benefits have been shown from insect-free or treated/fumigated seed being stored in insect-resistant containers. Most insects that infest seed have comparatively weak mouth parts and cannot cut through substantial wrappers used to package the seed. Many, however, can thrust their ovipositors through the meshes of cloth or jute bags and lay their eggs directly into the seed mass inside the bag. As they hatch, the larvae can also crawl through the fabric's meshes and through needle holes along the seams where paper bags are sewn. Closely-woven fabrics offer some resistance to insect penetration. Bags made of paper, paper laminated with plastic, or plastic bags are more resistant to insect penetration. However, if the bag is more or less vaporproof, they cannot safely store seed unless seed moisture content is very low (85 or less for starchy seed, 6% or less for oily seed).

Bags have also been treated with insecticides to help prevent insect penetration.

Preventing Damage by Birds and Rodents

An effective control program includes (1) an effective killing program, (2) good sanitation within and outside the plant, and (3) effective measures to keep birds and rodents out of the plant and its conditioning and storage areas.

Killing Program

Rodent and bird populations can be reduced by poison baits or water, trapping, and by fumigation. Pre-baiting with nonpoisonous food makes poison baits more acceptable. The use of anti-coagulants is safe and effective for ridding the plant of rodents.

Plant Sanitation

Good sanitation within the plant, and in outside areas around the plant, not only reduces fire and contamination hazards, but also reduces food and cover for rodents, and attractions for birds. A complete sanitation program should include means of protecting seed from high moisture; keeping the plant area dry; keeping the plant area free of trash or weeds which attract birds and rodents and provides cover or food; immediate removal of all waste products or storing it in covered metal containers; immediately cleaning up all spilled seed; storing bags on pallets; and avoiding stacking in corners and against walls.

Exclusion

Keeping birds and rats out of the plant id difficult. However, removing screenings and trash from the outside area, screening all openings (windows, etc.)into the plant, putting “semi-transparent door closing strips” over operating doors and entrances, and fumigating seed stacks helps control rodents and birds.

Exclusion measures should include using concrete floors, masonry walls or lower parts of walls, metal flashing on doors and sills, and covering all openings and potential rodent entryways with ¼ inch mesh hardware cloth or sheet metal. When a new plant or storage building is being constructed, include a 35-cm rat barrier ledge about 1 meter above ground level, at the top of the foundation, around the entire building. This should be smooth-finished concrete to help exclude rats and mice.

And, be alert for signs of rodents, birds and insects, and implement the control program as soon as signs are observed.

Construction of Conditioning and Storage Buildings

Good planning and proper construction techniques can help keep birds, insects and rodents out of the buildings, make them easier to keep clean and free of harbor for pests, facilitate fumigation, make seed handling easier, and facilitate maintaining complete identity of seed. Guidelines for building construction herein help improve management and profitable operations.

Preventing Loss of Seed Identity

If seed identity is lost, it is too risky to sell it; it is thus lost. To protect the enterprise's reputation for quality, any seed whose identity is lost, or is even questionable, must be discarded. Each bag, container and lot of seed must always have labeling which establishes its complete identity and ties it to records which provide full information on the seed.

Complete identity of seed requires:

1. Complete, up-to-date, and continuous records showing everything which happened to the seed and all necessary information on its identity, quality, condition, and origin.
2. Complete lot number identification, as discussed in the chapter on Lot Numbers.
3. Labeling each and every bag, bin, totebox, and any other container with identity of the seed, at least of a lot number of the current seed lot, and a permanent inventory number.
4. Stacking seed lots separately.
5. Using stack cards, with complete current information, on all stacks of stored seed.
6. Marking the floor of storages, including raw seed and all pre-conditioning storages, into clearly-defined small areas, and numbering or labeling each area.

Preventing Moisture Damage to Seed

Detailed information is discussed in the chapter on Seed Moisture. Seed must be kept dry, at or below the moisture contents which are safe for the particular handing of the particular crop seed. General procedures for maintaining safe moisture levels during seed conditioning include:

1. Sampling and testing moisture content of each seed lot immediately upon its delivery to the conditioning plant.
2. Immediately drying the seed to a safe moisture level.
3. Storing raw seed under dry, safe and protective conditions.
4. Cleaning the seed as soon as possible. As required (*e.g.*, re-drying maize seed after shelling, or dehumidified drying to put seed into vaporproof containers), re-dry seed during the conditioning sequence.
5. Bagging the seed at moisture levels safe for the type of package used.
6. Storing the cleaned seed in dry, safe storages.
7. Preventing moisture entry through floors/doors/walls/windows, protecting the storage from rain, having on water or water outlets in the storage (except emergency fire protection system), and keeping out animals, insects, unauthorized personnel, etc.
8. Regularly sampling seed in storage, and checking its moisture content and germination.

Preventing Admixture During Handling

Seed must be handled carefully during all drying, handling, conditioning, and storage operations. Management actions include:

1. Properly labeling every container of seed.
2. Have containers of only one seed lot in the conditioning area at one time. Do not bring in seed of another lot until the conditioning machines are cleaned up and ready to condition the new lot.
3. When laborers or the forklift operator brings raw seed from storage for conditioning, or takes cleaned seed into storage, he should have specific written description of the exact storage location from which, or to which, seed are moved.
4. At a prominent location in the conditioning area, have a chalkboard (blackboard) on which is shown the complete identity of the seed being conditioned, and all necessary information about its cleaning, bagging, lot number and labeling (Figure 17.1).

Current Conditioning Operations

Dates ______________________

Crop ______________ Variety ______________ Lot no. ______________

Total amount of seed ______________________________________

Raw seed storage location ___________________________________

Cleaning machines used _____________________________________

____

Clean seed bag ________________ Lot no. ____________________

Bag net weight ________________

Clean seed storage location __________________________________

Figure 17.1 Information to post on blackboard in conditioning area, for notice to staff, while a lot is being cleaned. Source: Gregg, Records and Reports, unpublished.

Personnel Allowed in the Plant

Persons who are not performing essential operations should not be allowed into the conditioning and storage areas. There is always a risk of accident or injury, and the resulting legal liabilities increase insurance costs. In addition, such personnel may accidentally or deliberately cause damage to the seed. Strict rules should be followed as to who can be in the conditioning and storage areas.

Chapter 18

Mechanical Injury to Seed

Seed are living entities, and must be handled carefully to avoid injuring or killing them. The ability of a seed to germinate and produce a normal seedling can be reduced or destroyed completely by mechanical injury received during harvesting, handling or conditioning.

Causes of Mechanical Injury

Any point in any operation where seed receive sharp blows or strike hard objects can cause mechanical damage to the seed. Any sharp blow to the seed, either from the seed falling and striking other seed or a hard surface, or a moving object striking the seed, can cause visible or internal seed damage. Impacts are received:

1. When the seed fall against a hard surface, as when they fall into a bin, or
2. When a moving hard object strikes the seed, as when the thresher cylinder strikes them.

Susceptibility to Mechanical Injury

Kind of Seed

All seed are susceptible to mechanical injury. Some seed kinds are more fragile and more easily damaged. Seed of soybean, pea, bean, onion, cowpea, etc., are highly sensitive to impact damage and must be handled with great care. For example, soybean seed are brittle and the cotyledons crack easily when impacted. The embryo may also be damaged, and damaged seed are more easily attacked by pathogens. The result is failure to germinate, or an abnormal seedling. In maize seed, the embryo is at the extreme end of the seed, so it is easily damaged in mechanical handling.

Other seed kinds are relatively resistant, and greater impact force is required to cause serious damage. Some seed, such as sudangrass, have a hard cover which protects them from damage. Cottonseed also has a protective cover, although it is more delicate and easier to damage. Most grass seed have structures which protect the embryo and endosperm from impact damage.

Regardless of their natural resistance, all seed should be handled gently to ensure minimum damage and maximum health of the seed, to ensure their ability to produce normal undamaged seedlings which are free of disease and rot.

Seed Moisture Content

Dry and brittle seed are more susceptible to mechanical injury. However, whether seed are dry and brittle, or moist and pliable, any sharp impact can cause injury to seed. Damage from impact or abrasion is more severe when seed are threshed or conditioned at moisture content below 10 percent. Excessively dry seed are easily damaged. Likewise, seed of very high moisture content (18% or above, with most seed) are pliable and are more susceptible to mechanical damage. In general, most seed kinds are most resistant to mechanical injury when moisture content is from 13% to 18%. To minimize damage, most mechanical operations on seed are done within this moisture range, while allowing for the effect of moisture content on physiological seed activity.

Damaging Force

Different degrees of impact force cause different degrees of damage, depending on the seed's natural resistance to damage. Research has shown that relatively mild impact force can cause mechanical injury. As seed fall, its velocity increases with distance until it reaches terminal velocity, which varies with seed kind. The greater the distance that seed fall, the higher is their impact speed and the greater is the mechanical damage.

While some seed kinds are relatively safe at this force level, the general recommendation is to avoid allowing seed to free-fall more than one meter and then strike a hard surface such as a steel bin wall or other seed.

Types of Injuries

Mechanical injuries that affect or destroy normal germination do not always appear as split or broken seed. Many impact injuries are visible only under magnification, and some may be internal injuries which cannot be detected except by germination tests or dissection. Types of injury include:

1. Broken seed, leaving seed in pieces.
2. Chipped seed, with the seed largely intact, but small portions broken off.
3. Cracked seed, with cracks across the seed endosperm, embryo or cotyledons.
4. Internal invisible damage, where seed appear undamaged but internally the embryo is damaged or the endosperm or cotyledons are cracked.

In addition to reduced germination or producing abnormal or damaged seedlings, mechanically damaged seed are more susceptible to chemical injury from treating chemicals, as the chemical can enter the damaged place and affect the embryo. This is especially noticeable when treated seed are stored for longer periods.

The seed coats of cottonseed are often cracked or chipped when the seed are improperly mechanically harvested or mechanically delinted too closely. When cottonseed with damaged seed coats are acid-delinted, the acid may enter the cracks and damage the embryos, especially if the seed are stored for longer periods such as carryover.

Effects of Mechanical Injuries

In addition to reduced germination or production of more abnormal or damaged seedlings, mechanically damaged seed also are more susceptible to chemical injury from treating chemicals, as the chemical can enter the damaged place and affect the seedling. This is especially noticeable when treated seed are stored for longer periods. Cracked seed are more susceptible to attack by pathogens such as fungi, either seed-borne or soil-borne.

The seed coats of cottonseed are often cracked or chipped when the seed are improperly mechanically harvested or mechanically delinted too closely. When cottonseed with mechanically-damaged seed coats are acid-delinted, the acid may enter the cracks and damage the seedlings, especially if the seed are stored for longer periods such as carryover.

Moisture Content and Mechanical Injury

Damage from impact or abrasion is more severe when seed are threshed or conditioned at a moisture content below 10 percent. Excessively dry seed are must more easily damaged. Likewise, seed of too high moisture content (18% or above, with most seed) are more susceptible to mechanical damage. In general, most kinds of seed are considered to be most resistant to mechanical injury when moisture content is in the 13% to 18% range. Mechanical shelling of maize seed is recommended when seed moisture content is 14–16%.

Seed Moisture Content Safe for Operations

While some compromise between the potential for mechanical damage and heating damage must be reached, seed moisture content should be within the safe range for as much as possible of the mechanical handling process. To minimize mechanical damage, seed operations are planned according to moisture content at which seed are most resistant to damage. Ensuring minimal damage to maintain high seed quality may require several drying operations.

For example, maize seed harvested on the ear at 22–35% moisture, rather than combine shelling, reduces damage. Then the ears are sorted and dried to 14–16% seed moisture and shelled. After shelling, the seed is dried to 10–12% for conditioning to prevent moisture or heating damage during bulk handling. Then, if seed are to be packed in vapor-proof bags, after treating they are dehumidified-dried down to 8% and then bagged. After bagging, the seed are handled on pallets, prevent impact damage to the seed due to handling or dropping individual bags.

Preventing Mechanical Injury

To protect the germination of seed and their ability to maintain viability during storage, eliminate all points in all operations where seed receive sharp blows or strike hard objects. Any sharp blow to the seed, either from seed falling and striking other seed or a hard surface, or a moving object striking the seed, can cause visible or internal seed damage.

Harvest: The combine or thresher cylinder is a major source of impact injury to seed. The harvester who prides himself on bringing clean seed out of the field may be unwittingly reducing seed germination. Conditioning should be used to remove undesirable materials. It is more profitable and safer for seed quality to bring in viable harvested seed with high trash content, than to bring in ready-cleaned seed with mechanical impact damage. Keep the combine or thresher cylinder speed as low as possible, and carefully adjust concave-cylinder clearance to minimize mechanical injury.

Even threshing floors cause significant mechanical injury. More damage can be done on a threshing floor than on a properly-adjusted thresher or combine. Use a thresher or combine if possible; if a threshing floor is used, keep a thick bed of straw or plant material on the floor, to cushion the seed.

Conditioning: Most severe impacts between seed and hard surfaces during conditioning can be eliminated or reduced by minimizing drops and padding points of impact.

Elevators: Elevators can be modified to reduce the number and force of impacts seed receive. Regular belt-bucket elevators operating at slower speed are used for seed; high-speed grain elevators should not be used. For easily-damaged seed such as beans and peas, only slow-speed bucket elevators, such as continuous-bucket or chain elevators, should be used. High-speed operation of standard belt-bucket seed elevators will increase capacity somewhat, but it significantly increases mechanical damage to seed. Elevator buckets are designed to empty properly at the recommended belt speed; even a slight increase in belt speed causes the buckets to sling seed out forcibly, and changes the angle at which seed strike the steel discharge chute. As the angle of impact approaches 90 degrees, seed strike the elevator head with more force and more mechanical seed damage occurs. Always operate elevators at exactly the speed recommended by the manufacturer.

Spouts and Pipes: For less fragile seed, a rubber or canvas baffle across the lower end of the spout filling the bin can absorb much of the speed and force of seed falling from the spout. The rubber baffle catches seed as they fall out of the spout, stops them, and then lets them fall more gently into the bin.

Avoid sharp bends in seed pipes or spouts, as falling seed strike the pipe in the bend at a sharp angle which will damage seed. This will also erode the wall of the bend in the pipe, ultimately causing a hole.

Long Drops: A common method of feeding a separator installed some distance from the elevator leg is to increase the height of the elevator, to get a 45-degree (or more) slope in the discharge spout from the elevator to the bin. This often requires a long discharge spout, and seed velocity can build up to damaging levels as the seed go from the elevator to the distant bin. This can be avoided by using a lower elevator combined with a vibrating or belt conveyor to bring seed horizontally to the elevator from the previous machine. Vibrating conveyors mounted on the floor, or trough-belt conveyors installed at elevator heads, or beneath separators, or at other locations, can move seed horizontally and eliminate damaging drops.

"Cushion boxes", with rubber-padded baffles which slow down seed without causing a damaging impact, can be installed in seed pipes (Figure 18.1).

Let-down ladders (Figure 18.2) which have rubber-padded baffle plates, can let seed down gently from a spout to a machine, elevator hopper, or bin location.

When seed drop more than one meter into a bin and strike other seed or the bin wall, seed damage may occur. Avoid long drops from elevator spouts to bins by placing the elevator closer to the bin to reduce the length of the spout. To move the elevator closer to the bin, a horizontal conveyor (usually a self-cleaning vibrating or belt conveyor) can bring seed from the previous machine to the elevator feed hopper.

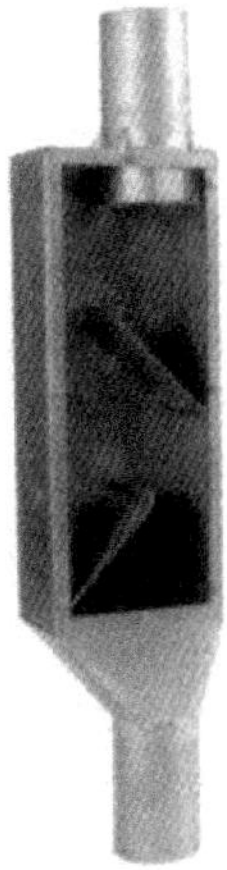

Figure 18.1 A cushion box, with two sloped rubber-padded surfaces to slow seed velocity, can be installed in long seed pipes to minimize mechanical injury caused by impacts. Source: Chris Ansel, Creamer Metal Products.

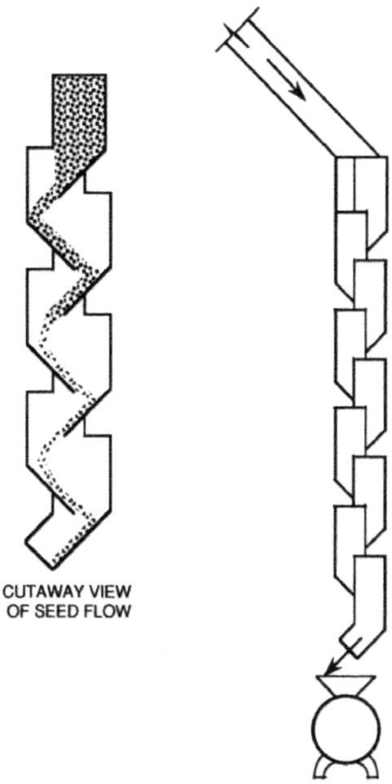

Figure 18.2 Diagram of "let-down ladders" which can drop seed to the feed hopper/bin of a separator, without causing mechanical impact damage to the seed. Source: Gregg *et al.*, 1970.

Holding Bins: When seed drop more than one meter into a bin and strike other seed or the bin wall, seed mechanical damage has been reported. Where possible, avoid long drops from elevator spouts to bins by placing the elevator closer to the bin and reducing the length of the down spout feeding the bin. To move the elevator closer to the bin, a horizontal conveyor (usually a self-cleaning vibrating conveyor) can be used to bring seed from the previous machine to the elevator feed hopper.

Where long down-spout drops cannot be eliminated and seed build up damaging velocity in falling down a long spout, or drop a long distance from the spout into the bin, several methods can decrease the force of the impact when seed fall into the bin.

Inside the bin, rubber-padded "seed step-down ladders or baffles" or rubber-padded spiral chutes, can carry seed safely from the spout down into the bin (Figure 18.3). They have open sides, and automatically discharge seed at a higher point as the seed level in the bin rises. Spiral slide let-downs (Figure 18.4) can also be installed in bins. One of these systems should be installed inside every bin.

In addition to rubber-coated step-down ladders, conditioners often line the inside of bins with rubber to cushion the shock of seed striking the sides of the bin.

Using smaller (about 1 ton capacity) toteboxes handled by forklift eliminates many deep seed bins, where seed can be damaged by falling long distances into the bins and striking the bin or other seed and being damaged.

For less fragile seed, a rubber or canvas baffle across the lower end of the spout filling the bin can absorb much of the speed and force of the falling seed. The rubber baffle catches seed as they fall out of the spout, stops them, and then lets them fall more gently into the bin.

A common method of feeding a separator installed some distance from the elevator leg is to increase the height of the elevator, so that a 45-degree (or more) slope can be obtained in the discharge spout from the elevator to the bin. This often requires a long discharge spout, and seed velocity can build up to damaging levels as the seed go from the elevator to the distant bin. This can be avoided by using a lower elevator combined with a vibrating conveyor to bring seed horizontally to the elevator from the previous machine. Vibrating conveyors mounted on the floor, or trough-belt conveyors installed at elevator heads, or beneath separators, or at many other locations in the conditioning plant, can move seed horizontally and eliminate damaging drops.

Sharp bends in seed pipes or spouts should be avoided, as falling seed will strike the bend in the pipe at a sharp angle which often causes breakage or internal damage to seed.

Other methods can be used to reduce the number and force of impacts which seed receive. Any reduction can reduce mechanical injury and improve seed quality.

Dryers: Larger bin dryers or continuous-flow dryers may damage seed falling into the dryer, especially the first seed fed into the dryer. The baffles, etc., in columnar dryers can be rubber-padded to minimize seed impacts. In bin dryers, step-down bean ladders can be used.

On bag dryers and small-capacity dryers, handle seed carefully so as to minimize the force and number of impacts.

Storage and Mechanical Injury

Storage Life

Mechanically-damaged seed deteriorate faster in storage, so lots with a high level of injury should not be carried over, especially for longer than one year.

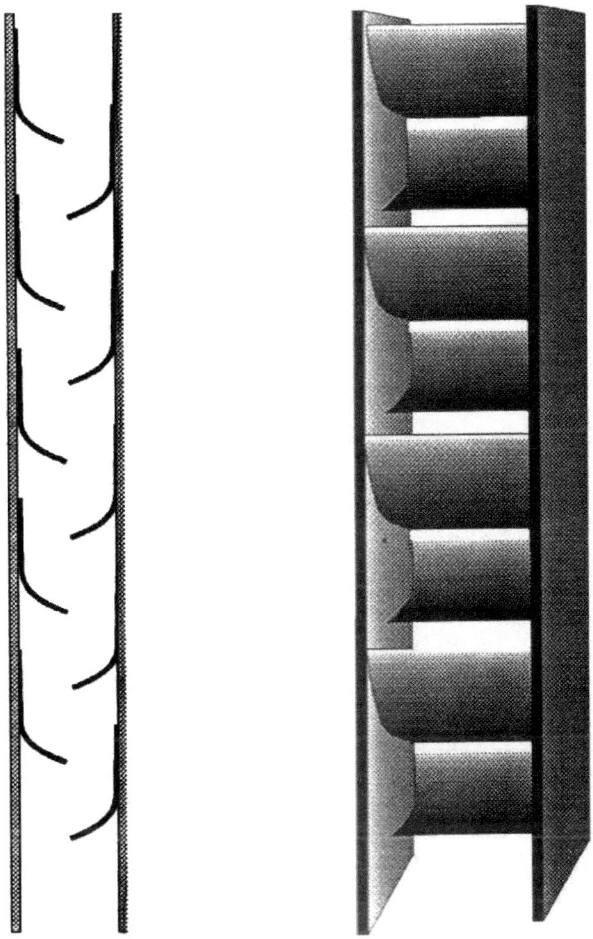

Figure 18.3 Let-down ladder which can be installed in a bin, to prevent long drops which cause mechanical damage to seed. Source: Crippen International descriptive literature.

Bulk Bin Storage

Larger storage bins, especially when they are first filled, have long drops which may cause significant seed damage. As an example, the first seed falling into a storage silo often are damaged so much that there is a layer of broken seed in the bin bottom. Larger storage bins should have step-down ladders or spiral chutes to load seed safely into the bin.

Using smaller (about 1 ton capacity) toteboxes handled by forklift eliminates many deep seed bins, where seed are damaged by falling long distances and striking the bin or other seed.

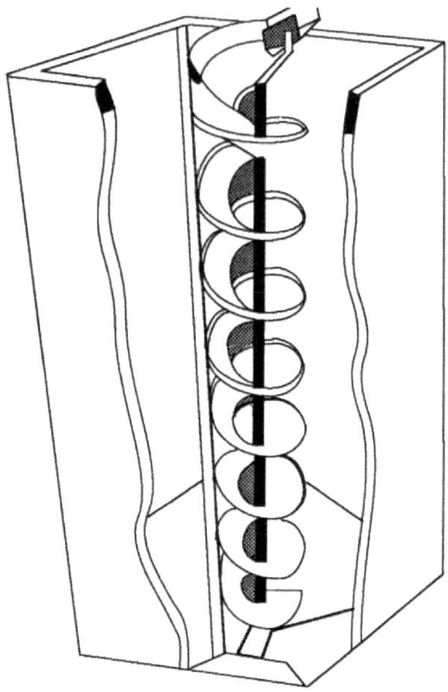

Figure 18.4 Diagram of a spiral let-down installed in a bin. The bin can be filled without long drops which cause impact damages to seed Source: Gregg *et al.*, 1970.

Bagged Seed Storage

Dropping or throwing bags off stacks causes serious mechanical injury when the bags strike the floor. Bags should be stacked and handled on pallets, to avoid falls and mechanical damage. If bags are handled carefully, there is no additional damage due to pressure in stacking.

Treatment and Mechanical Injury

If damaged seed are treated, they tend to lose germination faster during storage than do undamaged seed, or untreated damaged seed. Apparently, over a period of storage, the treatment chemical enters through damaged areas and affects the embryo.

Cottonseed coats may be cracked or broken in harvest, ginning or handling. If damaged seed are acid-delinted and then stored for longer periods, the acid which entered the broken seed coats has been shown to damage the embryo and reduce germination. There seems to be no adverse effect, if such seed are planted in the season immediately following harvest and handling.

Evaluating Facilities for Causing Mechanical Injury

IQC or specialists should carefully examine the entire flow, handling, separating, and storage system to identify locations where impact damage occurs. A major role of IQC is to sample seed and check for damage done at each point in handling and conditioning, and then recommend and help install solutions.

When a bin, elevator or machine is installed, IQC should sample seed going into and coming out of it. The samples are evaluated for mechanical damage, and the level of damage to seed coming out is compared to the level of damage in seed going in. Thereafter, a check once a year, or when any change in the installation or operation occurs, should help monitor mechanical damage.

Checking Seed for Mechanical Damage

The conditioner should carefully examine his entire flow, handling, separating, and storage system to identify locations where impact damage occurs. A major role of Internal Quality Control is to sample seed and checking the amount of damage done at each and every point in handling and conditioning, and then recommending solutions.

When the plant or any bin, elevator or machine is installed, IQC should sample seed going into and then coming out of each point. The sample is then carefully evaluated for mechanical damage, and the level of damage to seed coming out is compared to the level of damage in seed going into that point. Thereafter, a check once a year, or whenever any change in the installation or operation occurs, should help monitor mechanical damage.

Determining Extent of Injury

Several tests can be used to identify the presence and extent of mechanical injury in a sample of seed.

1. *Visual examination*: visually examine all surfaces of each seed, under magnification, to identify cracks, breaks and other external damage. This usually cannot show internal damage.
2. *Fast green test*: soaking seed in an aqueous solution of fast green helps visually identify damage. The fast green will enter cracks in the seed coat and dye the underlying starch a bright green color, which makes it easier to identify small breaks.
3. *Abnormal seedling check*: a germination test is conducted, and the seedlings carefully examined for damaged parts.

4. *Seed internal damage*: several methods of opening seed of different crops (*e.g.*, soybeans) to examine the cotyledons for breaks or damage have been used.
5. *Seed internal damage*: several methods of opening seed of different crops (*e.g.*, soybeans) to examine the cotyledons for breaks or damage have been used.

Chapter 19

Waste Products

Contaminants in Seed Lots

Every raw seed lot contains undesirable materials consisting of:

1. Trash, plant materials, leaves, stems, etc.
2. Broken seed and seed pieces.
3. Other crop seed.
4. Weed seed.
5. Immature, light, moisture-damaged, insect-damaged crop seed.
6. Seed appendages, husks, stems, etc.
7. Sometimes, extraneous materials such as rocks, mudballs, hand tools, broken pieces of equipment, bag strings, etc.

The purpose of seed conditioning is to remove all undesirable materials, leaving only pure crop seed which are mature, whole, undamaged, high-germinating, able to give the desired crop stand in the field, at the lowest planting rate.

Conditioning cleaners and separators remove specific undesirable materials from the crop seed. From one flow of seed going into the machine, it creates additional flows of various components of undesirable or waste materials and good seed which go out of the machine. These new waste product flows must be handled separately from the good seed flow. Often, some waste products must be handled separately, as they can be used as grain or feed, and must be kept separate from other waste products which have no value and will be burned or used for compost.

Pre-Cleaning to Reduce Waste Products

Modern plants have a separately-installed scalper or pre-cleaner, to pre-clean seed as they arrive at the plant, before going into pre-conditioning storage. This removes much of the waste materials, and reduces the storage required.

Many plants pass raw seed over a scalper or pre-cleaner before the seed enter the main cleaning line. This adds another machine to the conditioning "line", but it removes the bulk of the coarse trash, and sends it directly to a trash bin or container. This minimizes handling of waste products, increases overall conditioning capacity, and enables the air-screen cleaner to make a closer separation.

Handling Waste Products

Moving screenings, tailings, air liftings and other waste products out of the conditioning line to trash bins or for disposal is one of the more difficult problems in seed conditioning. These waste products must be removed from the cleaning line, but are often a major amount of material, and an adequate handling and disposal system is overlooked in planning the seed plant. Many headaches can be avoided by planning, installing and operating an adequate waste removal system.

In small operations of low capacity, waste products can be collected into bags mounted on bag holders beneath the discharge spouts of the machines. The bags are then changed by hand as they fill with waste products, and carried out of the conditioning area.

In medium to high capacity operations, however, hand handling of bagged waste products becomes costly and laborious. In addition to requiring more labor and bags, it requires more space around machines, to handle the bagged waste products.

In more efficient, or larger, conditioning plants, waste products are spouted to a conveyor, elevator or auger which moves them out of the conditioning plant to a trash bin. Suction fans are also used to pull light trash, such as air liftings from the air-screen cleaner, to a collector. The airlift elevator or vibrating conveyor makes an excellent system of waste removal.

Collecting Waste Products

Waste products can be collected in (1) a bin inside the conditioning plant, for emptying into bags or toteboxes, or in (2) an outside weatherproof overhead trash bin with a trapdoor bottom which permits dumping the trash into trucks (Figure 19.1).

Wastes Containing Weed Seed

Disposal of waste products which contain weed seed is often controlled by law. Burning such material in incinerators at the seed plant or elsewhere, is

Figure 19.1 An overhead outside bin to collect waste materials (in this case, burned linters from delinted cottonseed). An air conveying system carries the waste and drops it into the bin. As the bin is filled, a truck is driven under the bin, and the waste material is dumped into the truck. Source: Don Clark, Continental Eagle.

a common practice. A good disposal plan, if the wastes have no weed seed or treating chemicals, is to process them through a hammermill and pelletizer, and then sell the pellets as livestock feed. However, if wastes are used in livestock feed or in compost or mulch, be certain that weed seed are rendered non-viable by grinding or by lethal fumigation. In most cases, grinding will not be fine enough to destroy weed seed germination.

Separator Installation to Handle Waste Products

Waste products (and clean seed) are often discharged from a low level, or even bottom, of the cleaner or separator. If the separator is installed on the floor, it cannot feed waste products (or clean seed) into an elevator or conveyor. If a pit or trench is installed in the floor to handle elevators below the machine, it limits the flexibility of operations, is more expensive, may have soil moisture problems, and may create hazards to workers or potentially dropping contaminants into the seed.

This problem is eliminated by installing the cleaner/separator on a stand which supports the machine above the floor, at the required height so a conveyor or elevator can be installed to handle the discharged materials (Figure 19.2).

Before planning the installation of machines, identify all their discharge spouts, determine the height needed below the spout for conveying/elevating equipment, and then plan the machine's support stand. The stand should permit full operator access to the machine, sampling discharged seed and products, and spouting them into elevators or conveyors.

Storing Waste Products

Anecdotal experiences indicate that waste products are more susceptible to insects, and often harbor heavy insect infestations which spread to good seed. To prevent this, waste products should never be stored with cleaned seed.

Waste materials should be removed from the conditioning/storage plant as soon as possible. Properly labeled and identified, they may be stored for short periods with raw seed.

Waste products should be regularly checked for insect infestation, and fumigated as needed.

Waste Product Disposal

A definite program for regular and frequent disposal of waste products should be followed. Wastes should not be allowed to accumulate in the storage.

Wastes can be disposed of in several ways:

1. Put into livestock feed, if the wastes do not contain chemical treatments, harmful materials (such as ergot) or troublesome weed seed.
2. Used as food grain, if the wastes are untreated food grains which contain no harmful materials and have been handled carefully. (in most cases, seed are handled with greater care than are food grains).
3. Incinerated, to destroy trash and weed seed. Because of the fire hazard, the incinerator should not be located near to conditioning and storage operations.
4. Used as compost; in this case, the wastes should be decomposed properly and under controlled conditions, to ensure that it does not contain weed seed.

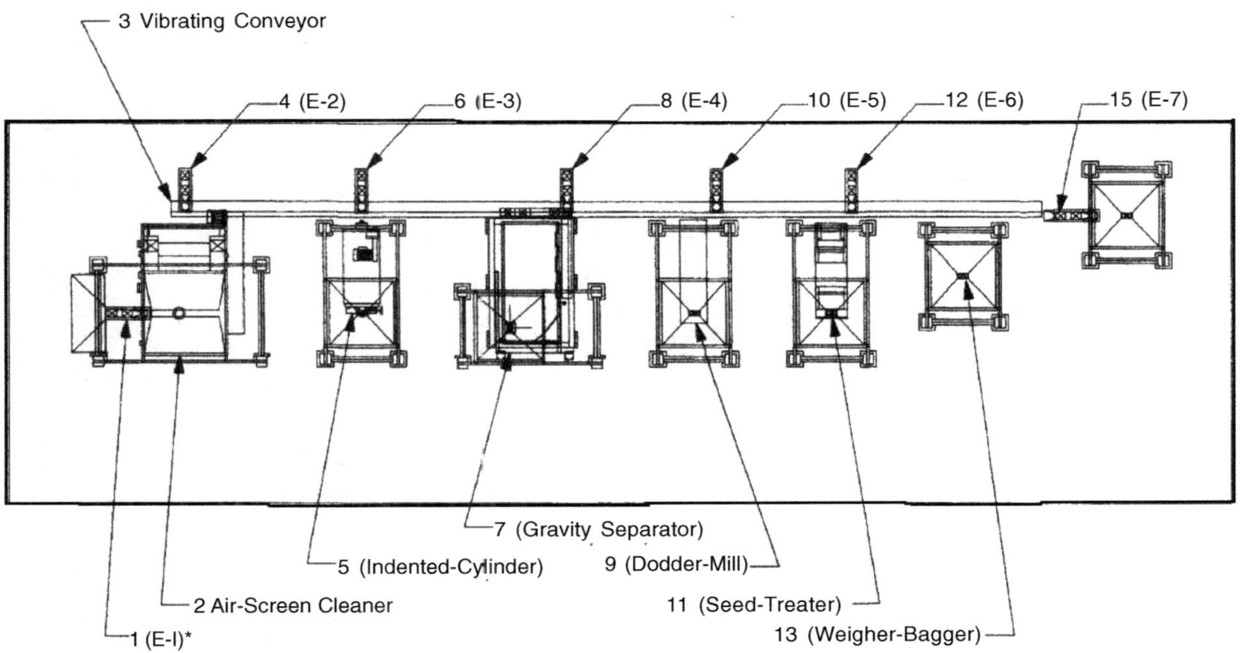

* E = elevator

Figure 19.2 Seed plant layout using a 2-channel vibrating conveyor, mounted on the floor, to move both seed and waste products. The first channel in the conveyor carries good seed from machine to machine; the second channel collects waste products from all machines, and carries them to a waste collecting bin at the end of the conditioning "line". Source: Billups unpublished designs.

Chapter 20

Dust Control

Dust in Seed

Dust consists of fine particles of soil, or equally fine particles of plant materials or materials rubbed off seed or grain. Rough, raw or uncleaned seed generally contain fine dirt, crushed plant material, and other matter which creates dust in moving the seed during conditioning. In addition, a fine dust is rubbed off of many kinds of seed as they pass through elevators and cleaners. These factors combine to create a dust problem in the conditioning plant which is unpleasant and dangerous to workers, and may be a fire or explosion hazard.

Dust Dangers and Explosions

Dust is unclean and unpleasant and often dangerous to workers, in that some kinds of dust cause breathing and respiratory hazards. Dust is also a problem for some machine parts such as bearings. More dangerously, dust from seed or grain is extremely flammable, and may catch fire so rapidly that it may cause a serious explosion. Even reinforced concrete storage silos have been destroyed by dust explosions.

Dust Control

For the safety of staff and the plant itself, dust must be carefully and completely controlled in each operation, and in all plant areas.

Dust will be generated in any operation of handling and moving grain or seed. Dust can be kept at manageable levels only by controlling where it goes, and removing it from the operations and operating area.

A dust collecting and handling system provides efficient trunking of dust-laden air to collectors where dust and light chaff settle out of the air without creating a hazard, and without being exhausted into the environment.

Most cleaners which have an air separation system have enough air pressure to move dust through pipes to dust-settling collectors. Quite often, auxiliary fans are needed to help move dust and chaff for the required distance to collectors, and to reduce inside air pollution. The additional fans are required to move dusty air from machines, elevators, bins, etc., which are not equipped with dust exhaust fans. Auxiliary fans are also needed to move dusty air from inside the plant, to reduce inside air pollution, prevent dust from settling inside the plant, and move dusty air safely to collectors. And, constant vacuuming, dusting, and cleaning are required to remove dust from the general plant area.

Scalper in a Separate Room

One means of reducing dust in the conditioning system and area is to install a scalper (with good air separation) in a separate room. As raw seed is received, it goes over this scalper which removes much of the dust and light dusty trash before the seed go into the conditioning system. This reduces the amount of raw seed in storage, and minimizes dust in conditioning.

Dust Collectors

The dust-laden air is piped into a collector, where the air flow moves into a chamber much larger than the cross-sectional area of the air pipes. As the air volume is increased, the air flow velocity is reduced, causing the dust to settle out. In the collection chamber, the lower-velocity air moves in a circular motion. Dust and light materials are no longer supported by the air flow, settle out of the air, and fall to the bottom of the collecting system. The excess air, now moving more slowly, is discharged into the environment, without carrying the dust.

Dust collecting systems are designed for specific cleaners, and dust collection systems for general plant dust control are available. Most equipment manufacturers either provide, at extra cost, a complete dust collecting system, or provide specifications for an effective dust control system.

Several types of dust collectors are designed for specific cleaners, and dust collection systems for general plant dust control are available. The cyclone type collectors are widely used, and the modern approach to complete removal of dust from the air is to add a filtering system after the cyclone removes larger pieces of light chaff.

Cyclone type collectors have been widely used for years. Cyclone systems usually do not completely remove all fine dust from the air flow, so they are installed outside, where the air can be exhausted without affecting the inside plant working area. Modern systems completely remove dust from the air with a filtering system, either alone or after the cyclone removes larger pieces of light chaff (Figures 20.1, 20.2, 20.3, 20.4). Filter systems can thus be installed indoors, inside the plant, without adversely affecting the environment.

Installing Dust Collectors

Dust collectors for machines such as the air-screen cleaner should be installed as near as possible to the cleaner they serve, to eliminate extended air travel which slows the air stream and increases the cost of ducting. The advantage of using a cyclone collector followed by an air filter system is that the system can be installed inside the plant, quite near the cleaner served.

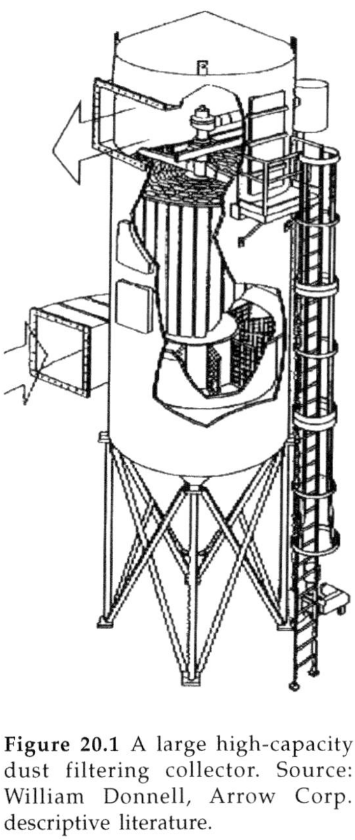

Figure 20.1 A large high-capacity dust filtering collector. Source: William Donnell, Arrow Corp. descriptive literature.

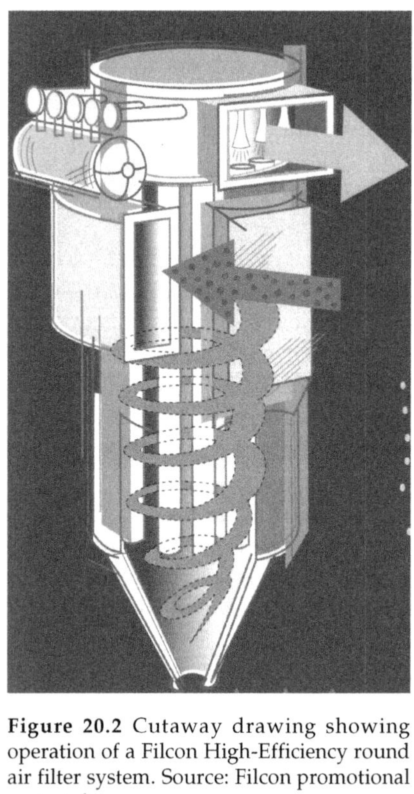

Figure 20.2 Cutaway drawing showing operation of a Filcon High-Efficiency round air filter system. Source: Filcon promotional materials.

Piping Dust-Laden Air

As the fan system collects the dust-laden air, it must be controlled and moved through airtight dust pipes to collectors.

Dust pipes between the separator or originating point and the collection system must ensure a smooth unimpeded air flow adequate to support the handling system. This requires pipes which are of the same cross-sectional areas as the exhausts on the separators or originating point, so that air flow in unimpeded. The pipes may be converted from square to round or from round to square, according to the pipe fittings present, so long as there is no change in cross-sectional area.

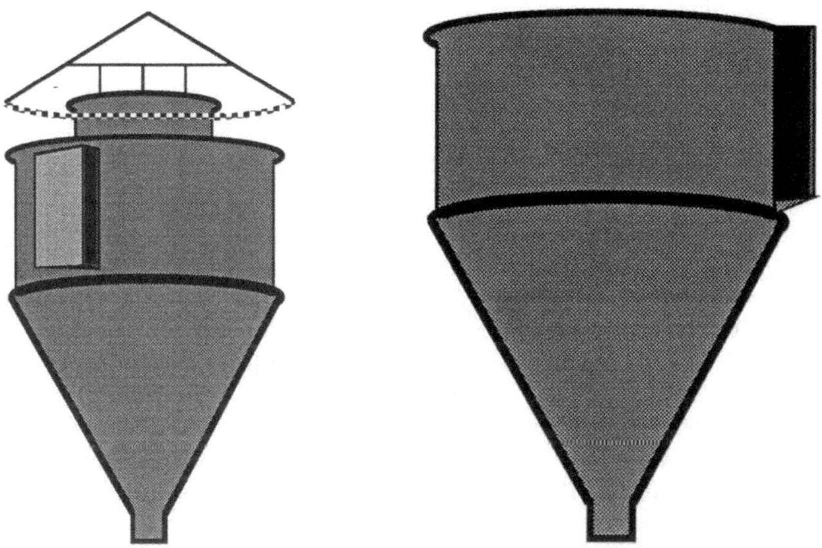

Figure 20.3 Cyclone dust collectors supplied by air-screen cleaner manufacturers. Left, left-hand cyclone with exhaust cover (top center) for outside mounting; right, a right-hand cyclone collector without exhaust cover. Left-hand or right-hand collectors can be selected to fit the machine installation. Source: Gregg *et al.*, 1970.

Dust collectors should be installed as close as is convenient and safe, to the points where the dust originates. The shorter the distance that dusty air must be moved, the less likely that there will be problems which interfere with movement of the dust-laden air.

Air-tight pipes must connect the fan moving the dust, and the collection system which removes the dust from the air flow. The air flow pipe system must be carefully designed, so as to ensure unimpeded air flow without dust settling inside the air pipes. This requires that the pipes permit smooth unimpeded air flow from the machines or fans, to the collector. Pressure of

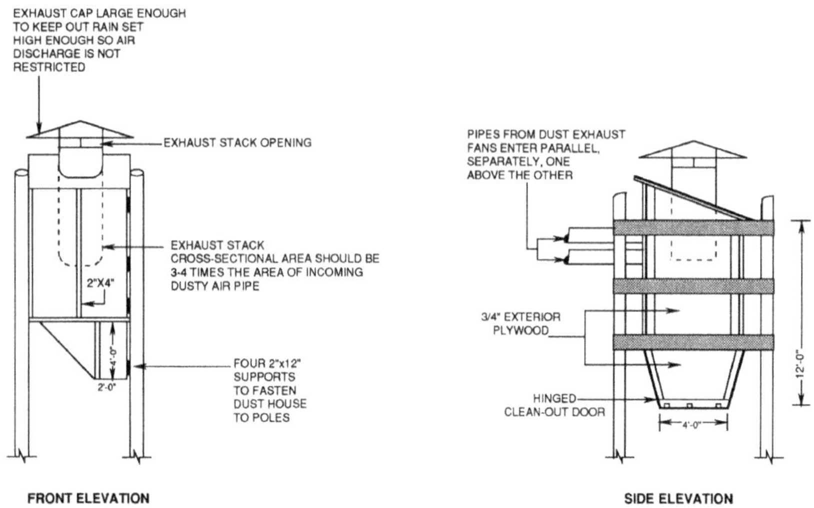

Figure 20.4 A cyclone dust collector for an air-screen cleaner, built by the plant staff. Source: Gregg *et al.*, 1970.

the flowing air must be uniformly maintained in the entire length of the pipes. This requires pipes which are of the same cross-sectional areas as the exhausts on the separators or originating point, so that air flow continues at an effective pressure and velocity. Pipes may be converted from square to round or from round to square, according to the fittings present, so long as there is no change in cross-sectional area.

All pipes must be smooth on the inside, to avoid frictional resistance to air flow. All joints must be tightly sealed, to prevent leaking dusty air. Edges of seams or joints must not face the air flow (*i.e.*, the edges must face in the direction of air flow). Allow an inside radius of the curve which is at least two times the diameter of the pipe. All bends should be of a wide arc, with a gradual curve, never a sharp or right-angle curve. All bends should be wide arc, smooth and present a gradual curve which does not interfere with air flow. Allow an inside radius of turns in air pipes of at least two times the diameter of the pipe (Figure 20.5). Sharp turns in pipes offer resistance to air flow, and cause dust to settle out in the bend; and, the sharp turns create back pressure to the separator or fan, and affect the air separation. Bends in pipes should also be in the same direction as the air currents in the settling collector chamber. If air currents inside the settling chamber move in a clockwise direction, avoid bends in air pipes which will move the air stream in a counter-clockwise direction. Both clockwise and counter-clockwise cyclone collectors are readily available.

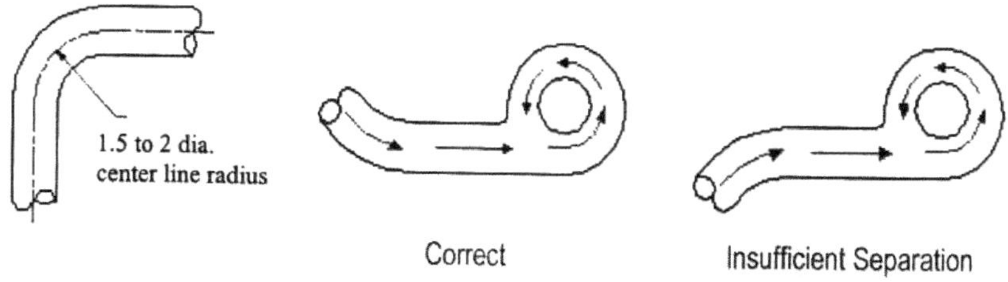

Figure 20.5 Correct bends in dusty-air pipes, to ensure unimpeded air flow and prevent particles from settling out in pipes. Left, minimum radius of a bend in pipes is 1.5–2 times the radius of the pipe; center, all bends in pipes should be in the same direction as air flow inside the cyclone collector; right, a bend in the opposite direction from the air flow in the cyclone will interfere with dust settling out in the collector. Source: Crippen International descriptive literature.

Bends in pipes should be in the same direction. Bends should be in the same direction as the air currents in the settling collector chamber. If air currents inside the settling chamber move in a clockwise direction, avoid bends in air pipes which move the air stream in a counter-clockwise direction. Both clockwise and counter-clockwise cyclone collectors are available.

Dusty Air From Cleaners and Separators

If a cleaner has a fan to create air flow and make an air separation, it usually generates dusty air which must be handled so as to collect the dust generated. Most manufacturers recommend a separate exhaust pipe and collector system for each fan. An individual pipe for each fan prevents back-blasts of air from one fan affecting the air separation made by the other fan. If it is necessary to use a single pipe and collector for two air separating fans, join the pipes from the two fans at a gradual angle, and install a divider inside the pipe at the point where the two exhaust pipes come together (Figure 20.6). The divider should be at least one meter long, and extend out toward the dust collecting system.

Collectors or collection systems recommended by the separator manufacturer should be used. However, a separate system with equal air handling capability may give equal results.

If auxiliary fans are used to boost air flow to distant collectors, they must not interfere with the air separation of the cleaner or separator; *i.e.*, they should not create negative or positive pressures on the cleaner fan, which will affect its air separation.

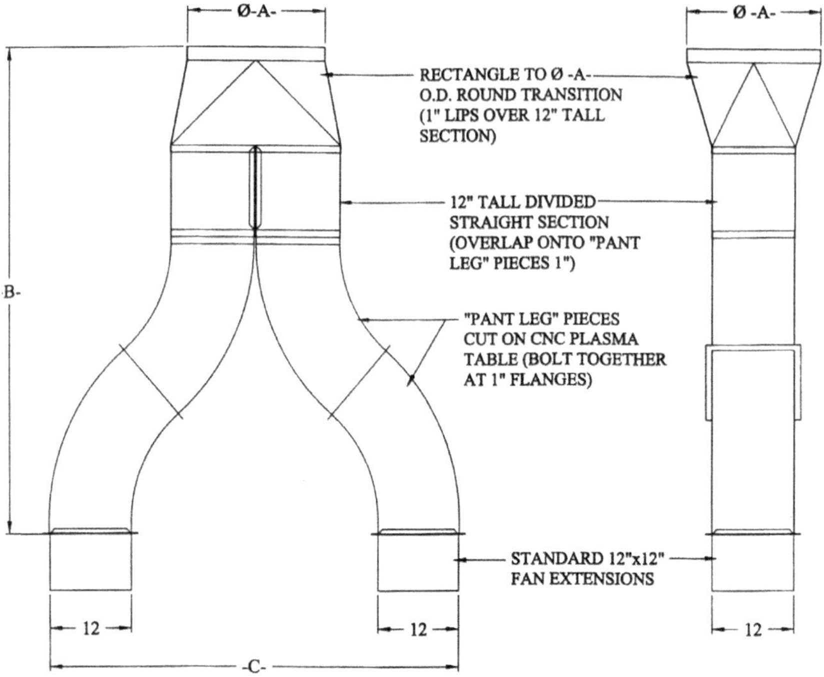

Figure 20.6 Curvature of the two dusty-air discharge pipes from a Crippen air-screen cleaner, to merge the two pipes into one exhaust pipe without causing back pressure to either of the cleaner's air exhaust fans. Source: Crippen International descriptive literature.

Air-Moving Fans

When auxiliary fans are used to help the built-in dusty air removal system of a cleaner, they must not interfere with the air separation of the cleaner or separator; *i.e.*, they should not create negative or positive pressures on the cleaner fan, which will affect its air separation.

For a complete in-plant dusty-air removal system, the fan must be powerful enough to handle the required volume of air at a useful velocity (Figures 20.7, 20.8). This usually requires professional installation.

Plant-Wide Dust Collection Systems

Most air handling and dust collection systems serve only one separator or cleaner. However, complete dust control systems can be installed to control dust throughout the conditioning plant. Dust collectors on all cleaning machines, exhaust hoods over dusty operations such as spouting and sacking points, removing dust from covered bins, and exhaust fans to

Figure 20.7 A typical fan system for moving dusty air. Source: Forsbergs Inc. descriptive literature.

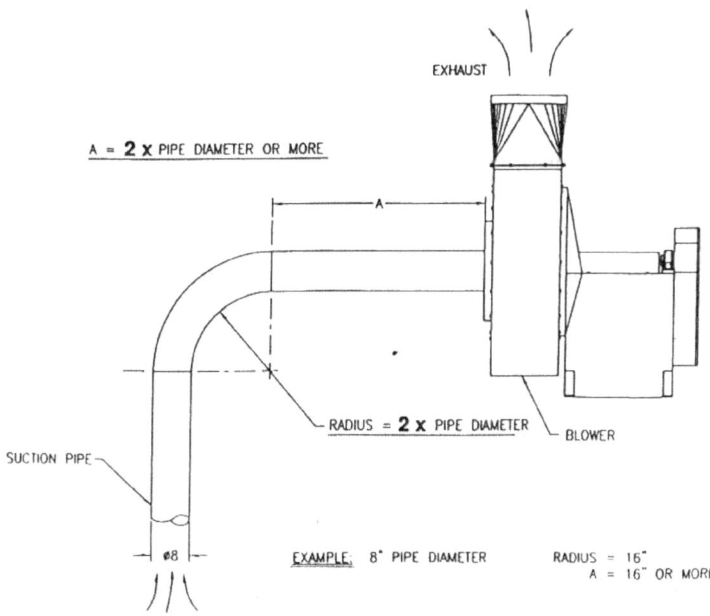

Figure 20.8 Connecting suction (intake) and exhaust (to collector) pipes to the fan in a plant dust control system. A similar system is used to remove dusty air from machines, including the vacuum gravity separator. Source: Forsbergs Inc. descriptive literature.

circulate clean air through the plant area can eliminate the fire and health hazards created by uncontrolled dust.

The complete plant dust control system uses a central fan to draw dusty air through pipes to a central collector. Not only dust from fans on cleaners can be handled, but hoods and fans can be installed to remove dust from other machines or treaters, from the air around dusty operations, from elevator heads or feed hoppers, from conveyor discharge spouts, from bins, etc. Individual collectors can be installed for each operation, or a central collecting system can be used to collect and handling dust from the entire system. Where dust particles are to fine for a single-cone collector to remove them from the dust-laden air stream, a double-cone collector or a cloth filter system can be used.

In a complete plant system, main dusty air removal pipes go through central plant areas, with smaller individual pipes going from this to each dust collection point. Often, the individual pipes will have shutoff valves, to close individual dust collection points when they are not needed.

At the dusty operation, a hood is used to collect air from the entire dust-generating area (Figures 20.9, 20.10). The dusty-air collection pipe is attached to the hood, so as to draw in the dusty air. The collection pipe is connected to the main air pipes, which feed air to the fan and on to the collection system.

Handling Settled-Out Wastes

Outside dust/chaff collectors, from which dust and light trash settle out, should be installed high enough for a forklift totebox or other container to be placed under the bottom outlet to collect the settled-out waste (Figure 20.11). Do not collect light trash in a bag hung onto the collector outlet. Always leave space between the container and the collector outlet, so that when workers forget to replace the filled bag, the waste material will not clog the collector.

Controlling Low-Level Dust

Seed kinds, or conditioning operations, which generate little dust may allow maintaining a safe relatively dust-free environment with only the air separation systems of the cleaners, combined with constant clean-up and adequate air movement through the plant areas.

Collectors or collection systems recommended by the machine's manufacturer should be used. However, a separate system with equal air handling capability may give equal results.

Figure 20.9 Dust collection hood over the discharge of a belt conveyor, to remove dust generated as the seed fall from the conveyor. The dusty-air removal pipe is connected to the top of the hood. Similar hoods and exhaust pipes can be installed over elevator feed hoppers and other points where dust is generated. Source: Darin Stutler, Bratney Companies descriptive literature.

Figure 20.10 A dust hood and suction exhaust pipe installed over the feed hopper of an elevator. The dust exhaust pipe is connected to the pipes leading to a central plant dust exhaust fan and collection system. Source: Crippen International descriptive literature.

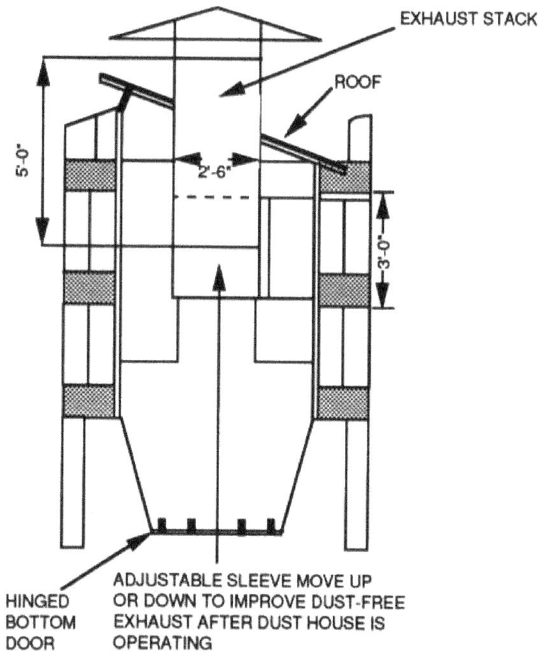

Figure 20.11 A "dust house" outside collector, built at a conditioning plant, with space at the bottom for dust and light trash to collect. When this area is filled with collected dust, a truck is driven under the dust house, and the bottom door is opened to dump the dust into the truck for transport to the disposal area. However, a problem with combined cyclone and collection area is that if the operator forgets to check and empty the collector, dust will pile up into the cyclone and interfere with dust collection or will blow into the outside area. Source: Gregg *et al.*, 1970.

High-Dust Operations

Some operations generate large amounts of dust. Sometimes, the dust can be harmful to workers, as the dust from sugar beet seed. Such operations require more complete dust collection systems.

This involves using:

1. Separating machines which have a complete covering, to keep dust inside the machine unit. Some modern machines are equipped with such a cover structure.
2. Auxiliary fans to remove dusty air from the machine. Some separators have built-in connections for dust exhaust pipes, and dust exhaust systems can be purchased as an additional component.
3. Hoods over dust-generating areas such as feed hoppers of elevators, bagging scales, gravity separators, etc. (Figure 20.12).

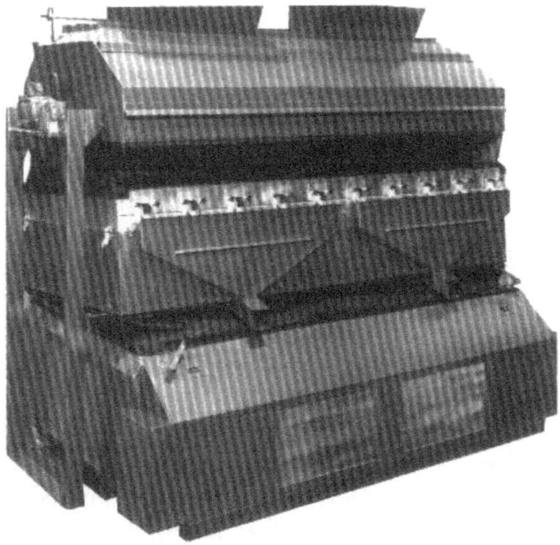

Figure 20.12 Crippen gravity separator with dust hood and exhaust connections for removal of dust generated by the gravity separation. Source: Crippen International descriptive literature.

4. General air movement system, with fans and screened windows to circulate outside clean air through the plant and move dusty air to the outside or to collectors. In colder climates, air inside the plant is sometimes sent through a filtering system to remove dust, and then the dust-free air is returned to the plant area.
5. General dusty-air removal system, with air pipes to all locations such as elevator heads, bins (which have top covers), etc., and auxiliary fans to pull dusty air from these locations and send it to a collector/filter system.

Chapter 21

Seed Bags and Containers

Bagging or packaging is the final step in making seed ready for marketing. The bag and its size are an important part of the seed marketing effort, and must be carefully managed to ensure that the bag and the seed are attractive to the farmer and provide what he wants.

Seed Packing

The bag or package holds the desired amount of seed as a specific and identifiable unit, and protects it from adulteration, loss, and to some extent damage.

It is easier for the conditioning plant to bag seed in large bags. However, the purpose of bagging is to provide what the farmer or ultimate consumer needs, not what is easier for conditioning. For example, if the typical farmer needs only 8 kg of maize seed, it is a serious marketing mistake to put the seed into 50-kg bags.

The closed and sealed bag identifies the seed brand and variety to the farmer, and eliminates the adulteration and quality problems which accompany sale of loose bulk seed. To protect both the using farmer and the supplying seedsman, seed should always be delivered to the using farmer in an unopened bag.

Bags

Bags are the most convenient means of packaging seed, except for very small quantities of vegetable and flower seed. As compared to boxes or cans, bags are relatively low-cost, easy and cost-efficient to fill and seal, are flexible, and can be stacked, handled, stored and delivered easily.

Bag Types

Material

Seed bags are constructed from several different kinds of material. There are cost differences, different materials are more readily available in different areas, and each has specific advantages. Most common bag materials are:

1. *Jute or burlap fiber*: Woven of relatively coarser fibers, this comparably low-cost material has been used for many years for making bags. Because of the space between threads, it is used for larger seed such as wheat, rice or maize. Not vapor-sealed, this bag material allows free exchange of air between the inside and outside of the bag. Its open mouth can be closed with a bag sewing machine or by hand stitching. It is used for larger amounts, up to 50 kg or more.
2. *Cloth*: Strong types of cotton-fiber cloth are widely used for making bags which range from small to the largest sizes in capacity. Cloth bags are readily vapor-porous. Because of their finer threads and closer weaving, these bags do not lose small seed through weaving pores, and are used for all kinds of seed, from small-seeded legumes to maize.
3. *Woven plastic threads*: The bag is of woven construction, using threads of a s strong plastic. These bags are often more readily available in local markets, and are used for handling seed in the same manner as jute and cloth bags. They are not vapor-proof.

   Bags of solid plastic or woven plastic threads do not stack well; they are smooth, will slide and fall off stacks. However, when a considerable area on both bag sides is printed, the bags will not slide and can be stacked normally.
4. *Woven plastic threads with a thin sheet of solid plastic*: Some woven-thread plastic bags have a thin solid-sheet plastic liner. These bags are suitable for both larger and smaller seed. While the sheet does restrict air and vapor exchange with the outside air, these bags are not vaporproof; both the thin sheet and the holes where the bag is sewn will allow air exchange.
5. *Solid sheet plastic*: Bags of solid sheet plastic are of a thicker ply, in order to provide the strength needed to hold seed. This construction is usually used for smaller bag sizes, down to 1 kg, or even smaller amounts as smaller packets. Solid plastic bags are strong, and are even used for fertilizers in amounts up to 50 kg. Closing this type of bag is usually done by heat-sealing it shut, to melt the plastic together to form a seal. This seals the bag and makes it vapor-proof; thus, seed must have very low moisture content at levels safe for vaporproof storage.

6. *Multiwall paper*: This bag construction is quite strong and is in common usage for all kinds of seed in amounts up to 25 kg. The bag is closed with a bag sewing machine, using care in spacing of stitches, to avoid cutting off the top of the bag.
7. *Multiwall paper with layers of aluminum foil and/or plastic*: This bagging material is often used, and is relatively vaporproof. It thus provides more protection for the seed, but seed must be of low moisture content.

Construction, Filling and Closing

There are two primary types of bag construction, which determines the method of filling and closing the bag:

1. *Open-mouth*: One end (the mouth) of the bag is completely open. The bag is suspended on a bagger-weigher machine, and held in place by a clamp. Seed is allowed to flow into the bag until the desired weight is reached. Seed flow is then stopped (usually automatically), and the bag is removed from the bagger-weigher and falls directly onto the bagging conveyor which is installed below the bagger-weigher. The conveyor carries the bag, at a coordinated speed, through the bag sewing machine which sews or stitches shut the mouth of the bag. The bag goes a further short distance on the bagging conveyor, and is then removed and stacked directly onto a pallet for forklift handling.
   In old plants, bags were filled by hand and weight was checked on a platform scale, and then stitched shut either by hand or with a hand-operated bag sewing machine. This method is slow and labor-consuming, and is no longer used in cost-efficient conditioning plants.
2. *Valve-pack*: Valve-pack bags are constructed with both ends already closed. In one corner of the top is a folding tube (the valve) of plastic with an open end. To fill the bag with seed, it is placed on a special bagger-weigher. On this specially-designed machine, the tube of the bag is placed over a pipe-like spout on the bagger-weigher, and is suspended from this spout. Seed flow into the bag through this spout, which is inside the tube on the bag. When the bag has the desired weight of seed, the bag is removed from the filling spout and the bag's tube is stuffed into the bag. The pressure of seed in the bag closes the tube, so seed do not fall out of the bag.

   The filled bag is thus self-closed, without sewing. It falls onto the bagging conveyor and goes directly to being stacked onto a pallet. This process takes little time and little labor.

A third method, heat-sealing the bag mouth, is used with bags of solid plastic sheeting. This method seals the bag completely vapor-proof, and is safely used only with low-moisture seed. An electric clamp-type heat sealer long enough to seal the entire bag mouth is commonly used. The heat level can be adjusted to provide a complete seal without burning the bag.

Vaporproof *vs.* Non-Vaporproof Containers

As described in a later chapter, vaporproof bags or packages prevent moisture-vapor exchange between the inside and outside of the bag/package. The vaporproof bag can be used only with low-moisture seed (with moisture content in equilibrium with less than 35% RH), as vapor-sealing a bag of higher-moisture seed will result in molding and death of the seed. However, when low-moisture seed are sealed into a vaporproof container, the seed maintain their low moisture content and can maintain viability longer under good ambient storage conditions.

Vaporproof containers are ideal for higher-value seed, and for all kinds of seed where environmental conditions are detrimental and seed may lose viability quickly, or where seed may be carried over.

Small Packages

Kinds and Uses

Smaller containers are used to pack high-value seed such as vegetable and flower seed, which are used in only small amounts by the final consumer. These packages range from a few seed or a few grams up to as much as 1–2 kg, depending on the needs of the consumer. Because of the high value of these seed, they are usually dried to very low moisture levels and then sealed into vaporproof packages so the seed will maintain high viability for longer periods.

Packages used include tin cans, small boxlike packets, and envelope-type packets. Tin cans are the usual package for larger amounts (1–2 kg, for example) of vegetable or flower seed, as required by large market-oriented farms or for retail dealers who re-package the seed into smaller packets bearing their proprietary brand name. Cans are sealed by a machine which fastens a lid onto the open mouth of the can. Not all sealers or lid machines will form an air-tight seal; in cans used for seed, the lid must form an airtight seal.

Small boxlike packets and envelope packets may be either:

1. Vapor-porous, of materials such as paper, if the seed are to be planted within a relatively short time; or,
2. Vaporproof packages for low-moisture seed, if the seed may be stored for some time before planting, or if the seed are of high value.

Increasingly, the more quality-conscious seed companies are using vaporproof packages to minimize seed deterioration and thus ensure high quality. Vapor-proof packaging materials include solid plastic sheets, paper packets with a lining of aluminum or vaporproof plastic, or similar materials. Using vaporproof packages requires drying systems which can dry the seed to very low moisture contents. Low-moisture drying often cannot be done in open-air drying systems; a special closed-system dehumidified drying system is required.

Filling Small Packages

Filling and sealing methods vary widely, according to the volume of seed handled and the investment which can be made in filling/sealing equipment. In small-scale operations, envelopes or packets may be filled by hand, measuring seed with small dippers which hold the quantity desired for each packet, which is then sealed by hand. For large-scale operations, completely automated machines measure the desired amount of seed, place it in packages, seal the package, and stamp the packets with required information. Some machines even form envelope-like packets from a preprinted roll of plastic material.

For special high-value seed handled in small volumes, a relatively low-cost seed counter is available which can count the desired number of seed and discharge them into small seed packets. The packets or envelopes are handled by hand; the counting machine ensures accuracy of fill and greatly speeds up the operation.

Selecting the Most Suitable Seed Bag or Package

Selecting a bag type is a significant part of the marketing effort. The following aspects must be considered.

1. *Weight of Seed Per Bag*: The primary concern is to provide farmers with the amount of seed they need, without selling loose seed from opened bags. To determine the best bag size for each crop seed, survey farmers to see how much seed they need and what bag size (and multiples of bags) would most efficiently fit their needs. Also, check the bag sizes already used by other seed suppliers. For example, if the average farmer needs 16 kg of maize seed and some farmers need only 8 kg, an 8-kg bag would take care of the needs of most farmers.
2. *Bag Material*: Check the materials available from local bag sources, evaluate their fit to your needs, and select the most useful and economic kind.

3. *Availability and Cost*: Not all kinds of bags are available in every locality. Check possible available sources, and compare costs of different sources and different bag materials.
4. *Kind of Seed*: Bags should be able to handle the crop seed safely. For example, jute bags could not be used alone for small seed.
5. *Suitability for Seed and Local Conditions*: Bags should be suitable for the seed kind handled (*i.e.*, small or large seed, etc.), local climatic conditions, drying facilities available, etc. For example, where seed must be widely distributed in a tropical climate where seed cannot be dried to safe low moisture contents, it would be disastrous to put high-moisture seed in vaporproof sealed plastic bags, unless the distribution system could keep the bags refrigerated.

Sampling and Seed Bags

Quality control during seed storage, handling and distribution requires periodic sampling and testing to ensure that seed maintain acceptable viability. To get a truly representative sample requires taking small samples from many bags; some trials using a single separate small bag containing the "testing sample" have not shown good results. When a trier or probe is inserted into a bag, it creates a hole which must be closed to prevent spilling seed. If the bag is of a loose-weave material such as jute, cloth, or woven plastic, careful sampling can separate the woven threads to insert the trier between the threads. When the sampling trier is withdrawn, the trier point can be used to put the threads back in position so that no problematic hole is left. However, solid plastic sheet bags, or even closely-woven materials, will leave a permanent hole in the bag when the trier is withdrawn. This hole must be closed to avoid losing seed. A special tape, suitable for the specific bag construction, can be obtained to cover the sampling hole. Some external QC agencies (seed law implementing agencies) use a special tape to seal these holes; the tape is pre-printed with wording such as "Sampled by the Seed Law Agency for your protection".

Chapter 22

Internal Quality Control (IQC)

External vs. Internal Quality Control

External quality control—Certification, Seed Law—is normally to ensure that your seed meets required standards. It gives an acceptable/non acceptable decision; it does not help your management efforts or prevent problems. Always ensure that your seed meet Seed Law standards, and as much as possible, use Certification.

However, external quality control does nothing to prevent problems or improve the cost/time-efficiency of your operations. Maintaining you own Internal Quality Control (IQC) is the only and best means to achieve consistently high seed quality and operating efficiency, prevent problems, and help improve cost-effectiveness and profitability.

Role and Operations of IQC

Internal quality control (IQC) is management's technical arm to prevent problems, ensure that seed meet required standards, minimize losses, maximize cost/time-efficiency, and improve operating profitability. No responsible seed program can operate most efficiently, in time, cost, and seed loss, without a strong IQC with competent personnel.

In this role, IQC can enter, sample/test/examine any operation and any seed or separated fractions. It can then modify or stop any operation, in order to achieve or improve seed quality and operating efficiency, or to prevent loss of good seed.

IQC's operations are not dictatorial, but supportive, in that it helps each operation, provides on-the-job training for staff, and helps the entire operations be more productive and profitable.

Internal Quality Control Activities and Responsibilities

IQC is a staff unit which works directly under the Manager; and acts as the quality and technical control arm of management. It is not subsidiary to any of the direct operating units such as production or conditioning. The IQC can enter, inspect, and supervise any operation of the enterprise. It is the only unit, short of the General Manager, which can stop and change any operation in order to achieve the desired seed quality or operating efficiency. Its aim is to ensure that all seed meets the required standards, and to prevent losses and excessive costs.

Responsibilities of this organizational staff unit include:

1. Formalize the standards, procedures and conditions required by the market, and as determined by management and external quality control systems such as Certification and Seed Law.
2. Set up, operate and maintain adequate testing facilities of lab and equipment.
3. Regularly observe each operation of the enterprise, evaluate the operation, sample and test seed, and stop, change, eliminate or adjust the operation until it fits the desired requirements.
4. Provide quality control supervision for all operations of the enterprise and all contracted growers, dealers, and others.

IQC does this by monitoring every operation of the seed enterprise, especially in conditioning and storage, to ensure that seed quality is achieved and maintained, and loss/damage of seed is minimized. To do this, IQC observes and evaluates every operation, samples and tests seed and waste products from every operation. It seeks to answer the following questions, in a way which can identify and implement improvements:

1. What must be done to the seed lot to achieve and maintain acceptable quality?
2. What can cause damage to seed/seed quality, and how can it be removed or prevented?
3. How did this get into, or happen to, the seed lot?
4. How can the undesirable condition be prevented in the future, as well as removed in conditioning the current and other seed lots? Is specialized handling, conditioning or treating required? What does it cost, in relation to the value of the seed lot?
5. Can the seed lot be accepted and meet standards, cost-effectively?
6. What can be done to minimize or prevent this problem in the future?

Preventing Problems

If a problem occurs, it is costly in both the loss of seed or efficiency, and in the cost of eliminating the problem. A major role of IQC is to examine and analyze operations so that potential problems can be identified and prevented before they occur.

Resolving Existing Problems

If a problem does occur, IQC should analyze the situation and develop the most efficient means of overcoming it. For example, if a seed lot contains a problem contaminant, IQC can either devise an efficient means of removing the contaminant, or else recommend discarding the entire lot before any expenses (in both time and money) are spent on a wasteful effort to salvage it.

Eliminating Undesirable Conditions

In its work, IQC should be able to evaluate operations and identify conditions which are, or which may be, detrimental to quality and efficiency. It should then devise changes which can be implemented to eliminate the undesirable conditions.

IQC Organization

IQC is a staff function, the quality and efficiency arm of management. It does not have "line" or direct responsibilities in producing and supplying seed. Instead, it assists production, conditioning, storage and marketing by monitoring their operations and seed condition, to ensure cost-effective operations and high seed quality, and minimum loss of seed.

IQC monitors these operations, and reports directly to the chief managing officer of the enterprise or program. Infrastructurally, its position in the management organization is as shown in Figure 22.1.

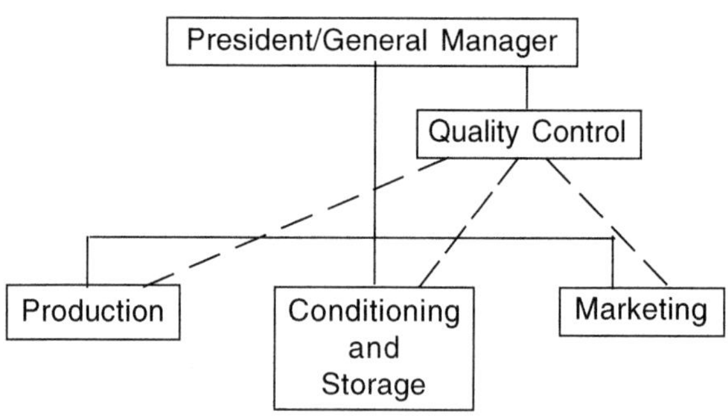

Figure 22.1 The place of IQC in the organizational structure of a seed enterprise or program.

IQC Staff

IQC staff should be the most highly-trained and experienced staff of the enterprise. They must be able to:

1. Guide growers in production planning and operations, train field men in inspecting seed fields, and generally provide technical and managerial guidance.
2. Inspect, evaluate, analyze and guide all conditioning operations: receiving, sampling, labeling, drying, raw seed storage, planning and scheduling conditioning sequences, operation and adjustment of each machine, etc.
3. Sample and test seed for all factors affecting quality, and to analyze samples to determine the efficiency of each conditioning machine and operation. This includes operating an internal seed testing laboratory, with at least the equipment and facilities required to make the necessary tests.

IQC staff usually consists of:

1. Senior IQC Specialist, who manages the IQC Unit.
2. IQC Specialists, who take samples, make tests, monitor field work and inspect fields, recommend conditioning requirements, etc.
3. Technicians, who carry out work such as sampling, sample and testing media preparation, cleaning and care of laboratory facilities and equipment, etc.
4. File clerk, who keeps records and reports up-to-date.

IQC Specialists with professional seed and quality control training should organize, supervise and conduct QC work. The Specialists' work can be expedited by well-trained but non-professional IQC Technicians. The numbers of Specialists and Technicians required depends on the amount of work, which in turn depends on the side of the seed enterprise's operations. As a minimum, one Specialist and two Technicians should be considered.

The Senior IQC Specialist must be in charge of the IQC Unit, with complete responsibility and authority for its operations, organization, facilities and staff.

IQC Facilities

IQC must establish and operate its own seed testing laboratory within the enterprise, plus offices or desk space, and computers to record data and communicate with other seed operations such as production, conditioning and storage. This lab must be able to do (1) sample bulk and bagged seed, (2) all seed analysis and testing procedures as required for the operations, (3) conditioning machine operational effectiveness testing, and (4) storage quality and condition testing. The lab facilities should, if at all possible, be air-conditioned to facilitate the work and help protect sensitive equipment.

An effective layout of IQC laboratories is shown in the chapter on buildings. IQC is often included in the administrative building, for economy and close working relations.

IQC must also have vehicles as required to travel to any production, conditioning, handling, or storage location, to be able to monitor all operations closely.

IQC in Conditioning

After conditioning organization, management and facilities have been organized and set up so as to improve seed quality and maintain its identity, an internal quality control system must be established to ensure constant high seed quality and operating efficiency. IQC helps organize and operate so that problems are prevented, and operations are time- and cost-efficient. IQC helps solve problems when they occur, but its greatest contribution is in preventing problems.

In conditioning, IQC constantly works to ensure handling seed to achieve and maintain high quality, maintain identity, and prevent contamination and loss. Its main work includes:

1. Sampling all incoming raw (field-run, uncleaned) seed immediately when it arrives at the conditioning plant facility.
2. Analyzing the samples to determine (A) moisture content and drying requirements; (B) insect infestation and need for fumigation, isolation, or special handling; and (C) seed germination and what is in the seed lot, to determine conditioning requirements and probable amount and quality of seed after cleaning. Based on these determinations, the seed lot is rejected immediately before money and time are spent on it, or it is accepted and the lot is immediately entered into the conditioning operations so as to maintain its quality.
3. Advising the conditioning staff on what each lot requires to clean it to the best quality with minimum loss. This includes cleaners and separators to use, approximate adjustments and settings (screen sizes, etc.), what must be removed, and how much loss can be expected.
4. Forming large "cleaning operations" lots from several smaller grower lots of the same kind and class of seed.
5. Determining conditioning sequence of lots, so that several lots of the same seed kind and quality can be cleaned in sequence, without losing time to clean-out the conditioning plant between these lots.
6. Then, at frequent intervals during conditioning IQC samples and analyzes the clean seed and all waste products coming from each separator. In some efficient plants, clean seed is sampled and analyzed

every 1–2 hours, and waste products are sampled and analyzed every 3–4 hours. If the clean seed does not meet standards, or too much good seed is lost in waste products, IQC advises the conditioning staff to stop the operation and re-set the machines so as to correct the situation.

7. Forming final clean seed lots as the seed come out of conditioning and are bagged (packaged).
8. Ensuring sampling and testing of each final lot in accordance with Seed Law and/or Certification requirements.
9. Advising on storing and handling the seed after conditioning, to maintain its quality and identity.
10. Advising on lots to market first, and lots for potential carryover.
11. Monitoring storage conditions to prevent loss or damage.
12. Regular germination testing, according to the pre-arranged schedule, to ensure that seed quality does not deteriorate.

Sampling

Samples for analysis to evaluate seed and operations are taken:

1. When the lot first arrives at the conditioning plant, to (1) determine seed moisture content and drying needs; (2) determine insect infestations and need for fumigation; (3) determine if the seed lot can be economically cleaned up to standards; and (4) conditioning requirements.
2. From the clean seed and all waste fractions leaving each separator, to determine if the separator is removing all of the undesirable materials, and is not losing too many good crop seed.
3. From each machine, elevator, bin, conveyor, etc., to determine the amount of mechanical damage done at this point, and to identify means of minimizing the damage done here in handling the seed.
4. From the final clean seed lot, for testing to ensure that the seed meet standards, and to obtain the lot's quality information for labeling (tags).

Moisture Testing

A seed moisture test is made immediately of all raw seed received, and its need for drying is determined. Conditioning is then assisted in planning and implementing the required drying, as to immediacy of drying, drying conditions, and final moisture content required.

Insect Infestation Testing

Samples of seed are analyzed for insects, and the conditioning plant and storages are examined for signs of insect presence. Then, fumigation, treatment, cleaning, etc., are recommended to control insects, eliminate infestations, and prevent damage.

Purity Testing

The standard purity test is one of the first and most important tests made. The sample of specified size is examined in a standardized procedure, seed by seed, by the trained IQC specialist to identify the impurities in the seed lot.

Physiological Quality Testing

The most common test for viability, conducted by IQC, is the standard germination test. IQC often runs this test, which requires several days to complete, and provisionally accepts the incoming raw seed lot based on its quality appearance as evaluated by the experienced IQC specialists.

If more definite information is needed quickly on the seed's viability, the TZ (tetrazolium) test is often used. Other methods, such as the electronic test made on seed leachates, are sometimes used. However, these are more complex and expensive than tests normally used in IQC, and are not common.

Seed Identification and Labeling

Exact identity of every bin, totebox, and bag of seed is essential. IQC should help supervise the complete use of a seed identification system and labeling procedures.

Storability Testing

Often, a certain amount of seed is carried over; a small amount, such as 5–10%, above the expected market demand will be produced to enable the enterprise to meet changes in market demand. Seed to be carried-over should be identified as soon as it arrives, and should be moved into good carryover storage as soon as possible after it is cleaned. Leaving it in ordinary ambient storage (used for seed to be sold and planted within a relatively short time) may initiate deterioration which will cause the seed to be below standard by the time they are carried over to the next year.

Testing for Conditioning Requirements

IQC identifies the conditioning requirements of raw (uncleaned) seed lots, and advises conditioning of the machines, screens, settings, etc., to use. This saves a tremendous amount of operating time and labor for conditioning.

To test for conditioning requirements, a sample of the raw (uncleaned) seed is examined, with hand screens, lab model seed cleaning machines, and visual examination, to determine the physical differences between the good seed and materials which should be separated. Then, the cleaning machines required, approximate screen and cylinder sizes, and approximate machine adjustments and percentage cleaning loss, are determined by the following methodology:

1. Identify all materials (inert, weed seed, other crop seed, damaged or immature crop seed, etc.) which must be removed.
2. Evaluate physical differences between the good seed and the undesirable materials, and select the proper separating machines. For example, if there are length differences, the properly-set-up length (cylinder or disk) separator is selected; if there are apparent weight differences, the gravity separator is also selected.
3. Use hand testing screens to determine approximate screen perforation sizes needed in the air-screen cleaner. Use small testing machines, visual and other evaluations of the seed, combined with experience, to determine the required sequence of cleaning machines and their approximate settings.

Accepting or Rejecting Incoming Lots

As soon as each lot of raw seed arrives at the conditioning plant, IQC samples and tests it, as described above. If the lot appears to be good, it is accepted and entered into the conditioning system. If it is doubtful that the lot can be brought up to quality standards, or if it appears that too high a percentage of the lot must be taken out, the lot is rejected and moved out of the conditioning plant immediately, before time and operations are wasted on it.

Testing During Conditioning

At regular intervals during conditioning, IQC samples and tests:

1. The good seed leaving each cleaner and separator, and does a quick purity test to ensure that all undesirable materials are properly removed.
2. All waste product fractions separated by each machine, to be sure that excessive amounts of good seed are not lost.

If too much undesirable material is left in the good seed, or too much good seed is lost in a particular waste product, IQC advises the conditioning manager, the machines are stopped and re-adjusted to correct the problem.

Determining Purchase Price of Lots

Many seed enterprises pay growers for seed on the basis of "final clean seed estimates" which are based on analyses made by IQC as soon as the lot is received. IQC (1) determines moisture content and adjusts weight for drying weight loss; (2) analyzes the sample and identifies what must be removed; (3) does "conditioning analyses" to determine how to remove the undesirable materials and the accompanying good seed reduction; and (4) calculates the final weight of the cleaned seed lot.

This allows immediate payment to the farmer-grower, who often needs money immediately at harvest time. It also avoids losses by the seed enterprise due to un-accounted-for cleaning weight reduction. In the rare cases where the actual final yield of good clean seed is higher than estimated, an additional payment is made to the grower after the seed is cleaned and the actual amount of clean seed is determined. However, experience has shown that a good IQC estimate of cleaned-seed turnout is usually quite accurate.

Disposal of Waste Products

IQC plans safe disposal of waste materials removed in conditioning. If the waste products contain troublesome or noxious weeds, they may be burned; or, if possible to fumigate or grind them to destroy weed seed germination, this may be done. If nothing seriously harmful is present, the waste products may be used in livestock feeds. Wastes containing weed seed should not be returned to the growers, as they seldom are able to destroy the seed effectively.

Identifying Lots to Carryover

Most seed operations usually produce about 5% or so, more seed than the marketing plan calls for. This is produced to allow for fluctuations in market requirements, and to be sure that new customers or additional demands can be supplied. If this extra seed is not sold, it should be carried-over and sold in the following year.

Testing During Storage

Seed age and die, as do all living things. The objective of storage is to keep the seed alive and high-germinating until they can be sold. The objective of marketing is to sell the seed while their quality is still high.

IQC assists by regularly sampling and testing the germination of all stored seed lots, to determine:

1. Which lots still have high germination and can be kept for marketing.
2. Which lots have lost germination and should be discarded (or, if possible and economic, re-conditioned to bring germination of part of the lot back up to the required level).
3. What is the relationship between conditions in the storage with the maintenance or loss of germinability, and thus the need to improve storage conditions.

IQC should maintain a regular schedule of germination testing at specified intervals of all stored seed lots.

Examining Storages and Storage Conditions

IQC regularly monitors storage temperature and dampness (Figures 22.2, 22.3), and inspects storages for signs of rodent, bird, insect, or other damage. As soon as any potential problem is identified, IQC recommends immediate actions and helps implement preventive measures.

Figure 22.2 A hygrothermograph used to monitor storage conditions. Placed in the storage area, this machine makes a continuous recording of temperature and relative humidity at the point where it is located. Source: Kathy Reading, Seedburo catalogs.

Follow-up on Field Supervision and Inspection

IQC is also active in ensuring quality in seed production. IQC specialists train the field men who supervise seed growers. The field men are taught how to identify potential and actual problems, and eliminate them in the field. This includes isolation, roguing, diseases, etc. Good field management, assisted by IQC, significantly reduces loss of fields, improves quality of seed delivered for conditioning, and reduces conditioning work and problems.

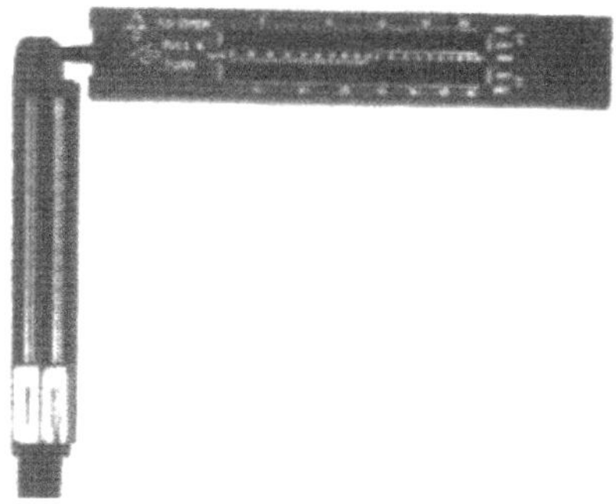

Figure 22.3 A sling psychrometer, used to measure relative humidity of the air. A small portable device, it can easily be used in storages, in conditioning, in the drying area, or in outside or on-farm areas. Source: Kathy Reading, Seedburo catalogs.

Guiding Operations

IQC is closely involved in helping other units (Production, Storage, Marketing, Conditioning) plan their operations so they are cost/time-efficient, with high seed quality and minimum losses.

IQC helps them plan their season's operations, and constantly monitors their operations to ensure efficiency.

Reviewing Performance

At the end of the year or season, the seed enterprise, and the conditioning unit, should review its performance. This ongoing activity should include analysis of the successful operations and the problems encountered, so that the next season can be improved. IQC is a major help in such reviews, improving operations, cost-efficiency, seed quality, and loss prevention.

Managing Quality Control

Follow-up and Supervision

Every manager, from the top manager/owner down to every manager of operations or personnel who must manage fields/machines/operations, must follow-up to see that the necessary activities are carried out properly at the right time.

In addition, Internal Quality Control (IQC) is a special management tool to implement follow-up and supervision of all technical operations. Its benefits are to:

1. Ensure that all seed are of high quality, and will enhance the enterprise's stature with customers.
2. Minimize losses and unnecessary expenses.
3. Prevent problems by foreseeing them and advising on ways to avoid them. When problems do occur, to identify the most cost-effective solution (which is sometimes to discard the seed or abandon a field or operation) to them.
4. Keep operations as cost-effective and efficient as possible.

IQC may be a special unit, one specially-trained person, or a part of the duties of the single manager-operator. Regardless of how many are involved, IQC exercises follow-up and supervision to guide, control and advise every operation of the enterprise: production, harvest, drying, cleaning, treating, bagging, storage, handling, identifying seed, testing, etc.

Records and Reports

IQC records the details and results of analysis of every operation, field inspection, and seed sampling and testing within the enterprise. This includes causes, effects, conditions, evaluation and recommendation for improving the operation, field or seed.

Analyzing and Using Records as Management Tools

Regular reports are compiled with an overview of operations and quality/efficiency aspects, in a format which helps management improve and redirect the operations. Management should always ensure that IQC has the authority to change or stop any operation at any time to improve it, and can recommend the disposal of substandard fields or seed. The recommendations of IQC should always be reflected immediately in decisions and actions by management.

Equipment for IQC

IQC must have the basic equipment required to take samples and conduct tests. Table 22.1 shows basic equipment required. Additional equipment, such as mechanized samplers for incoming bulk raw seed, should be added as needed and as possible.

Table 22.1 Examples of equipment specifications for Internal Quality Control.

Note:

1. Example equipment specifications are shown; other brands/models which do the same work are equally suitable.
2. Designed for IQC of 2 purity analysts plus sampling and other testing.
3. Some listed equipment is not required, when crops involved are not handled, or volume is low.
4. Some equipment (*e.g.*, microscopes in purity and pathology) can be used in more than one place, but is listed here in both.

Item No.	Number Required	Description
Sampling, Receiving, Environmental Testing:		
01	1	Probe, sampler, bag, thief type; 13" (33.3 cm) peanut and rice trier, nickel-plated steel; 1-1/2" (3.8 cm) outside diameter at large end; slot 7-3/8" (18.9 cm) long, tapering to 7/8" (2.3 cm) wide. Brand name: Seedburo No. 76 or replacement or later-technology model or equivalent.
02	1	Trier, double-tube, 40-inch (102 cm) long, 6 openings; open-end, without partitions; constructed of drawn and tempered seamless heavy-gauge tubing; 1-3/8" outside diameter; heavy bronze point. Brand name: Seedburo No. 20-OH or replacement or later-technology model or equivalent.
03	1	Trier, double-tube, 63-inch (160 cm) long; 10 openings; heavy bronze point; open-end, without partitions; constructed of drawn tempered seamless heavy-gauge tubing; 1-3/8" outside diameter; heavy bronze point. Brand name: Seedburo No. 22-OH or replacement or later-technology model or equivalent.
04	1	Trier, double-tube, 72-inch (183 cm) long; 12 openings; open-end, without partitions; constructed of drawn seamless tempered heavy-gauge tubing; 13/8" outside diameter; heavy bronze point. Brand name: Seedburo No. 16-OH or replacement or later-technology model or equivalent.
05	1	Trier, brass, partition, 10-ft. (3.05 m) long; 20 openings, extra heavy bronze point. Brand name: Seedburo No. 230 or replacement or later-technology model or equivalent.
06	1	Probe/trier, deep bin cup, w/T-handle and four (4) 3-ft. extensions; brass cups, 1.5-inch outside diameter, 15-inch deep, 265-gram capacity; cup remains closed as probe is pushed down, and opens when probe is pulled out. Brand name: Seedburo No. 234 deep bin cup probe w/ No. STD-3 extensions & STD-T T-handle or replacement or later-technology model or equivalent.
07	1	Thermometer probe, for taking temperature of bulk seed in bins; Centigrade; w/standard T-handle and four (4) 3-foot-long handle extensions; 81/4" (21.1 cm) brass cover case over a 53/8" (13.8 cm) glass thermometer protected by a plastic cover; temperature range 0–60°C w/1°C divisions. Brand name: Seedburo No. 213 or replacement or later-technology model or equivalent.

Item No.	Number Required	Description
08	1	Sling psychrometer; w/36 extra wicks, 2 sets extra thermometers; Centigrade; two 5" etched mercurial lens magnifying thermometers attached to aluminum back with metalhandle for field use; scale range –5 to 45°C. Brand name: Seedburo No. 317-SC or replacement or later-technology model or equivalent.
09	2	Thermometer, dial; w/12-inch stem; Centigrade, range –10 to 100°C; stainless steel construction; 5-inch dial; bimetallic system, 1% accuracy; recalibrator. Brand name: Seedburo No. 5000R or replacement or later-technology model or equivalent.
10	2	Hygrothermograph, weekly recording, w/extra drum, 6 extra pens, 1,000 weekly recording charts; Centigrade; easily-viewed chart, pens and mechanism, protected by a clear plastic window; ready access to moving parts; recording mechanism w/8-day spring-wound clock; external pen lifter; gear for 7-day (172 hours) chart rotation; operates on one D-cell battery; slip-on felt-tip pen cartridges, no ink required; 11-1/2" × 7" chart paper; 0 to 100% relative humidity range in 1% increments, accuracy ± 3.0%. Brand name: Seedburo No. 8368-00 w/8368-40 charts or replacement or later-technology model or equivalent.
11	1	Numbering stamper, hand-used, dating stamp type w/consecutive numbering. Brand name: Bates No. 6A or replacement or later-technology model or equivalent.
12	1	Tag/label printer, stencil-printing type similar to machine used for printing library cards; w/all required components, supplies adequate for 100,000 seed bag tags. 220–240 v, 50 hz. Brand name: Weber Table-top Tag & Label Printer or replacement or later-technology model or equivalent.
Moisture Testing:		
13	2	Moisture tester, portable electronic; Dole 400 w/charts for all crop seed; transistorized electronic circuits, low power use; operates on a 9-volt transistor battery; with AC/DC converter, temperature/moisture correction indicator; direct dial reading for common grain/seed, charts for all crop seed available; balance meter for adjustment; built-in sample weighing scale. Brand name: Seedburo Model PB-70-21 or replacement or later-technology model or equivalent.
14	1	Moisture tester, Motomco; w/calibration charts for handling all crop seed available; seed in test samples not damaged/affected; wide test moisture range for testing low- and high-moisture seed.Complete, ready to operate. 220–240 v, 50 hz. Brand name: Seedburo Motomco No. 919/A or replacement or later-technology model or equivalent.
15	25	Dish, analytical, for oven moisture determination; aluminum, w/straight edge and flat bottom; diameter 2-1/2", depth 5/8"; w/loose-fitting lids. Brand name: Seedburo No. 25 or replacement or later-technology model or equivalent.

contd. . . .

. . . Table 22.1 contd.

Item No.	Number Required	Description
16	1	Tongs, for removing sample containers from oven; general purpose, nickel-plated steel; 8" long (20 cm), 1" (2.5 cm) stirrup jaws with teeth. Brand name: Fisher Scientific Catalog 208, No. 15-202 or replacement or later-technology model or equivalent.
17	1	Oven, heated-air, electric; double-wall construction w/corrosion-resistant aluminum interior, NLT 1" fiberglass insulation, enameled steel exterior; bimetallic oven temperature thermostat w/indicating pilot light; mercury thermometer, w/mount in oven top; pre-set high-limit safety thermostat; NLT 3 adjustable shelves; chamber NLT 46 × 34 × 56 cm, 3.1 cu. ft. volume. 220–240 v, 50 hz. Brand name: Seedburo Model 3512-1 or replacement or later-technology model or equivalent.
18	25	Can, grain/seed sample, for moisture-test samples; polyethylene plastic, w/tight-fitting enameled plastic cap and gasket insert for air seal; neck sized for No. 10 rubber stopper w/glass thermometer; capacity approx. 1 liter. Brand name: Seedburo No. 630 w/No. 10 rubber stopper or replacement or later-technology model or equivalent.
19	1	Desiccator; heavy annealed glass; sealing w/greased internal edge of flanges; rim in lower portion to hold sample plate over desiccant; 200 mm inside diameter; w/cover; w/190-mm unbreakable polypropylene plate w/o feet, w/flanged outside rim. Brand name: Fisher Scientific Catalog 208, No. 08-595D w/No. 08-643-2B plate or replacement or later-technology model or equivalent.
Dividing:		
20	20	Sample pan; aluminum, rectangular w/pouring spout; 1-1/2 qt. capacity; approx. size 8-1/2" × 12" × 1-1/2". Brand name: none.
21	20	Sample pan, triangular; heavy tin or aluminum; approx. size 10" × 10" × 2-1/2". Brand name: none.
22	1	Divider, Boerner; w/pans and extension; w/hopper over center separating cone, 38 sections for separating seed, alternately feeding into two streams which empty into separate pans; heavy brass and copper construction; entire bottom section removable; brass-plated steel legs and supports. Brand name: Seedburo No. 34 or replacement or later-technology model or equivalent.
23	8	Sample catch pan, to sample flowing stream of seed from dividers; mouth 6-1/2" (16.5 cm) × 1" (2.5 cm); 14" (36 cm) long × 9" (23 cm) wide. Brand name: Seedburo No. 23 or replacement or later-technology model or equivalent.
24	1	Test weight apparatus; in kg/hectoliter, complete w/filling hopper, 1-liter dent-proof bucket, adjustable stand, triangular pans for overflow, kg/hl-calibrated weigh beam; filling hopper w/sliding gate valve and standard 1.25" opening; hopper height adjustable to ensure standard 2" above bucket; cast iron base. Brand name: Seedburo No. 26M w/No. 4184 1-liter bucket, No. 151 filling hopper & 2 No. 64P pans or replacement or later-technology model or equivalent.

Item No.	Number Required	Description
Purity Testing:		
25	2	Purity workboard, w/diaphanoscope and light; work area of smooth opal glass surface for seed analysis; light w/off-on switch under the glass to shine through when required for seed examination; removable sample drawer. Brand name: Seedburo No. 135 or replacement or later-technology model or equivalent.
26	2	Forceps, analyst, 5.25-inch, medium-sharp point for manipulating medium and small seed. Brand name: Seedburo No. MM or replacement or later-technology model or equivalent.
27	2	Forceps, analyst, 6-inch, blunt point; for manipulating large seed. Brand name: Seedburo No. 62 or replacement or later-technology model or equivalent.
28	2	Lamp, Dazor, w/magnifier; 6 extra light bulbs, desk base; floating arm magnifier lamp to provide optically-correct fluorescent lighting; 5-inch magnifying lens providing + 75% magnification at 13" focal length; 18" adjustable arm, 10" long light fixture w/3 60-watt tubes. 220–240 v, 50 hz. Brand name: Seedburo No. M 270-25 or replacement or later-technology model or equivalent.
29	2	Magnifier, analyst's, w/heavy metal base; adjustable, may be set at any angle or easily removed from base for hand use; removable, adjustable reflection shade to shield lens from side reflections; ground, polished optical glass lens, 3-power, 4" diameter, 7" focal distance. Brand name: Seedburo No. 4-3x-17 or replacement or later-technology model or equivalent.
30	1	Binocular microscope, w/light; stereo-zoom, magnification NLT from 7x to 30X; illuminating system suitable for 220 v 50 hz. Brand name: Seedburo No. 4E1422 or Olympus model VMZ-001 w/VM-LSG illuminator, or replacement or later-technology model or equivalent.
31	30	Purity test sample dishes, metal or plastic, about size of petri dish (approx. diameter 4", depth 5/8"); shape may be round or square. Brand name: none.
32	1	Seed blower, Bates laboratory aspirator type for aspirating purity samples; separates light foreign materials by passing a closely-controlled air stream through the sample as it passes over a spreading cone in a closely-regulated flow; valve and slide to regulate air flow; centrifugal collector; NLT 2-qt. seed receiving hopper w/flow control for instantaneous start/stop of seed flow, and control dial to set flow rate; transparent cup for instant viewing lifted materials; swinging shelf for pans, vertically-adjustable for different pans; locking screw to set pan under feed cone; w/triangular seed pan; solid-state rheostat for variable fan speed. 220–240 v, 50 hz. Brand name: Seedburo No.95BGV or replacement or later-technology model or equivalent.
33	1	Rice sheller; tests rice samples to indicate red rice. Complete with hopper, dust collector, flexible hose, all components and spare parts.

contd. . . .

. . . Table 22.1 contd.

Item No.	Number Required	Description
		220–240 v, 50 hz. Brand name: Seedburo McGill Sample Sheller for Rice No.580DC/C or replacement or later-technology model or equivalent.
Weighing:		
34	1	Balance, Torsion, precision; 120-gram capacity, 1 g × 0.01 g (10 mg) graduations; 2 mg dial readability, 5 mg accuracy; scoop size 4-3/8" × 3-1/4" × 1-1/16", of polished stainless steel; dimensions 11-3/4" × 6" × 8-1/4"; w/weight loader, 9 g total × 1 g increments; w/Class P brass weight set, from 50 grams to 1 gram in hinged box, velvet-lined cover w/forceps. Brand name: Seedburo No. DWM2-1 w/No. BC5DO3 weight set or replacement or later-technology model or equivalent.
35	1	Scale, gram, triple-beam, metric system, w/seed pan; aluminum alloy beam w/high-grade polished agate self-aligning bearings; lithographed aluminum beam graduations for clear readings and to remain legible over long period of use; stainless steel scoop seed pan 12"× 2.75", shaped for easy pouring of sample; stable sliding-type poise, w/ tongue which fits into accurately-milled notches on beam; capacity 610 g, w/No. 707 weight set increases to 2,610 g; sensitivity rated 0.1 g; magnetic dampening to speed up weighing. Brand name: Seedburo No. 99-S or replacement or later-technology model or equivalent.
36	1	Balance, electronic; w/top weigh pan; digital read-out; dual capacity of (1) 3,000 g at 0.1 g readability, and (2) 300 g at 0.01 g readability. 220–240 v, 50 hz. Brand name: Seedburo Ohaus Brainweigh No. B3000-D or replacement or later-technology model or equivalent.
Germination and Vigor Testing:		
37	1	Germinator; cabinet, w/lights; cabinet capacity NLT 0.6 m3 (20 cu. ft.), w/supports for NLT 30 trays; all trays provided; single or double chamber; fluorescent lighting for each chamber; exterior finish stainless steel or rust-resistant enamel baked on; interior finish stainless steel; can reach and maintain temperatures 2°C to 40°C w/in 2°C; maintain 95% RH; meets ISTA specifications. 220–240 v, 50 hz. Brand name: (1) Hoffman SG-2-22 Chamber germinator, (2) Conviron Model G-30 seed germinator, or (3) Seedburo Microprocessor Germinator Model MPG 1000/c, complete w/all required accessories or replacement or later-technology model or equivalent.
38	1	Germination counting boards, set for all available crop seed; counting and placement tray for 100 seed; constructed of acrylic and brass to allow washing or sterilization; tray size approx. 17" × 19". Brand name: Seedburo Nos. 715B, 715C, and others for other crop seed or replacement or later-technology model or equivalent.

Item No.	Number Required	Description
39	200	Petri dish, standard size; plastic, w/cover fitting loosely to provide ventilation during germination tests; diameter 4", depth 5/8". Brand name: none.
40	2	Thermometer, lab; Centigrade, red alcohol, white back; temperature range –20°C to +110°C, 305 mm long, immersion 76 mm. Brand name: Fisher Scientific Catalog 208, No. 14-998 or replacement or later-technology model or equivalent.
41	30	Pan, aluminum, rectangular, approx. 10" × 20" × 3" deep, for rolled and folded paper towel tests. Brand name: none.
42	1	Germinator, single-chamber, for use in vigor, TZ and/or accelerated aging tests; clear plexiglass top for light, angled to drain condensation; stainless steel interior; insulated w/high-density foam, non-absorbent and insect-proof; chamber designed for air flow for uniform temperature and humidity; NLT 8 trays NLT 15" × 21", anodized aluminum; water pan; thermostatically-controlled heating elements; temperature levels to 40°C; thermometer; rubber feet. 220–240 v, 50 hz. Brand name: Seedburo No. 548R-2 or replacement or later-technology model or equivalent.
43	1	Germination room/conversion unit for existing masonry-walled and concrete-floored room in building; walk-in; w/thermal insulation and moisture/vapor barrier on all surfaces; one entry door in one end, insulated and vapor-barriered w/rubber gasket seals on all surfaces, and door; inside push opener. Temperature and humidity controls, air circulation fan(s) to prevent temp. stratification; fluorescent lights to provide light required for germinating seed, with day/night timing switches; corrosion-proof wall, ceiling, floor, and door surfaces; metal adjustable shelves for germination tests; capacity NLT 2,500 simultaneous tests; capable of establishing and holding 10, 15, 20, 30, alternating 10–20, 20–30°C, or other desired 24-hour germination temperature regime within ISTA standards. 220–240 v, 50 hz. Brand name: (1) Conviron Model CG 96 plant growth chamber; (2) El Chark Co. (Egypt) built-in 4m × 3m × 3m germination room modified for air circulation, or replacement or later-technology model or equivalent.
Conditioning Testing:		
44	1	Hand testing screens, complete set of 160 different perforation sizes and shapes; w/storage racks, 12 blank screens; 12" screen frame size, wood frame w/design to permit secure stacking of screens for hand-shaking. Brand name: Westrup or replacement or later-technology model or equivalent.
45	50	Sample pan; aluminum, rectangular w/pouring spout; approx. 1-1/2 qt. capacity; size approx. 8-1/2" × 12" × 1-1/2". Brand name: none.
46	20	Sample pan, triangular; heavy tin; approx. size 10" × 10" × 2-1/2". Brand name: none.

contd. . . .

. . . Table 22.1 contd.

Item No.	Number Required	Description
47	1	Lab-model air-screen cleaner; must be continuous-flow to process either samples or small lots; w/160 screens of different perforations for cleaning field crop and vegetable seed from small to large sizes; air aspiration separation before and after screen separating; dust collector; quick-change screens; NLT 3 screens in sequence flow, 2 scalping and 1 grading/sifting; rubber ball screen cleaning system; mounted on work table of standard lab work height; adjustable feed hopper; variable-speed screen shake with tachometer or similar system to indicate shake speed to facilitate reproducing research/testing; seed pans to collect all discharged fractions; w/all required motors, drives, switches. 220–240 v, 50 hz. Brand name: Westrup LA-LS or replacement or later-technology model or equivalent.
48	1	Lab model indented cylinder separator; quick-change cylinder, variable-speed cylinder drive; adjustable liftings discharge; seed pans for discharged fractions; mounted on work table of standard lab work height; adjustable feed hopper; w/10 common-grain indented cylinder sizes; all motors, switches, and drive. With complete conversion kit for sizing (width and thickness, perforated cylinders) separations, including commonly-used perforation sizes and shapes for major field crop seed. 220–240 v, 50 hz. Brand name: Westrup LA-T or replacement or later-technology model or equivalent.
49	1	Lab model width-and-thickness separator; motor-driven; w/feed hopper, pan to catch throughs, motor, switch; w/20 commonly-used perforation sizes; quickly-removed cylinders. 220–240 v, 50 hz. Brand name: Carter-Day Laboratory Model Precision Grader or replacement or later-technology model or equivalent. (FOR SIZE-GRADING MAIZE SEED, ETC.)
50	1	Lab model seed scarifier/huller; w/brush-type adjustable feed hopper; adjustable scarifying mechanism to provide different degrees of polishing/hulling/scarifying action; seed pans for all discharged fractions; dust aspiration and collection system; mounted on work table of standard height; w/a range of 20 interchangeable mantles; all required motors, switches, and drive. 220–240 v, 50 hz. Brand name: Westrup LA-H or replacement or later-technology model or equivalent.
51	1	Lab model gravity separator; w/precise air control system; interchangeable deck surfaces to permit separating small to large seed; includes 5 often-used deck surfaces; deck slope adjustable from end-to-end and from side-to-side; adjustable deck speed with tachometer or similar system to indicate exact speed to allow reproducible research/ testing; continuous-flow operation for small seed lots; seed pans for all discharged fractions; adjustable feed hopper; mounted on work table of standard height; all required motors, switches, and drive. 220–240 v, 50 hz. Brand name: Westrup LA-K or replacement or later-technology model or equivalent.

Item No.	Number Required	Description
52	1	Lab model roll mill; hand-operated, adjustable feed rate, removable tray for lifted fractions. Brand name: WARSCO Hand-operated Test Model Roll Mill or replacement or later-technology model or equivalent. (FOR SMALL LEGUME SEED)
Operations:		
53	1	Air compressor, heavy duty; w/NLT 7.5 gallon tank; portable, on semi-pneumatic tires, w/pull handle; 3/4-hp motor; oil-less twin-cylinder air pump; one 15-m hose w/quick-snap connectors, trigger-operated air gun; pressure regulator; automatic pressure-actuated on/off compressor switch; 2 extra complete (male and female units) quick-snap connectors. 220–240 v, 50 hz. Brand name: Grainger Catalog No. 371, p. 872, No. 3Z921 w/accessories or replacement or later-technology model or equivalent.
54	1	Hand tools, complete set, for small repairs to mechanical and electrical equipment; w/socket wrench set, ratchet wrench set, key wrench set, screwdrivers, open-end wrench set, pin punch set, ignition wrench set, hammer set, plier set, pinch bar, philips screwdriver set, torque wrench, tin snips, hacksaw, etc.; w/steel toolbox and lock. Brand name: Sears Craftsman or similar set of general tools or replacement or later-technology model or equivalent.
55	1 set	Air conditioners, split unit preferred, to air-condition all internal QC work area. 220–240 v, 50 hz. Brand name: none.
56	1	Vacuum cleaner, industrial, on casters for easy rolling; tank to collect liftings; bag or filter to collect/control exhaust of dust; flexible hose of NLT 3 m length; set of nozzles to permit vacuuming seed from corners and small spaces. 220–240 v, 50 hz. Brand name: none.
Seed Health (Fungi) Testing:		
57	1	Compound microscope, DKK 33,600, w/attachments required to mount camera on microscope. W/luxury 10x eyepiece and 4x, and 10x, 100x objective lens; 220–240 v, 50 hz; tube w/ 10x eyepieces for camera. Brand name: none.
58	1	Binocular microscope, w/light; stereo-zoom, magnification NLT from7x to 30X; illuminating system suitable for 220 v 50 hz. Brand name: Seedburo No. 4E1422 or Olympus model VMZ-001 w/VM-LSG illuminator, or replacement or later-technology model or equivalent.
59	1	Ultra-violet light; hooded w/cabinet, for fluorescing seedlings; w/2 sets spare bulbs; cabinet to minimize natural light on sample; aluminum reflector, two 15-watt ultraviolet bulbs (365 nanometers); side door for easy sample placement. 220–240 v, 50 hz. Brand name: Seedburo No. 229 or replacement or later-technology model or equivalent.

contd. . . .

. . . Table 22.1 contd.

Item No.	Number Required	Description
60	1	Ultra-violet light, portable; high-intensity for detecting low-level fluorescence or for use in brightly-lighted areas; w/standard 100-watt spot bulb, 5" glass absorption filter to screen visible white light while transmitting maximum ultraviolet light intensity; range 365 nanometers in the ultraviolet spectrum; lamp head removable from base for hand use, or swivels 360o on the stand. 220–240 v, 50 hz. Brand name: Seedburo No. B-100F or replacement or later-technology model or equivalent.
61	1	Balance, 0-1200 g with 1 dec graduations, to four decimal places. Brand name: none.
62	1	Refrigerator, NLT 16 cu. ft.; w/stainless steel trays; both refrigeration and freezer compartments; automatic defrost; 220–240 v, 50 hz. Brand name: Sears Kenmore or replacement or later-technology model or equivalent model.
63	1 set	Glassware, standard laboratory, each set including the following: 4 Conical flask (2 liter) 1-P-5662 5 Conical flask (1 liter) 1-P-5652 12 Conical flask (100 ml) 1-P-5617 2 Measuring cylinder (1 liter) 1-M-0527 4 Measuring cylinder (250 ml) 1-M-0454, graduated w/tip 4 Measuring cylinder (25 ml) 1-M-0451, graduated w/tip 5 Beaker (250 ml) 1-B-1420 100 Glass petri dish 9 cm 1-P-0180 100 Test tube (160 × 16 mm) glass Pyrex 1-P-4460 6 Bottle with cap. (100 ml) Brown 1-M-0211 and 1-M-0235 12 Bottle with droppers, Brown 1-P-0575-A 5 Glass funnel 1-P-8260, 15 cm diameter Brand name: Pyrex or replacement or later-technology model or equivalent.
64	20	Haemocytometer 1-T-2060 (Burker Turk), w/cover glass 1-P-0710; 1/10 mm deep, for counting spores. Brand name: none.
65	1	Magnifying glass, 10X, DKK 150. Brand name: none.
66	4	Metallic spoon, of different sizes. Brand name: none.
67	6	Plastic bowl, approx. 20 cm diameter, 6 cm deep. Brand name: none.
68	1	Box for permanent slides, 1-P-4770; 17 cm × 25 cm × 3–5 cm, for 50 slides; of plastic material. Brand name: none.
69	1	Box for temporary slides, 1-P-4765; 10 cm × 25 cm × 3–5 cm, for 20 slides, carton material. Brand name: none.
70	1	Test tube rack for incubation (50 holes each) 1-T-1967; hard metal wire covered w/plastic; 50 holes × 1.5 cm diameter for standard test tube. Brand name: none.
71	1	Sieve, 3.15 mm, 200 mm diameter, stainless steel. Brand name: Struers 611200-24 or replacement or later-technology model or equivalent.

Item No.	Number Required	Description
72	1	Sieve, 2.0 mm, 200 mm diameter, stainless steel. Brand name: Struers 611200-22 or replacement or later-technology model or equivalent.
73	11	Sieve, 1.0 mm, 200 mm diameter, stainless steel. Brand name: Struers 611200-19 or replacement or later-technology model or equivalent.
74	10 m	Rubber tubing; 6.5 mm inside diameter × 16.5 mm outside diameter. Brand name: none.
75	50	Stopper 1-K-390, rubber. Brand name: none.
76	10	Stand w/holder for fixing funnel; w/boxwood adjusting screw, double, 15/15 cm each (one meter). Brand name: none.
77	120	Grooved plate, for dishes; 9 cm diameter; 3 grooves. Brand name: none.
78	1	Hand-operated counter, 1-S-3775; count up to 10,000 by single units. Brand name: none.
79	1	Brush, 1-B-1140. Brand name: none.
80	2	Test-tube brush, 100 mm. Brand name: none.
81	1	Paint brush, 5 mm. Brand name: none.

Chapter 23

Sampling

Sampling is Key to Quality Determination

The value of seed depends on their quality. The entire concept of seed quality is based on the assumption that tests conducted on seed from a very small sample will give accurate information about the entire seed lot. For this to happen, two things are of paramount importance:

1. The seed lot must be homogeneous, of the same quality throughout its entire mass of seed.
2. The sample must be taken ("drawn") in such a way that it includes parts from different areas in the lot, and is truly representative of the entire seed lot.

To help make tests on seed samples more relevant to the lot's actual quality, external quality control agencies limit the maximum size of lots (see ISTA, etc., lot requirements), and require that a lot be a "homogeneous quantity of seed", and if not, that a non-homogeneous lot be subdivided into two or more new lots.

To ensure that the sample is taken in a manner which makes it more representative of the lot, most certification and external quality control agencies require that samples be drawn by a trained sampler-inspector.

One of the most important things that the seedsman or the seed IQC specialist does is to draw samples in such a manner that it presents an accurate "representative" picture of the seed lot. To ensure that sample are as representative as possible, and to establish uniform procedures, external seed quality control agencies have established standardized seed sampling procedures. The following guidelines are based on ISTA procedures, which are quite similar to other standardized procedures such as AOSA.

In seed conditioning, samples are taken to evaluate the work being done—drying, cleaning, separation, treatment, etc.—and/or determine the quality or condition of the seed. It is important in all cases to have a true picture of the seed; if the sample does not accurately represent the seed lot, the seedsman/conditioner may make mistakes and lose seed or, worse, give customers poor-quality seed and lose customers or go out of business.

Sampling in Internal Quality Control

IQC, Internal Quality Control, is the seed quality/operating efficiency monitoring and control are of the seed enterprise's management. Frequent and ongoing sampling of seed and waste products is the basic means of checking to ensure that complete cleaning is being done. It samples seed at all stages in the conditioning operations, and tests the samples for different aspects (Table 23.1). Based on results of these tests, the enterprise conducts its conditioning operations, modifies/changes/ eliminates/adds operations, and accepts or rejects the seed lot.

IQC checks on seed condition and conditioning operations is a critical part of conditioning, and is essential to turning out high-quality seed and operating at minimum cost and seed loss.

Representative Samples

If the sample is not representative of the entire lot, the information obtained from testing the sample is worthless. Sampling is a statistical process in which a small amount of seed represents the entire lot. To ensure that the sample represents the entire lot, external quality control agencies such as ISTA have intensively studied the statistical relationships between samples and the larger lots. The procedure used today to ensure that the sample is representative, involves taking small samples ("preliminary samples") in several places throughout the seed lot. These are then blended together to form the larger sample ("composite sample") on which tests are conducted to evaluate the lot or the operation.

Sampling Bulk Seed

Raw, uncleaned seed often arrive at the conditioning plant in bulk, in trucks or in "toteboxes" handled by forklift. Seed may be stored in bulk, in bins or toteboxes. The bins may be large silo-type or smaller holding-type bins. Seed are also sometimes transported from place to place in bulk, either in trucks or in toteboxes or large bulk baglike containers handled by forklift. These seed must be sampled and tested at different stages, to measure or determine several conditions.

Sampling Equipment

The sleeve-type trier is the most commonly-used, and officially accepted, sampling device for both bulk seed and for bagged seed. Different sizes, with similar construction, are used for sampling seed in bulk or in bags.

Triers used for bulk seed are long, of varying lengths up to a few meters, and have larger diameters. To accommodate the longer length, there are more openings or slots in the sides. A typical bin sampling trier may have 6 to 9 slots.

The sleeve trier (not to be confused with the French word "Trieur" or cylinder separator) consists of a hollow brass tube which is inside a closely-fitting but loose outer sleeve which has a pointed end at the end which is inserted into the seed mass. The handle of the inner brass tube extends out at the operator's end, so that the operator can turn the inner tube within the outer sleeve.

Both the outer sleeve and the inner tube have slots, of the same size and position, in one side. The sampler can turn the inner tube so that the slotted openings are closed (seed cannot enter the trier), or he can turn the tube so that the slots in the tube and sleeve are matched together to create an opening, so seed flow into the cavity of the trier's inner tube.

Large triers used for bulk-stored seed (or grain) may have an entirely open inner tube, so that seed can flow through the tube, or the tube may have inner walls which separate each opening into a separate compartment. The advantage of the compartmentalized tube is that it keeps seed from each depth separate, so the sampler can examine seed from each depth to determine its condition individually.

The trier may be used either horizontally or vertically; when used vertically, the trier must have these inner tube walls which divide the tube into different compartments; without these, seed will drop into the trier from the upper layers when the slots are opened to take the sample. This over-represents the upper layers and under-represents the lower layers of seed in the bulk mass. This does not give a truly representative sample, and may cause the sampler and testing to miss conditions taking place at lower levels in the seed mass.

Bulk Sampling Machines

Seed companies which receive bulk seed in large trucks, or even rail cars, use large mechanically-operated probes which extend out over the surface of the truck or rail car to the desired spot, then insert the trier part down into the seed mass (Figures 23.1,23.2). The preliminary sample is then taken and moved through a vacuum tube to the sample receptacle in the sampling office beside the sampling probe. The probe can be controlled so that it takes preliminary samples from different depths and at different locations.

Table 23.1 Purposes of sampling seed in conditioning cperations.

Where Sample is Taken	What is Sampled	Test Made on Sample	Tested to Determine
When raw, uncleaned seed arrive at plant	Raw seed	1. Moisture 2. Purity analysis 3. Germination 4. Insect presence 5. Storability (AA) 6. Presence or susceptibility to pathogens	1. Drying needs, possible moisture Damage. 2. Cleaning needs, approx. % cleaning Loss. 3. Seed acceptance or rejection. 4. Fumigation needs. 5. Potential carryover. 7. Need for seed treatment, and kind.
During drying	Raw seed	1. Moisture 2. Temperature	1. If drying is completed. 2. If drying heat is safe.
When forming large "cleaning lots" or sequences	Raw seed	Purity, germination, class of seed	If small grower lots are of the same quality, so they can be conditioned as a large "cleaning" lot in sequence for forming clean seed lots, and without clean-out between them.
After each separator	Clean seed	Purity analysis	If desired separation is made completely; re-adjust machine.
After each separator	All waste fractions	Purity analysis	If too much good seed is removed; re-adjust machine. If the machine setup is causing mechanical damage to seed.
Before & after each elevator and conveyor	Seed	Mechanical damage	If excessive mechanical damage is done to seed.

Before and after seed flow into each bin	Seed	Mechanical damage	If excessive mechanical damage is done to seed.
After treating	Seed	Treatment coverage	If treatment is uniformly spread over all seed.
At bagging of clean seed	Clean seed	Purity, germination	If seed meet standards, and can be marketed, or need recleaning or rejection.
Periodically, in storage	Bagged seed	1. Moisture 2. Germination 3. Insect presence	1. If seed gain too much moisture, storage needs improving, or seed need redrying. 2. If seed maintain germination, if storage is good, or needs improving, or if seed lots must be discarded. 3. If fumigation is needed.

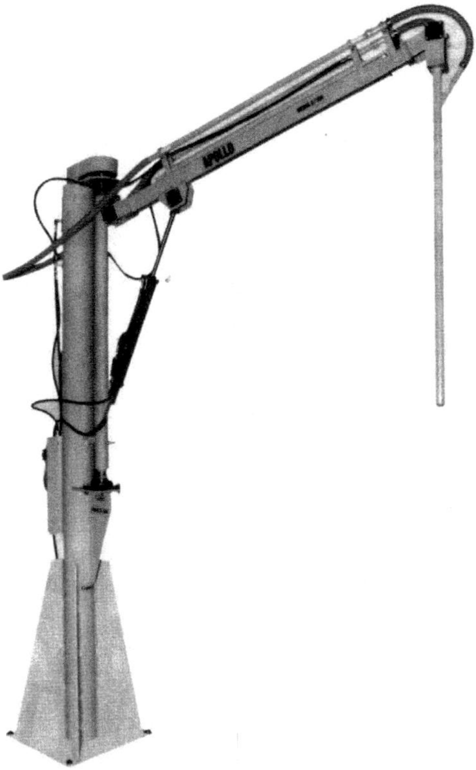

Figure 23.1 Gamet power probe for sampling bulk seed in trucks or rail cars. Source: Kathy Reading, Seedburo catalogs.

Trier Sampling Procedures

The trier is a long, slender tube-like hollow device which is inserted into the seed mass, to collect a small amount of seed from different positions within the seed mass, and bring these seed out of the mass so they can become part of the sample which is tested.

For seed in bulk, vertical insertion of the trier (the sampler working on top of the seed mass, and pushing the trier vertically down into the seed mass) is more practicable.

The inner tube is turned so that the slots in the side are closed. The trier is inserted into the seed mass at the selected place. The entire trier is then turned halfway around, to dislodge seed which may have been dragged along as the trier was inserted.

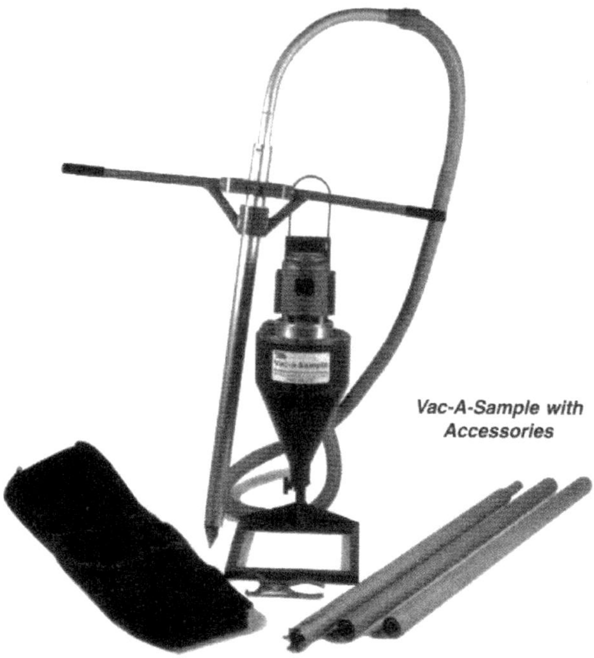

Figure 23.2 "Vac-a-Sample" vacuum probe used to sample bulk seed in trucks and bins. Source: Kathy Reading, Seedburo catalogs.

The inner tube is then turned so that the side slot openings are open, and seed can flow into the trier. Gently agitate the trier from side-to-side for a short distance, for several times, to ensure that enough seed flow into the trier to fill it.

Then, turn the inner tube to close the side slots, and withdraw the trier, *i.e.*, pull the loaded trier out of the seed mass. When closing the slots, try to be careful and avoid damaging any seed caught in the slot openings.

Have a piece of canvas, plastic sheet, heavy paper, etc., as long as the trier. Lay the trier down on this sheet, and gently turn the inner tube so the slots open and the seed flow out of the trier, onto the sheet.

Because of the compartment walls inside the inner tube, seed taken from different depths are in different parts of the trier, and thus are in different small piles on the sheet. Quickly examine the seed withdrawn by the trier, to be sure that all the seed lot sampled is homogeneous, and that do spots of different composition were encountered.

If the seed all seems similar in quality, lift up the sheet and pour the entire "preliminary sample" into a bucket or other container to work toward forming the composite sample. Continue taking as many preliminary samples as required, in the same manner.

Large Bins: To sample seed stored in large quantities in very large silo bins or the so-called "Butler" bin, several procedures may be used:

1. The seed are conveyed out of the bin into another bin or to an operation, and samples are taken from the flowing seed stream.
2. The bin may have openings (with adjustable covers) at intervals on its sides, so that a trier may be inserted into the openings to draw samples.
3. If the bin is not too deep and inside head space is adequate, the sampler enters the bin, and moves around over the top of the bulk seed, inserting the trier (which should be long enough to reach nearly to the bottom of the seed mass) into the seed mass at the required number of randomly-selected points about the surface. This is the most commonly-used approach.

Smaller Holding Bins and Toteboxes: For toteboxes and smaller bins such as those installed above cleaners to provide uniform feeding, methods of sampling which have been used include:

1. The sampler stands on top of the seed mass in the bin or totebox, and inserts the trier vertically into the seed mass at several points. This is the most common method.
2. To examine conditions such as mechanical injury to seed which first fall into the bin, the sampler may enter the bin and take preliminary and composite samples when the bin is only partially filled.
3. To examine condition or cleanliness of seed being fed to the bin/machine, samples may be taken from the stream of seed feeding into the bin.

Sampling Frequency

Each probe with the trier brings out a small amount of seed which makes up one preliminary sample which represents only a part of the bulk seed mass. A number of preliminary samples are taken from different parts of the seed mass, and these are combined to form the composite sample which represents the entire seed lot.

ISTA recommends the number of primary samples shown in Table 23.2, for bulk seed lots with different amounts of seed.

Table 23.2 ISTA recommendations on sampling frequency in bulk seed lots. Source: ISTA Rules.

Lot Size	Number of Primary Samples to Take
Up to 500 kg	At least 5 primary samples
501–3,000 kg	One primary sample for each 300 kg, but not less than 5
3,001–20,000 kg	One primary sample for each 500 kg, but not less than 10
20,001 kg and above	One primary sample for each 700 kg, but not less than 40

For example, for a totebox holding 1 ton (1,000 kg), 5 probes would be made to take 5 primary samples. For a bin holding 2 tons (2,000 kg), 6–7 probes would be taken for 6–7 primary samples.

Size of the Composite Sample

The total amount of seed taken in the several preliminary samples is often more than the sample size needed for analysis and testing. After all the preliminary samples are combined and thoroughly mixed together, the needed composite sample is taken by reducing the total sample in the approved manner. The remaining seed can be returned to the lot. CAUTION: be sure that the leftover sample is returned to the proper lot!

Cleaned Seed: The composite sample need only be large enough to contain the amount need for testing, as shown by the seed testing agency (for example, ISTA calls it the submitted sample) plus a file sample or other need of the IQC of the enterprise.

Raw Uncleaned Seed: The composite sample should contain enough seed to permit all the required internal IQC tests. For example, a certain amount would be needed for moisture testing, and several kg may be needed to conduct the pre-conditioning tests to determine conditioning needs.

Sampling Flowing Seed

Flowing streams of seed and waste products are sampled and tested during conditioning, to determine if (1) the required separation is completely made, (2) a machine is adjusted and performing properly, (3) excessive amounts of good seed are not being lost with the separated waste materials, and/or (4) if excessive mechanical damage is inflicted on seed as they move through the mechanical handling system.

Equipment

Sampling "cups" to catch seed as they fall out a spout are used. These are of many different sizes and forms, but all do the same job—collect seed or waste products leaving a machine. In general, the sampling cup should be long enough to reach fully across the flowing stream. It may be narrower, and the sampler can move the cup across the face of the flowing stream.

Facilitating Sampling From Machines

Too often, machines, flow spouts and discharge pipes are installed so that it is difficult or impossible to sample the discharging material. At the time of installation, provisions should be made to ensure quick and effective sampling of every stream of products discharging from the machine.

There should be an opening adequate for inserting the cup or hand and withdrawing the sample. This should be in a position which the operator can reach easily and safely. Usually, a hinged flap or lid is closed over the sampling opening when it is not being used, to prevent materials from leaking out.

Procedures

There are two different methods of sampling and testing seed flowing through the conditioning plant:

1. The IQC takes good representative samples, analyzes them, and advises the conditioning operators if the seed are acceptable, if the machine needs re-adjusting, or if operations need changing.
2. An experienced seed conditioning operator may take samples by inserting a sampling cup or his hand, and examine the seed to determine the efficiency of a machine or the operations.

In each case, the procedure is simply to insert the cup into the flowing stream of seed or waste products, catch the required size of sample, and withdraw the cup so the sample can be analyzed.

Clean Seed from each separator should be given a quick purity analysis, to be sure that the desired foreign seed, trash, and low-quality or damaged crop seed are removed. The sample taken should be large enough to permit a good evaluation of the separation.

Waste Products samples should be large enough to enable a purity-type analysis to determine if an unacceptable amount of good seed are being lost.

Sampling Frequency

There may be minor variations within a seed lot, the vibration of a machine may change its work, or something may go wrong within a separator. These all can affect the performance of the separator. Constant and close monitoring of each machine is required. To determine the performance of a separator, the quality of seed after the separator does its work, or the effectiveness of conditioning, samples are taken at frequent time periods. For example, the IQC of larger and better-organized seed companies regularly sample and analyze seed and wastes from each separator at intervals of one to three hours.

When the operator is sampling and checking separation performance, he should constantly monitor the machines. He should move back and forth to check all machines, their operation, supply of seed to them, flow of materials from them, and the quality of separations. He should never leave the conditioning "line" unattended for more than a few minutes. And, he should observe the machines themselves; watching a control panel in the office does not provide good management of the conditioning operation.

For obtaining a representative sample, ISTA recommends the same sampling frequency for flowing streams of seed, as for bulk seed. The numbers of primary samples shown in Table 23.3 are recommended to ensure a representative sample from a seed lot as it flows through the operations.

Table 23.3 ISTA recommendations on frequency of sampling from a flowing seed stream. Source: ISTA Rules.

Lot Size	Number of Primary Samples to Take
Up to 500 kg	At least 5 primary samples
501–3,000 kg	One primary sample for each 300 kg, but not less than 5
3,001–20,000 kg	One primary sample for each 500 kg, but not less than 10
20,001 kg and above	One primary sample for each 700 kg, but not less than 40

To determine the occurrence of mechanical damage to seed, intensive sampling and analyzing should be done when the plant, or any new equipment, is first installed. The plant or equipment should be modified as required to minimize mechanical damage. After the plant has been modified, precursory examinations of mechanical injury can be made at infrequent intervals.

Sampling Bagged Seed

Raw un-conditioned seed often arrive at the conditioning plant in bags. They are sampled to obtain the samples used to determine the lot's acceptability, need for drying, conditioning requirements, homogeneity, and potential for combining with other similar-quality small raw seed lots to make larger economic conditioning lots.

After seed are cleaned and bagged in their final or marketing bags or packages, they are sampled and tested to obtain the quality information required on the label. Also, cleaned and bagged seed in storage are regularly sampled to be sure their germination and moisture content are still at acceptable levels.

Equipment

Triers similar in construction to those used for bulk seed are used (Figure 23.3). However, the bag trier is much smaller in diameter, as less seed are taken from each bag. It is also shorter, as it need only reach through the bag. Since bags are usually lying flat and the trier is inserted horizontally into the bag, seed will flow into the trier only from the area around each slot. Thus, the compartmental dividers in vertically-inserted triers are not required.

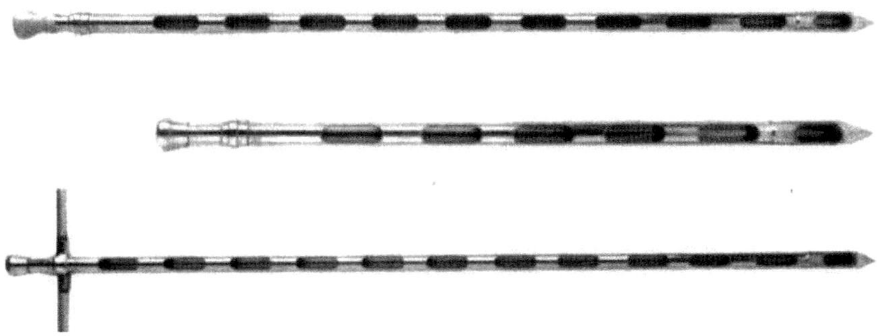

Figure 23.3 Typical triers used to sample seed in bags, forklift "toteboxes", small bins and shallow loads in trucks or trailers. Source: Kathy Reading, Seedburo catalogs.

Procedures

Bags are usually lying flat on their sides when they are sampled. Bags are selected at random throughout the lot, in a manner which ensures the proper number of bags to be sampled. Bags should be selected at the top, middle and lower part of stacks.

Seed should be taken from the top, middle and bottom of the bag. Insert the trier into one corner of the bag, so that the trier extends diagonally through the inside of the bag, toward the opposite corner of the other end of the bag. Insert the with the openings closed and facing down; when the trier is fully inserted, gently agitate it to shake loose any seed dragged along by the side slot openings, and turn the trier so that its openings face up. Open the side slot openings, and slightly agitate the trier so that seed will fall into it. When the trier should be full of seed, close the openings and withdraw the filled trier. The removed seed are then emptied onto a tarp or plastic sheet for examination, or into a pan or container.

This is repeated until the required number of bags has been sampled.

Closing Holes Created by Sampling Bags

When the trier is inserted into bags, it leaves a hole which is large enough for seed to spill out. With jute or cloth bags, this hole can be closed by using the point of the trier to rake the fabric threads of the bag back together over the hole. If the bags are of paper or other material without loose-woven fabric threads, a self-adhesive tape should be used to seal the holes. Many agencies cover sampling holes in bags with a specially-printed tape bearing a message such as "**sampled and tested by ____ for your protection**" (Figure 23.4).

Figure 23.4 A roll of tape used to seal holes in paper/plastic bags caused by inserting the probe to sample the seed. After sampling, a section of the tape is torn off the roll (the tape is perforated at intervals, for easy tearing) and the tape is placed over the hole in the bag. Source: Kathy Reading, Seedburo catalogs.

Access to Stacked Bags

The bags in a lot must be stacked in a manner which allows the sampler to reach the required number of bags. If the sampler cannot reach the required number of bags, the stack must be dismantled and re-stacked.

A useful way to stack bags is (1) have only one lot in each stack; (2) stack bags on pallets and handle them by forklift, with the stack or pile only one pallet side (*i.e.*, do not stack pallets side-by-side); (3) have adequate aisle space on both sides and both ends for the sampler to work. To protect the seed, do not stack bags against the wall or directly on the floor!

Sampling Frequency

External quality control agencies establish standardized rules for taking samples, including how many bags from which the sampler must take preliminary samples. ISTA requirements for frequency of sampling bags are shown in Table 23.4.

Table 23.4 ISTA requirements for sampling frequency of bagged seed. Source: ISTA Rules.

Number of Bags or Containers	Number of Bags From Which to Take Preliminary Samples
Up to 5 bags or containers	Sample each bag, and always take at least 5 preliminary samples.
6–30 bags	Sample 5 bags, or at least one in every three, whichever is the greater.
31–400 bags	Sample 10 bags, or at least one in every five bags, whichever is the greater.
401 or more bags	Sample 80 bags or at least one in every seven bags, whichever is the greater.

File Samples

External quality control labs keep a file sample for approximately one year, to provide a reference in case doubt arises about the seed lot. This is also a good practice for the seed enterprise's IQC. The file sample can be composed of some additional seed from the composite sample, and/or the fractions separated in the purity analysis.

Other Sampling Tools

Other tools have been developed and used for specific testing uses. Some of the more common ones are described below.

Hand Sampling

Some chaffy seed do not flow well enough to enable sampling by triers. In this case, the seed must be sampled by hand. In hand sampling, the hand is inserted into the seed mass with the palm and fingers open and flat. When the desired position is reached, the fingers are closed around a handful of seed, and it is withdrawn to form the preliminary sample.

The hand is inserted different times to different levels in the opened bags. However, the hand cannot be inserted into large bags which are deeper than the arm is long. In such cases, bags may be opened and emptied onto a clean surface so that preliminary samples can be taken of seed in the lower parts of bags.

Thief Probe or Trier

The thief probe or trier (Figure 23.5) is a short version of the seed trier. It is composed of a single pointed hollow tube with an opening in its side. When inserted into a bag, it allows seed to flow through the tube to the outside, where the sampler catches the seed in a container.

Seedsmen use the thief probe to take quick samples from bagged seed. However, it is to short to take seed from all parts of the bag, and is not considered a valid means of taking representative samples.

Cotton Seed Sampling

Auger trier: Gin-run or machine-delinted cottonseed still have enough linters to prevent them from flowing readily. These seed are usually stored in bulk in large bins or piles, and are sampled in bulk before preparing for planting (further machine delinting, acid delinting, cleaning, treating). Because they are not free-flowing they cannot be sampled by standard triers, and the

Figure 23.5 "Thief probe" used for checking seed in bags. While this type of sampling device does not sample seed across the entire bag and is not recommended for official sampling, it is often used to check seed during operations. Source: Kathy Reading, Seedburo catalogs.

large bulks preclude hand sampling. A hand-operated cottonseed sampling auger, an open corkscrew-like auger which is 5–6 feet long, is used to bore down into the pile and withdraw a 6–8-inch wide column of seed. The seed in the auger are examined and then the preliminary sample is taken from the seed in the auger, from the desired depth.

Bulk vacuum samplers: Cottonseed in large bulk quantities are often sampled with vacuum probes (Figure 23.6). The hollow open-ended probe, attached to a hose which connects it to the vacuum, is inserted to the predetermined depths in the seed mass, and the preliminary sample is sucked in through the hose to the collecting tank. The sampling pattern, similar to the standard bulk sampling procedure, covers different parts and different depths of the seed mass.

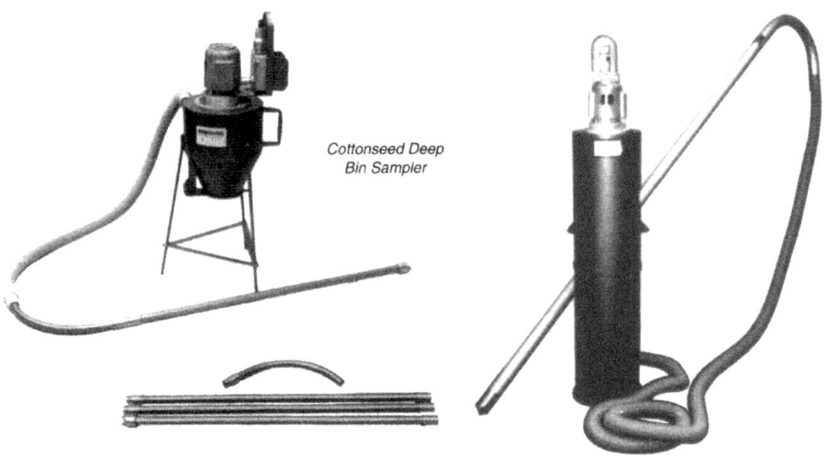

Figure 23.6 Vacuum samplers for bulk cotton seed. Left, the vacuum deep-bin sampler with extra sample probe tube sections; right, the vacuum truck sampler. These illustrations are not on the same scale; the truck vacuum is smaller than the deep bin sampler. Source: Kathy Reading, Seedburo catalogs.

Nobbe Trier

The Nobbe trier is a bag trier composed of a pointed tube, long enough to reach the center of a bag, with an oval hole near the pointed end. A typical unit (as described by ISTA) is about 500 mm long including a handle of about 100 mm and a point of about 60 mm. This leaves about 340 mm to penetrate into the bag. Nobbe triers for cereal seed are about 14 mm in inside diameter, and about 10 mm for clovers and other small seed.

In use, it is inserted into the bag, pointing upward at about 30° to the horizontal, with the hole on the bottom side. When the trier end reaches the center of the bag, is is turned until the hole faces upward. It is then slowly withdrawn, at a decreasing speed, so that the amount of seed from successive locations increases progressively as the hole is moved toward the side of the bag. As the trier is withdrawn, gently agitate it so that seed will flow into the open hole. Alternatively, if the trier is long enough to go almost all the way diagonally across the bag, it is withdrawn at a constant speed.

Automatic Samplers

Automatic samplers are available, to sample seed in flowing streams such as at bagging-weighing of cleaned seed. These can be pre-set to take preliminary samples (of sub-samples) of desired size at desired intervals.

Chapter 24

Testing Seed

Tests of cleaned seed for official labeling of seed before marketing are usually made by an approved external quality control laboratory, often operated by the external government quality control agency. In some cases, there are privately-operated independent seed testing laboratories. In other cases, the seed enterprise's own quality control lab may be approved, monitored and/or bonded to conduct official tests.

In any case, within the seed conditioning enterprise's own operations, there are constant and ongoing needs for testing and analyzing different aspects of seed quality. To ensure high seed quality and efficient operations, the enterprise must have an IQC (internal quality control) program which can sample and test seed quickly and accurately.

When and Why Seed Are Tested

Seed are tested at various points in the receiving, conditioning, storage, and marketing process, to determine:

1. Seed conditioning and separation requirements.
2. Performance of conditioning, in removing undesirable materials and saving good seed.
3. Initial seed quality, for labeling and for accepting/rejecting/re-cleaning the lot.
4. Maintenance of seed quality in storage and handling, and suitability of storage conditions.

Raw Seed Arriving at the Plant

Raw seed arriving at the conditioning plant is tested:

1. For moisture content, to determine drying needs.

2. For "conditioning" purity, to determine the impurities which must be removed, to determine if they can be cost-efficiently removed in conditioning, and if the lot should be accepted or discarded.
3. For estimating germination and purity after conditioning, to determine if the lot should be accepted or rejected at this stage.

Evaluating Raw Seed Lots

Raw seed lots are analyzed to determine their purity, undesirable materials which must be removed, germination, certification level, etc., to determine if several lots are sufficiently similar in conditioning requirements and quality for them to be cleaned in sequence without requiring clean-out of the plant, or if they can be combined to make a larger "conditioning lot".

During Conditioning

Clean seed from each machine is subjected to a rough or "conditioning" purity analysis to determine if all undesirable materials are being adequately removed. This analysis is made at regular frequent intervals, to ensure that all seed will be of the desired quality. Some conditioning plants make such analyses every two hours; they should be made at least once daily.

Each waste product from each machine is analyzed at the same time, to determine if too much good seed is being lost.

If required, the machine settings and adjustments are changed to prevent the loss of too much good seed (a small amount of good seed is expected to be lost at each separator), or to remove all undesirable particles. In some cases, an additional separator may be required, or the "middling waste product" may be re-cleaned to salvage the lost good seed.

Cleaned Seed

As the cleaned seed are bagged, a sample is taken for official testing to ensure that the seed meet the required quality standards. This testing usually involves a standard purity test and germination test. In some cases, health testing and evaluation of treatment coverage may be done.

During Storage

Seed are living beings, and deteriorate over time; if storage conditions are unfavorable, seed will lose germination faster. Seed Laws generally specify that seed must be tested for germination within a certain period (usually 6 months or one year) before they are sold.

A testing schedule, showing dates each lot must be re-tested, is usually maintained for all seed kept in storage. Each lot is re-tested appropriately before it is sold, to ensure that it is still of high quality.

Testing for Seed Quality

Seed quality determination basically involves testing (1) genetic or varietal purity, (2) physical purity, and (3) germination/viability or life; however, health and conditioning quality are also tested.

Tested Sample

The sample tested must be drawn (taken) by recommended methods, to ensure that it is statistically representative of the entire lot. If the sample does not accurately represent the lot, the testing effort is wasted.

For official testing, specific sample sizes or weights are specified for each crop seed by ISTA or AOSA, by properly reducing or sub-dividing the drawn sample. These sample sizes should also be used for internal IQC testing. However, for tests to determine conditioning requirements, a few kilograms of seed may be required to make a specific test, such as hand screen selection, cylinder selection, or gravity separation. With careful handling, this test sample can be kept pure and then returned to the lot after it has been tested. Seed sample preparation equipment is shown in Figure 24.1

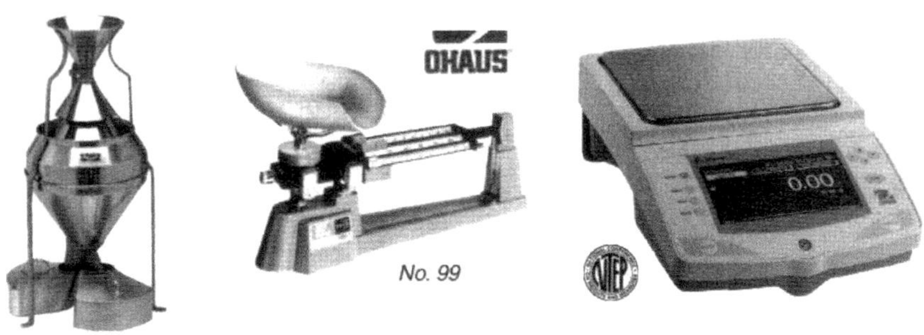

A B C

Figure 24.1 (A)Boerner divider, used to reduce the size of seed samples while still maintaining the sample's accurate representation of the larger lot. (B) Ohaus no. 99 balance, used to weigh different fractions of a relatively larger seed sample in lab testing and in IQC evaluation of raw seed lots (C) electronic balance used to weigh small fractions separated in the purity test. Source: Kathy Reading, Seedburo catalogs.

Records of Tests

Results of tests are kept in several records:

1. Raw seed testing results are kept by IQC and Conditioning, by crop variety, and in operations and "large conditioning lot" formation records.
2. For cleaned seed, the test report is kept in the file on testing, or sometimes in the file related to the crop variety or lot.
3. Test results on cleaned seed are entered into the "lot permanent record", and in the records on quality monitoring of stored seed.

Moisture Testing

Seed moisture content is critical to maintaining seed quality, so it is tested at many stages in receiving, drying, conditioning and storing. Most accurate test for moisture content is the oven drying method, but, for most work in seed conditioning, one of the electronic moisture testers is sufficiently accurate (Figure 24.2). However, some electronic moisture testers are less accurate at very low seed moisture contents. If dehumidified drying is required for vapor-proof seed packaging, an electronic moisture meter which is accurate at low seed moisture must be used.

Figure 24.2 Some of the most popular electronic moisture testers. Left to right are the Dole, Seedburo, and Steinlite. Source: Kathy Reading, Seedburo catalogs.

Additional equipment used in moisture testing include ovens to dry seed (also used in the most precise moisture test, *i.e.*, drying the sample) and sealed dessicators to hold samples without exposing them to humidity in the air (Figure 24.3).

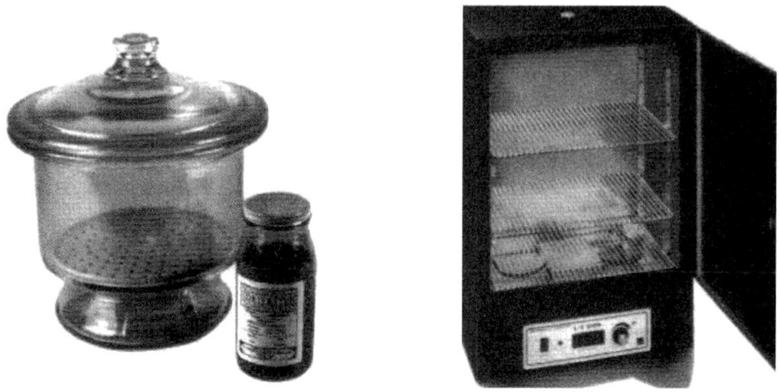

Figure 24.3 Desiccator (and silica gel desiccant) used to hold samples during moisture testing, and a drying oven used in the drying method of determining seed moisture content. Source: Kathy Reading, Seedburo catalogs.

Purity Tests

The purity test determines the physical purity of the seed sample, and identifies what kinds of materials are in the seed sample. It is made by a trained analyst going through the sample, visually examining each seed or piece of material, and separating the sample into good crop seed, other crop seed, weed seed, and inert material.

To prepare the sample for analysis, the analyst repeated divides the total sample until it is reduced to the sample size required for analysis. ISTA and other agencies establish minimum sample sizes for purity tests for each crop seed. The sample may be larger than this required minimum, but should never be smaller.

To make a purity analysis, the analyst places the working sample on a purity workboard or other working surface (Figure 24.4), and uses forceps

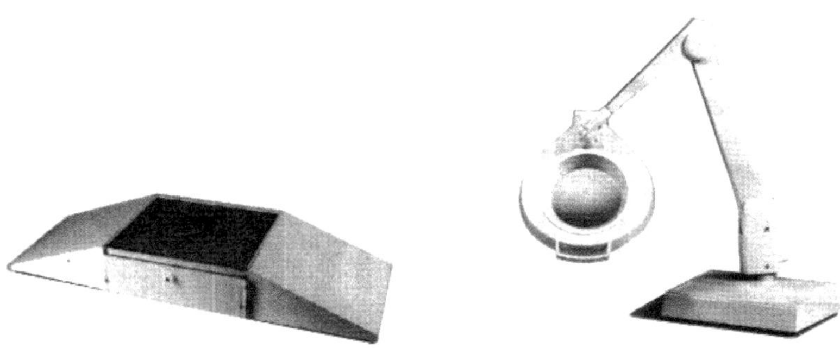

Figure 24.4 A seed purity testing workboard and an analyst's worklight with magnifier. These are used in making purity tests, noxious weed seed determinations, and to identify undesirable seed and particles which must be removed in conditioning. Source: Kathy Reading, Seedburo catalogs.

or a pointed object to move a few seed at a time toward himself from the sample. He separates each particle or seed into the proper fractions, and places each separated fraction into a separate Petri dish, plastic container or other dish, until the entire sample is separated into the required fractions of different materials.

After the entire working sample has been separated into the desired fractions, each fraction is carefully weighed. The percentage of each fraction is computed as a percentage of the total weight of the working sample. The number per kilogram (or pound) of each weed seed and other crop seed is also calculated.

Germination Tests

Standard Germination Tests

Germination tests indicate the potential ability of seed to produce normal seedlings under favorable growing conditions in the field. A germination test involves planting several small samples on a moist substrate under the proper temperature and humidity conditions. ISTA, AOSA, etc., list the proper testing conditions for each kind of seed. Different crop seed require different temperature regimes and different times for producing seedlings.

Small germinators are suitable for germination tests in the conditioning plant (Figure 24.5). The only requirement is that the germinator provide consistently accurate testing conditions, and be large enough to handle the required volume of tests. With germinators set at the proper temperature regime, the enterprise's IQC can make germination tests to (1) determine the viability of incoming raw seed, (2) check the efficiency of conditioning in removing immature or damaged non-viable seed, and to (3) determine the germination of the cleaned seed or pure seed fraction of the sample.

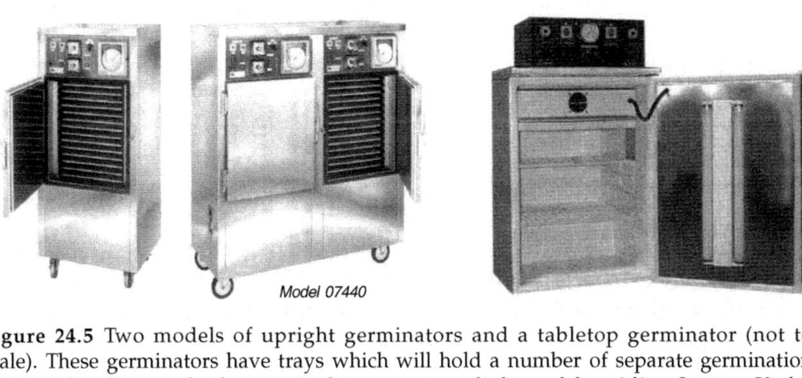

Figure 24.5 Two models of upright germinators and a tabletop germinator (not to scale). These germinators have trays which will hold a number of separate germination tests, and will provide the required temperature, light and humidity. Source: Kathy Reading, Seedburo catalogs.

Correct interpretation or evaluation of seedlings is very important. The IQC specialists should be well-trained in all their work, and conditioning staff should also be trained in rough evaluation of seed for quality and its likely viability. To assist in evaluating seedlings, photographs of normal and abnormal seedlings are available for a number of crop seed.

Quick Germination Tests

The standard germination test requires days or even weeks to grow out seedlings to a stage adequate for evaluating viability. Too often, an immediate evaluation of seed viability is needed when raw seed is delivered to the conditioning plant. Depending upon the skill and experience of the IQC specialists, reasonably accurate estimates can be made in a shorter time. Some methods used include:

1. *Sampling and testing before the seed is delivered to the conditioning plant*: In some cases, the seed will be stored or held in other places, or even in storage controlled by the conditioning plant, for some time before it is delivered to conditioning. In this case, advance samples can be taken and tested to determine if the seed should be accepted or rejected and what conditioning will be required.
2. *Tetrazolium (TZ) testing*: TZ tests can indicate within a day or so, the seed's germinability. Little extra equipment is required, but the IQC testing specialist must be extremely well-trained and have special training in conducting this specialized test and evaluating the test results. However, the TZ test, when properly used, is a very valuable tool in rapid evaluation of seed quality. In some cases, TZ test results are acceptable for seed labeling.
3. *Seed appearance*: Many seed conditioners and IQC specialists have years of experience and can judge by the appearance, color, texture, odor, etc., the general condition of the seed. They can tell if the seed is good and high quality, or if it has been damaged and will be of low quality. This, within rough limits, can help in the process of accepting or rejecting incoming seed lots but is not accurate for determining germination percentage.

Determining Conditioning Needs

Much time, cost and effort can be saved by analyzing and pre-testing a sample from each raw seed lot (or grower sub-lot) to determine which separators will be needed to clean the lot or make a specific separation, and if the seed can be cleaned cost-effectively and brought up to the required quality standards.

The conditioner who runs a raw seed lot into the "cleaning line" and then looks to see whether or not the separation is being made will find his operating costs higher and his cleaned seed of lower quality, which will ultimately ruin his enterprise.

The key to efficient conditioning is to identify the undesirable or contaminant materials which must be removed, compare their physical characteristics with those of the good crop seed, and then select and use the separator which can separate two seed which differ in the noted physical difference.

The following methods have been used to pre-test the likelihood and cost of cleaning raw seed lots, and the machines which are likely to be needed to clean them.

Conditioning Analysis Test

Using the same procedures as are used in a purity test, the analyst can easily identify what is in a seed lot and can determine the cleaning required to bring the lot up to the required standard. This can be used to:

1. Determine what conditioning operations are needed to bring the seed up to the required quality standards.
2. Identify what must be separated out, and which separators will be required.
3. How much material and weight will be removed in conditioning.
4. The final weight and percentage of good seed, and its quality.

This test is conducted in the same manner as a purity test, with some additional aspects:

1. Damaged, immature, etc., crop seed which are likely to be of low quality are also separated into a separate fraction.
2. (A) Specific identification is made of all seed/particles which must be removed, (B) comparison of their differences in physical characteristics from the good crop seed, and (C) based on this, selection of the separators and adjustments/components which are required to separate them.
3. Based on this information, a determination is made of the estimated amount and percentage of cleaning loss, final cleaned seed, time required for cleaning, and other similar lots which could be cleaned in the same sequence without equipment clean-out between lots are identified.
4. From this, a decision is made to either accept the raw seed lot for cleaning, or to reject it on the basis on excessive cleaning loss, need for a separation which cannot be made, or inadequate estimated quality of the final seed lot.

Records of Previous Cleaning

The efficient conditioner keeps complete records on each lot of seed which is conditioned, including the contaminants which must be removed and the separators and adjustments which removed them.

By making a quick-reference file of what successfully removed specific contaminants from different seed varieties, the conditioner has a ready guide for conditioning a new lot which presents the same contaminants.

Screen Selection

The air-screen cleaner is the basic cleaning machine, and is used on all raw seed lots. However, screen perforations and adjustment of air blasts differ from lot to lot. Inexpensive hand screens are available with the complete range of perforation sizes and shapes which are available for the machine. By using hand screens, it is possible to select the approximate screen sizes needed, without wasting time

Length Separation

1. A close visual examination of seed, combined with experience, will often tell a competent conditioner whether or net he can properly separate specific contaminants by differences in length.
2. Also, a single-disk hand-operated disk separator is available for test use. And, a lab-model of the indented cylinder separator is available, for testing samples from large lots or for cleaning small lots such as vegetable or flower seed.

Seed Coat Texture Test

1. Visual examination of the seed will generally differences in seed coat texture which may be sufficient for separation in a roll mill or magnetic separator.
2. Also, small-scale lab models of the velvet-covered roll mill and the magnetic separator are available for testing samples or cleaning very small lots.

Gravity Separation

The gravity separation is often needed to improve both the physical and physiological quality of seed, by removing lighter/heavier materials and separating light poor-germinating seed.

A small table model of the gravity separator is available for cleaning small lots or to show the gravity separation possible on large lots by cleaning samples.

A simple procedure for determining the degree of specific gravity difference between crop seed and contaminants, and thus estimate the ability of the gravity separator to separate them, is the waterglass method.

Other Tests

Some other lab-model equipment is available to the seed conditioner, and may be used to determine separation efficiency on samples from large lots, or may be used to clean small lots of valuable seed.

However, the most useful technique is for the conditioner or his IQC carefully and systematically examine the seed before starting to clean it. Based on observation of the seeds' physical differences and the personnel's operating experience, a careful examination will indicate possible separations and the machines to make them. The will always save time and reduce cost of seed conditioning.

Test Weight: Test weight per bushel, or test weight in kilograms per hectoliter, are often used in seed conditioning, to determine:

1. The test weight (density) of different fractions discharged from the gravity separator, especially as test weight is often related to seed maturity or condition, and thus to germination potential.
2. Test weight of raw seed lots, as a measure of maturity, chaffiness, possible quality, etc.

The test weight apparatus has many uses in IQC in seed conditioning; some plants install a test weight apparatus beside the gravity as an aid to adjusting the gravity and getting good separations.

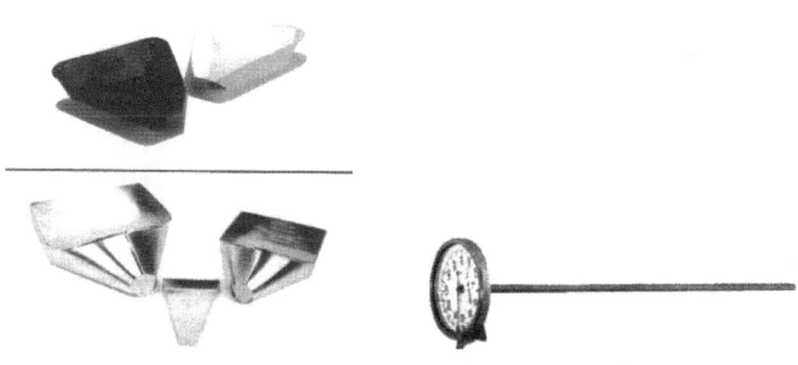

Figure 24.6 Sample pans and probe thermometer which are essential in IQC operations. Source: Kathy Reading, Seedburo catalogs.

Figure 24.7 One of the time/labor-saving items which should be used in an IQC lab is a cart which can move, handle and protect test samples during operations. Source: Kathy Rreading, Seedburo catalogs.

Lab Operations

An IQC lab is an important part of ensuring quality and profitability in a seed enterprise and its seed conditioning. It should be operated efficiently and in accord with internationally-established procedures and standards. It should also be equipped as required to make all useful tests, and so that it can save time, labor and costs in making tests (Figures 24.6, 24.7).

Chapter 25

Lot Numbers

Maintaining Seed Identity

The most important factors of seed value are (1) quality, to be sure that they will give a good stand of the desired plants, and (2) identity, to be sure that the seed is what is wanted. This requires:

1. Positive identification of each bag or package of seed.
2. Being sure of the kind of seed in the bag.
3. Knowing the quality of the seed in each bag.
4. Knowing where the seed came from, what happened to it, what was done when, to determine what may have affected the seed's quality, either in a good or bad way. This requires being able to trace a bag of seed back to its lot, tracing the lot back to its production field, this field back to the seed which planted it, and ultimately back to its original Breeder seed stock.

An effective lot numbering and seed identification system, to provide this information, is essential to protect:

1. The farmer, by giving him good seed of the right kind.
2. The seed enterprise, by always being sure of which seed is sold to a farmer.
3. The seed enterprise, by always being sure of the quality of each bag of seed.

A seed enterprise handles many lots of seed of different varieties and crops. There must be a good system for maintaining the identity and purity of each lot, identifying each lot separately, and identifying each bag in each lot. The easiest, most efficient way is to have a comprehensive field/lot numbering system which definitely identifies each field/lot, and is logical so that staff can learn it easily and work without constantly referring to records. This can be done only if the enterprise can identify each lot and bag of seed, in a manner which allows:

1. Being sure that deteriorated, low-quality seed is identified and discarded, and not sold to customers.
2. Monitoring the quality of seed, so it can be sold and planted before its natural deterioration causes it to lose germination.
3. Managing operations cost-effectively, by being able to identify and eliminate growers/conditions/ operations which result in lower seed quality.
4. Knowing which farmer purchased seed of which exact lot.
5. Being sure that the proper seed is delivered to each customer.

Most efficient and easily-used is a structured numbering for growers and/or fields. Then, use the same numbers, with necessary modifications, for each lot of seed from the fields.

Information From the Lot Number

The easiest way to identify fields and seed lots is to have a numbering system which provides a completely different number for each, and the field and lot number is the same or similar. Managing and operating the enterprise is made even easier if the numbering system is organized so that each part of the number provides some information about the field and seed it identifies.

Applying the Numbering System

The number assigned to a seed production field must be shown prominently on each record which applies to that field or seed. Field records should also show the lot numbers of seed used to plant the field.

The number assigned to a seed lot must be shown clearly and permanently on each bag or container of seed, and on each record which applies to that seed lot. Appropriate seed lot records should also show the number of the field which produced the seed.

An easily-managed numbering system should:

1. Provide a separate number for each seed lot, over at least 10 years.
2. Give basic information on seed without studying records.
3. Easily enable staff to learn the meanings of numbers.
4. Facilitate use of "stack cards" in storages.

The lot number is placed on:

1. Stack cards which identify what is in stacks of seed in the storage.
2. Each bag or container of seed in each lot, on the tag on the bag, and stenciled on the bag/package.

3. Each bin, box or bulk container of seed.
4. All records, delivery receipts, sales slips, etc., pertaining to a lot.
5. All summaries and reports of seed in storage and/or sold.

Existing Systems

If the Certification system or Seed Law implementation agency applies an organized system of seed field and lot numbers, follow their system, in order to ensure a numbering system which is consistent.

Setting up a Numbering System

A systematic and distinct numbering system can be set up easily. Points to consider in setting up a numbering system include:

1. No two fields or lots should have the same number.
2. The numbering system should—to persons familiar with the system—give some information about the field/lot.
3. The number should be on all records and each seed container.
4. The numbering system should be easy to apply and use.
5. A specific number is permanently assigned to its one field or lot; it is not duplicated in different years or places.

A numbering system meeting these requirements can be set up as follows. Each element of the numbering system signifies specific information about the field or lot.

Before setting up a lot numbering system, make a separate and complete list of all crops, all varieties within each crop, all growers and fields in current use. The lot numbering system will always provide room for new additions.

Then, assign a specific, unchanged identification (number or letter) to each crop, variety, grower and field, according to the plan of the numbering system.

Structure of a Numbering System

For quick and easy, the numbering system should use both letters and numbers, in a simple sequence which identifies the field/grower/seed lot and gives several items of important information on it. For example, a lot number could be A7BF2-2; it identifies the lot as a distinct unit of seed, and gives important information about the seed.

Basic Lot Numbering System

In a simple but effective system, the basic identification number consists of 5 parts. For fields or growers, only the first 4 parts are used. For seed lots, all 5 parts are used.

First Part—1 or 2 letters: example **A**

The letter (or 2 letters) identifies the crop. Using the English alphabet, one letter allows including 26 crops; two letters allow including over 600 crops. To prepare the number:

1. Make a list of all crops.
2. The list can be alphabetical or according to importance.
3. Assign each crop a permanent letter identification.

Second Part—a number: example **7** (number is now **A7)**

The number represents the variety within that one crop. 0 to 9 allows 10 varieties. If you have more varieties in one crop, use 2 numbers, and be sure to allow for 2 numbers in all lot numbers; *e.g.*, for number 1 use 01. To prepare the numbering system:

1. List all varieties within each crop, alphabetically, by year of release, importance, type, area of adaptation, or other characteristic.
2. In each crop, give each variety an ID number. Start with 1 (or 01) in each crop, as each is a separate group.

Third Part—2 letters: example **BF** (number is now **A7BF**)

These 2 letters identify the grower or field number. Using 2 letters allows over 600 growers or fields. To start the numbering system:

1. List all growers or fields, alphabetically or by size/location. Assign each a letter identification.
2. Start with 1 letter, then use 2 letters for more growers. There should be no confusion caused by using either one or two letters, as a number is on each side of the letters.

Fourth Part—one number: **2** (number is now **A7BF2**)

This number indicates the last digit of the year the seed was produced (2002). This number will be repeated every 10 years; however, a seed enterprise seldom can keep a seed lot for 10 years. To prepare the numbering system, each year, change the numbers accordingly.

2000 = 0
2001 = 1
2002 = 2, etc.

Fifth Part—a dash and a number: **-2** (number is now **A7BF2-2**)

This indicates the consecutive seed lot from this field in this year. Large fields may produce several final seed lots, which are usually limited in size by seed law or Certification requirements. To use the numbering system:

1. As seed are cleaned and bagged, form consecutive lots of the allowed size/number of bags allowed or desired.
2. Number the lots consecutively, starting with 1 in each crop and variety.

The Final Lot Number: **A7BF2-2**

This lot number indicates:

A: the crop—for example, maize
7: the variety—for example, Obatanpa
BF: the grower or field, for example, Mohammad Ali or his field
2: the last digit of the year of production, for example, 2002
-2: the second lot from this field in this production year.

Adding More Information

If desired, additional information can be included in the lot number. For example, the class of certified seed can be indicated by adding a letter at the end of the lot number:

A7BF2-2 becomes A7BF2-2C for certified seed, or A7BF2-2R for registered seed.

Or, other information, such as area of adaptation, can be added by adding to the number, as **A7BF2-2** can be modified by adding a letter after the year of production, as A7BF2A-2 (to indicate adaptation to area A). Letters and numbers should be alternated, to avoid confusion.

Alternative Numbering System

If the alphabet of the local language makes it difficult to use letters in the lot number, a separate system using only numbers can be structured. To avoid confusion, the numbers representing different factors should be separated by a dash. For example, instead of A7BF2-2, using only numbers, the lot number would be 1-7-2-6-2-2.

Using the Numbering System in Management

Using the numbering system in managing a seed enterprise and seed conditioning involves:

1. Keeping records of numbers used for crops, varieties, growers, and all operations involving the fields and seed.
2. Key staff should learn numbers which specify important crops, varieties, and growers.
3. When a variety/crop/grower is dropped, do not re-assign that number. Just stop using the number.
4. When a new variety/crop/grower is added, assign a new number.
5. In all offices and work areas, put up a poster showing basic numbers.
6. Show numbers on all bags, bins, records, and sales. Put the number on the tag, and stencil it on the bag.
7. Put the number on every record, sales slip, etc., on the seed and field.
8. Use "stack cards" in storage warehouses, and chalkboards on bins, to show the lot number.
9. When delivering or selling seed, double-check numbers on bags and records to be sure the proper seed is sold.
10. Advise farmer customers to keep the tags from their seed bags in their permanent records.

Chapter 26

Labels and Seed Identification

Seed Identity is Critical

A critical part of the quality of seed is to know with certainty what the seed is, and what its quality is. This requires complete and exact identity; identity thus is really a component of seed quality.

Complete identity of every seed lot and every seed bag at all times is essential. Since both the condition and the containers of a seed lot are changed during conditioning, it is essential to have each container completely labeled at all times. To maintain correct identity of each seed lot and seed bag requires 3 things:

1. An adequate label on each container (bag, bin, totebox) of seed, and especially on seed bags, because of their large number.
2. Complete records, which carry complete and up-to-date information on the seed (see chapter on records).
3. A comprehensive lot numbering system, which gives each lot a distinct number which ties each bag of the lot to all records, all activities, and the field where it was produced.

A comprehensive lot numbering system, described in another chapter, is essential to maintain distinct identification of each lot of seed, and each bag, bin or totebox within the lot. The record system described in another chapter, is essential.

Labels

Each container of seed must bear a label describing the seed, for two major reasons:

1. To identify the seed, as distinct from all other seed lots, to ensure that it is of the proper kind and the labeled quality. In doing this, identity actually becomes a part of the quality of seed.

2. Describe the important aspects of its quality, so the farmer knows what he is getting. This label also serves as the seller's guarantee of certain quality standards.

Raw Seed

During conditioning and handling, the exact identity of each container of raw seed must be maintained, which requires a label. To identify raw seed, bags, forklift toteboxes, bins, etc., must have facilities for attaching at least one label and/or lot number which fully identifies the seed. Normally the seed's identity can be established by labeling it with only a lot number, but often the label also includes variety.

Stacks in Storage

Stacks of bagged raw seed and conditioned seed in storage normally bear a "stack card" which carries information about the seed, including crop, variety, lot number, bag weight, number of bags in the stack, and date of last count of the bags.

Bags of Cleaned Seed

Labels required: Bagged or packaged seed offered for sale are required by law to be labeled with certain information. This includes the information shown below as to seed quality, where the seed was produced, and the person/company labeling the seed.

Attaching labels: The conditioner usually attaches the labels to bags as the cleaned seed is bagged. Because a number of days are required to complete germination tests to determine seed quality, seed bags are sometimes tagged in a separate operation after the quality tests are completed. However, this requires an extra operation, so tags are usually attached to the bags when the seed is bagged, even though the quality tests are not yet completed. To accommodate this situation, the conditioner is usually allowed by the Certification and seed law implementing agency to:

1. Label the bags with the minimum required germination percentage, with the understanding that real germination will meet or exceed this. In fact, many seed lots germinate higher than the labeled percentage.
2. The labeled bags of seed must be held in the conditioner's storage until official quality tests are completed and show that the seed meet or exceed standards. Then, the seed can be marketed.

Kinds of labels on cleaned seed bags: The identity and quality information on a seed bag are shown by (1) a lot number stenciled or printed on the bag itself; (2) a quality analysis tag attached to the bag; and (3) if required or desired, one or more other tags with information such as seed treatment.

Form of labels on seed bags: Labels on bags are usually in the form of a tag. The tag may be of either or a combination of several forms, such as (1) a separate tag(s) sewn onto the bag when the bag is sewn shut; (2) separate tag(s) placed with a seal onto the end of the thread with which the bag was sewed shut; (3) adhesive label/tag(s) which is stuck onto the seed bag; of (4) printed onto the bag.

Certification tag: Sometimes the Certification tag is separate from the quality analysis tag information. In some cases, the quality analysis and Certification data are all included on one tag (Figures 26.1, 26.2 and 26.3).

Required quality information: The seed laws in most countries require specific information on the seed label or tag. However, additional information may be shown on the tag, so long as it does not mislead the farmer or contradict the legally-required information. This information on seed quality includes:

1. Kind of seed and variety or type when the seed is sold as a variety or type or this information is necessary to the farmer. The kind, variety and type must be that which is officially recognized.
2. Lot number or other identification to establish the identity of a particular lot or bag.
3. Origin (where produced) is required for some seed, or if unknown, the words "origin unknown".
4. Percentage of pure crop seed, as listed as the "kind, variety and type".
5. Percentage of weed seed, and the names and rates of occurrence (number of weed seed per pound or kg) of all noxious weed seed occurring in the crop seed lot.
6. Percentage of other crop seed, crops other than those listed as the "kind, variety and type".
7. Percentage by weight of inert matter.
8. Germination percentage of the crop seed.
9. The date—month and year—in which the labeled germination test was completed. Germination tests are valid only for a specified period, and must be kept up-to-date.
10. The name or code designation of the producer or seller.

Required seed treatment information: If the seed are treated, information on the treatment materials and handling the seed, as shown in the chapter on seed treatment, must also be carried on the bag, usually as special tags/labels.

Farmer keeps labels: Seed promotion agencies, such as Agricultural Extension, encourage farmers to keep records on their crops, and to save and include seed tags as part of their records. This helps the farmer correlate seed quality with crop performance, and is the basis for purchasing seed next year, and for resolving problems with seed quality.

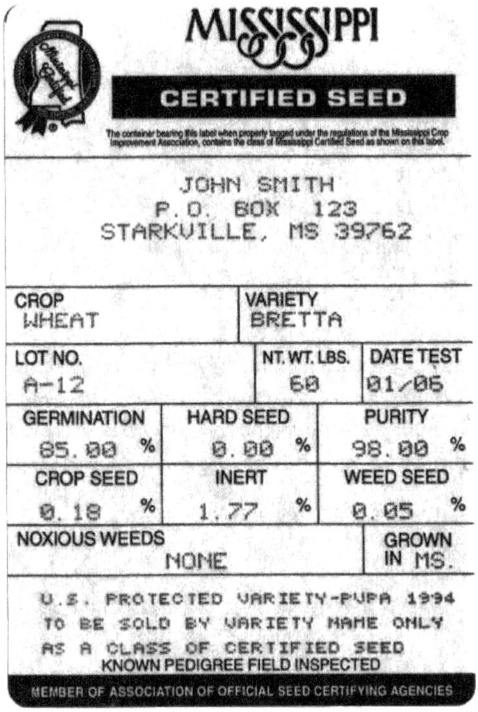

Figure 26.1 A seed certification tag which also bears quality analysis data required by law. This tag is blue, to indicate that it is of the Certified generation in the limited-generation sequence of Certification. Source: Bennie Keith, Mississippi Seed Improvement Association.

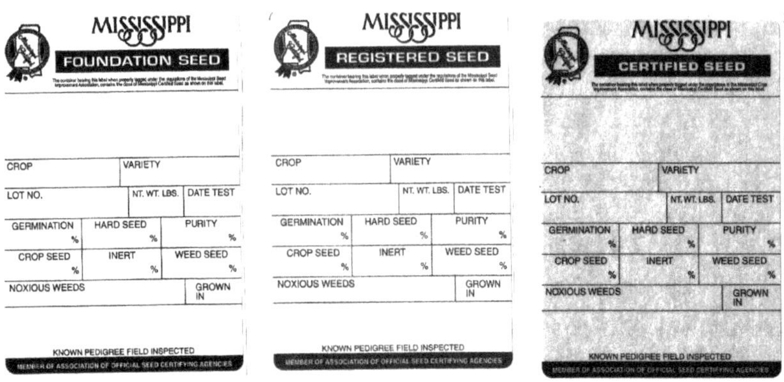

Figure 26.2 Tags for the three generations of Certified seed. Left, Foundation seed tag is white; center, Registered seed tag is purple; right, Certified seed tag is blue. These tags also contain legally-required seed quality information. The style of these tags is as an adhesive label attached to the bag. Source: Bennie Keith, MS Crop Improvement Association.

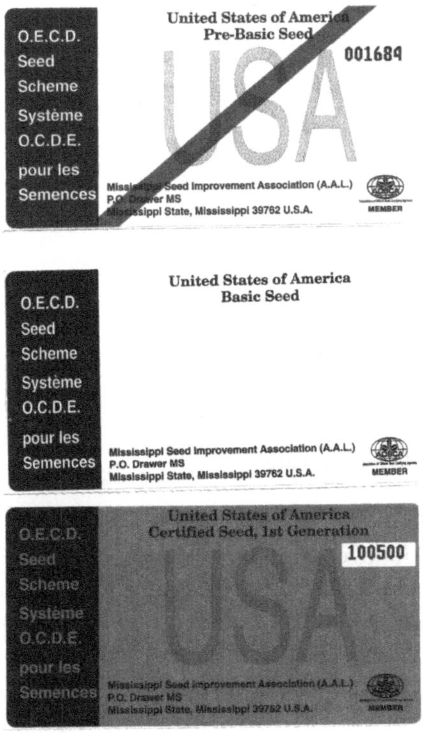

Figure 26.3 OECD tags for pre-basic, basic and certified seed. Source: Bennie Keith, Mississippi Seed Improvement Association.

Chapter 27

Blending

Blending in Seed Conditioning

Blending is the mixing or intermingling of two or more different seed kinds to produce a final lot of overall uniformity.

Blending of seed, either within the same lot or from two or more lots, is a common practice. A single lot of seed will have areas of dissimilarity in test weight, germination, concentration of weed seed, other crop seed, or inert matter. This is especially true if the lot is made up of seed from several different fields. Blending a lot, to mix the different parts of the lot thoroughly with each other, makes it more uniform in quality.

Reliable seedsmen make every effort to market seed of uniformly high quality. It is often necessary to blend two or more lots to produce a final lot of the required uniform quality. A lot of seed with slightly substandard germination can be blended with a lot of unusually high germination, to produce one large lot of acceptable germination. In the same way, small amounts of crop seed or common weed seed that are not troublesome in the field but are difficult to remove by conditioning, can be diluted to give a product which equals or exceeds the quality required.

Blending should never be used to enable the sale of seed with diseases, troublesome or noxious weeds or crop seed. This is not a recommended, nor morally acceptable, practice. Blending, however, is an excellent tool to ensure uniformly high quality. If there is ever any question, blending should not be practiced and the offending seed should be discarded.

Managing Blending for Uniform Quality

Proper blending is difficult and requires careful planning, management and operation. The theoretical idea of blending seed to complete uniformity is an unattainable ideal; in fact, few conditioning plants have sophisticated blending equipment. When two different seed are blended, the final result is a three-dimensional shuffling process, and must be handled carefully to ensure uniformity throughout the seed lot.

The seedsman must decide how uniform his final product must be, to satisfy seed law and certification requirements, ensure giving all farmers high-quality seed, and to zealously protect the seedsman's reputation, which is the key to continued farmer acceptance of his seed. The seedsman must establish a high standard of uniformity and quality for his seed, and constantly work toward it.

He must then have a means of testing uniformity of his seed, based on a reliable sampling technique. Blending introduces a certain amount of chance, and a series of samples taken from the same lot will naturally vary in composition, around a mean value. The sampling technique must follow statistical and mathematical approaches. ASOSA or ISTA sampling rules and procedures should be carefully followed; in the case of difficult blends, the recommended sampling and testing frequencies, etc., should be exceeded.

For the actual blending process, blending equipment must be selected, and used properly, which will produce the desired results at the desired capacity.

The degree of blending obtained depends on the physical properties of seed size, shape, specific gravity, surface texture, size distribution, relative flowability, and other factors which affect the random flow and dispersion of seed. Blending two lots of the same kind of seed, such as to produce uniform weight or germination, presents fewer problems, since both seed are similar in physical properties and will flow similarly. However, blending different kinds of seed to produce a uniform seeding mixture is more difficult. For example, it is almost impossible to get a uniform blend of a small legume seed and a larger, non-free-flowing grass seed, or similar mixes of seed which vary widely in size, shape, or flowability.

Blending Dimensions

Blending involves either (or both) of two dimensions:

1. *Horizontal blending*: blending different parts of a seed lot which are near the same level in the lot; for example, blending a seed mass as it flows through a continuous-flow system blends "side by side" parts of the lot in a horizontal blending process.
2. *Vertical blending*: blending different parts of the lot, such as from the "front" of the lot with the "back" of the lot as it flows.

Areas of non-uniformity may occur in horizontal or vertical arrangement within the seed mass, or the seed may be non-uniform in both directions. Ideally, blending should include the mixing together of parts of the seed lot from the front, back, center, and all sides. However, this is not always possible and requires certain types of batch blending equipment.

Equipment

Blending may be done in batches or in a continuous flow. A choice between the two methods would depend on the quantity of seed to be blended, nature of the seed to be blended, and the blending equipment available. The blending method used must produce the desired uniformity throughout the entire lot.

Batch Blenders

Equipment for batch blending can be divided into the following categories:

1. *Rotary Drum Blenders*: The working part of these blenders is a revolving drum which may have baffles on the inside walls or center to stir the seed mass. As the drum revolves, it forces seed to move within the drum and intermix with other parts of the enclosed seed mass. Baffles scoop up seed and cascade them down in the seed mass. The barrel may revolve about its long axis or it may turn end-over-end about an off-center shaft.
2. *Conical Mixers*: One or both ends of the mixing chamber is conical. A common conical mixer resembles a drum with both ends shaped like a cone. As the drum turns end-over-end about a central shaft, seed pout into and out of the conical ends. This spreading, tumbling, and re-blending action blends the seed mass. Baffles or paddles may be mounted inside the blender.
3. *Vertical Spiral Conveyor Blenders*: This blender consists of a relatively tall (taller than wide) mixing chamber with a vertical spiral conveyor or auger mounted in the center. Seed are lifted up by the conveyor; when they reach the top, the seed fall outward and downward. This continuous circulating motion blends different parts of the seed mass (Figure 27.1).
4. *Planetary Vertical Batch Mixers*: Off-center paddles rotate through a mixing chamber to fold and circulate seed into a uniform blend.

Continuous-Flow Blenders

When the quantity of seed to be blended is too great for the capacity of a batch blender, a continuous-flow blender can be used. Continuous-flow blenders do not accomplish vertical blending of a seed lot, since seed at the "front" of the lot will not be blended with seed at the "back end" of the lot. This may result in non-uniformity.

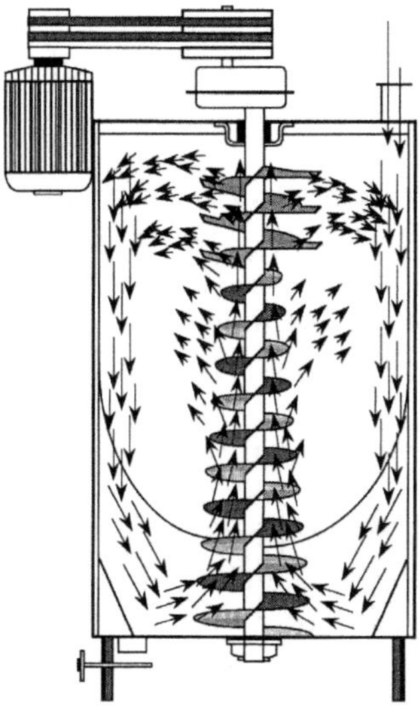

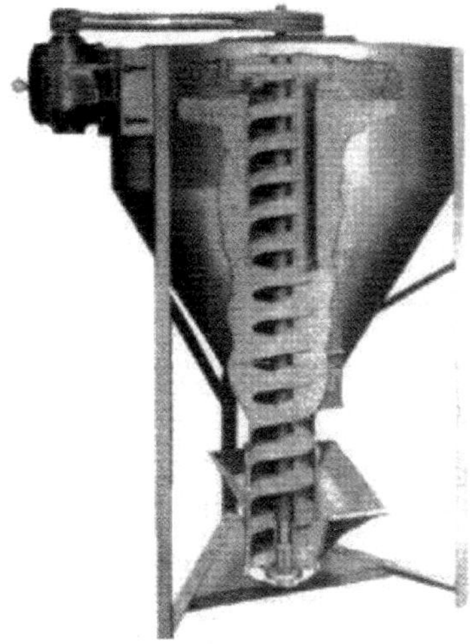

Figure 27.1 Cutaway drawings of two types of vertical auger blenders which blend small batches. Sources: A.T. Ferrell descriptive literature, Kathy Reading, Seedburo catalogs.

Continuous-flow blending occurs, to some degree, in the conditioning sequence. Also, a continuous-flow blender can be placed in the conditioning "line" to blend horizontally within each single lot as it is conditioned.

If several different lots are to be blended together into a single lot, they must be metered together into the batch or continuous-flow blender at rates to give the proper ratio. If a single lot is blended within itself, it should be split into several streams which are then poured back together and re-blended.

Equipment for continuous-flow blending includes the following.

1. *Continuous Centrifugal Blender*: This blender is similar to the seed analyst's Gamet divider. The seed fall onto a spinning rotor which separates the seed flow into different streams. The different streams are then re-blended and fed out a single discharge spout.
2. *Horizontal Blender*: Spirals, ribbons, blades, or other devices move seed through this blender, which resembles an auger conveyor. Seed are blended as they move through the conveyor-blender.
3. *Bin-Feeding Devices*: A wide range of devices can be used to feed seed uniformly into a blending bin. The seed stream is fed into the blending bin in uniform or calculated amounts which are then withdrawn (discharged) from several outlets in the sides and/or bottom of the bin. This method effectively blends seed which have "vertical" differences.

4. *Multiple Bin Arrangements*: Non-uniform seed are fed from several bins into an auger or mixing conveyor which blends them into a uniform single stream. This stream, in turn, may be re-divided into several bins and then blended a second time.
5. *Simultaneous Feeding*: The calculated amounts of each of several different lots are simultaneously fed into the hopper of an elevator. Flowing through the elevator into a bin, then repeating this process with one or more additional elevators and bins, performs some blending of the different lots fed into the system.

Computing Proportions for Blending Different Seed Lots

The Pearson Square is an easy way to determine the proportions of different lots to blend together to give a final lot of a desired quality. This method is based on the fact that the final seed lot will always be intermediate in quality (the quality factor blended for, as germination, purity, test weight, etc.) between the two original lots.

The following examples show how the Pearson Square can be used. It is equally suitable for blending for germination, purity, test weight or weed seed content; the principle is the same in either case.

Example 1: Blending 2 lots to get a specified final quality

Find proportions of two seed lots that can be blended together to give a final lot of a given germination percentage. You have a lot of 64% germination, and another lot of 96% germination; what proportions of these two lots can be blended together to give a final lot of 86% germination?

Procedure:

A. Draw a square and put final germination percentage—86%—in the center of the square.
B. Put the germination percentages of the two original lots on separate left-hand corners of the square. (Figure 27.2).

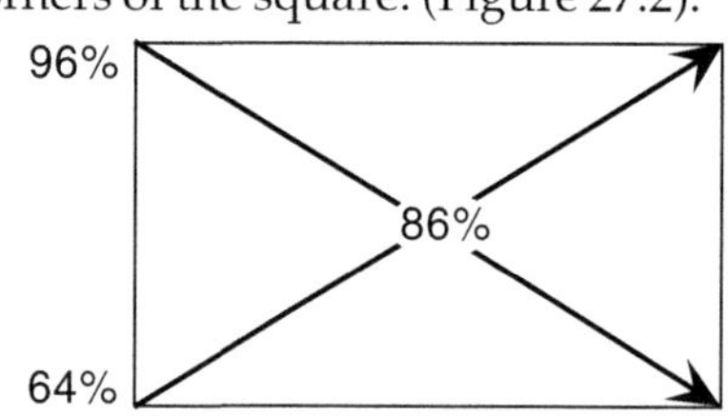

Figure 27.2 Starting Pearson Square chart for blending two lots of known quality to get a final lot of specific quality.

C. Draw lines diagonally across the square, and subtract diagonally (along the lines) the percentage in the center and the percentages on the left-hand corners. It makes no difference which figure (center or left-hand corner) is larger; just subtract and find the difference.

D. Place this subtracted difference on the right-hand corners of the square, at the end of the line with which you made the subtraction, diagonally across the square from its corresponding lot percentage on the left-hand corner. Do this with both lines and their corresponding figures. (Figure 27.3).

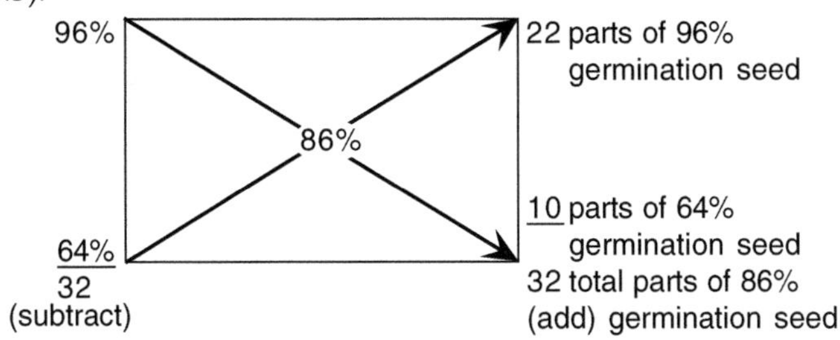

Figure 27.3 Pearson Square for blending 2 lots into one lot of final specific quality: subtracting to get the number of parts of each original lot.

E. The figure placed on each right-hand corner is the number of parts of the seed directly across on the left-hand corner (NOT along the line of subtraction) that should be blended to produce a final lot of the germination percentage shown in the center of the square.
Thus, for a total of 32 parts of 86% germinating seed, 10 parts of 64% seed must be blended with 22 parts of 96% seed.

F. Proofs:

(1)
22 parts × 96% = 2,112 (ignore the "parts, %" etc.; just use the numbers)
10 parts × 64% = 640
32 parts × 86% + 2,752

(2)
32 parts × 86% = 2,752

The two final figures are equal; therefore, your parts of each lot of seed are correct.

Example 2: To use a given amount of seed of one lot in a blend

You have 500 pounds of seed that germinates 60%. How much 90%-germinating seed must be blended with this 60%-germination seed to give a final lot whose germination is 80%?

Procedure:

A. Draw a square (as before), and put the percentages in position. Put the known germination percentages of initial lots on left-hand corners, and the desired final germination in the center. Draw diagonal lines, and subtract along the diagonal lines. Put the difference in the right-hand corners at the end of their respective lines.

B. Then, determine the amount of 90% seed by ratio; *i.e.*, known pounds are to parts as X (unknown weight) pounds are to corresponding parts. 500 lb. of 60% seed : 10 parts :: X lb. of 90% seed : 20 parts
X lb. of 90% seed = 500x20/10 = 1,000 lb. of 90% seed (Figure 27.4).

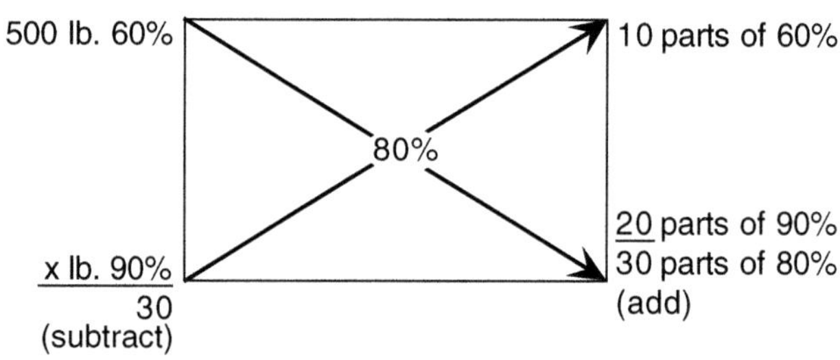

Figure 27.4 Pearson Square for determining number of parts of each of two original lots.

C. Proof:

Add:

500 lb. × 60 = 30,000
1,000 lb. × 90 = 90,000
1,500 lb. 120,000 lb.

Multiply:

1,500 × 80 = 120,000 lb.

The two figures are the same, so your parts of each lot are correct.

Example 3: To produce a definite amount of the final blended lot:

You want to make a total blend of 1,000 pounds of seed with a test weight of 16 pounds per bushel (lb/bu) from two lots that have test weights of 18 lb/bu and 13 lb/bu, respectively. How much of each lot will be needed?

Procedure:

A. Draw the square and put in the figures as before (initial test weights on left-hand corners, final in the center). However, use X to represent the weight of one original lot, and 1,000 minus X (1,000-X) to represent the weight of the second original lot. (Figure 27.5).

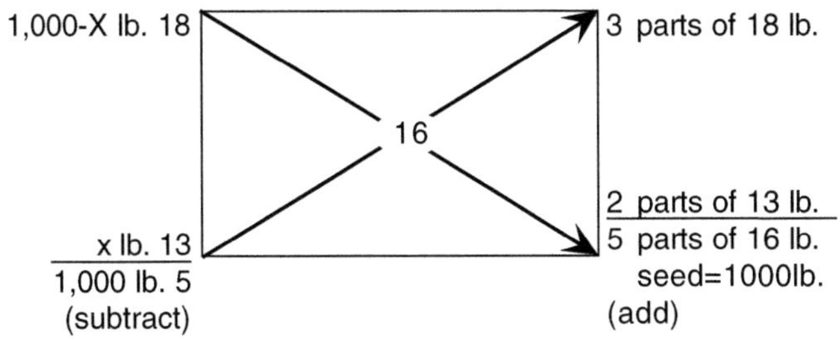

Figure 27.5 Pearson Square for determining parts used to produce a specified final lot amount.

B. Again, convert this to simple ratios:
   X lb. : 2 parts :: 1,000 lb. : 5 parts
   5X = 2 × 1,000
   X = 2 × 1,000/5 = 400
   = 400 lb. of 13-lb/bu test weight seed
   1,000 lb. – (minus) 400 lb. – 600 lb. of 18-lb/bu test weight seed
   OR,
   5 parts total – 1,000 lb. total weight of final lot
   1 part = 1,000/5 = 200 lb.
   2 parts of 13-lb/bu seed = 400 lb. weight of seed of this initial lot
   3 parts of 18-lb/bu seed = 600 lb. weight of seed of this 2nd initial lot.

C. Proof:
   (1)
   400 lb. × 13 = 5,200
   600 lb. × 18 = 10,800
   1,000 lb. 16,000 lb.
   (2)
   1,000 lb. × 16 = 16,000 lb.

The 2 figures are the same, so your proportions are correct.

A rapid check of the accuracy of the number of parts is:

1. Subtract the 2 percentages on the left-hand corners.
2. Add the numbers of parts obtained on the right-hand corners.
3. The products are equal if the number of parts is correct.

Example 4: Blending more than 2 lots

The Pearson Square can also be used to determine proportions when more than two lots are to be blended. The only requirement is that there must be the same number of original lots whose quality is HIGHER than that of the final lot as there are lots whose quality is LOWER than the desired final lot. If there is an unequal number of either, then one lot must be represented twice on the square, so that there are exactly the same number of HIGHER and LOWER lots.

You have lots testing 98%, 96%, 94%, 72%, 76%, and 80% germination. What proportions of these lots must be blended together to produce one final lot of 86% germination?

Procedure:

A. Draw the square and place the initial lots (sub-lots) germination percentage figures along the left-hand side, and 85% (final lot) in the center.
B. Draw a diagonal line from each left-hand figure (original lot germination) through the center number (final lot germination) to the opposite or right-hand side.
C. Subtract along the diagonal lines, and put the difference on the right side. Then, read straight (horizontally) across. This shows the number of parts of each lot (represented by its germination percentage) on the left-hand side, to blend together to make a final lot of the germination shown in the center (86%). (Figure 27.6).

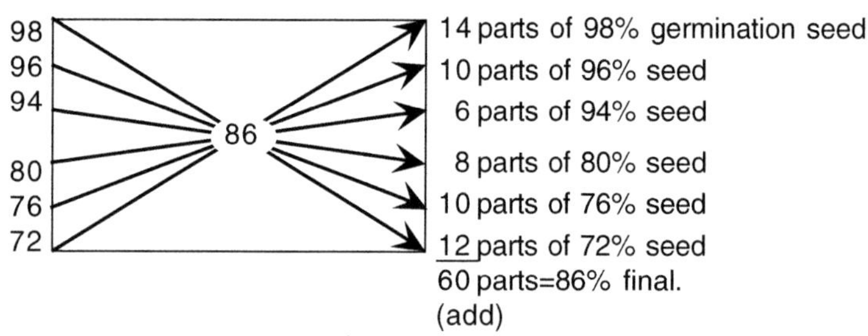

Figure 27.6 Using the Pearson Square to blend several lots into one final lot.

D. The proportions, or number of parts, of each original lot (shown on the left-hand side) can be varied by re-arranging the position of the different lots (shown by their germination percentage) on the left-hand side. For example, to use 10 parts of 98% seed, move the 96% (or 76%) lot figure to the lower left-hand corner. This would result in the numbers of parts shown in Figure 27.7.

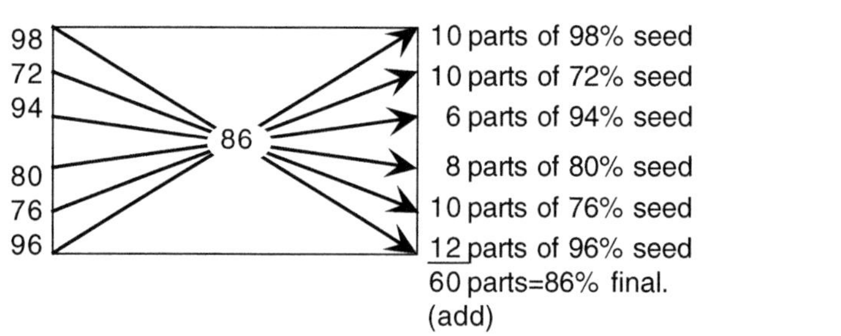

Figure 27.7 Using the Pearson Square to determine original lot parts, using a specified number of parts of one original lot.

Chapter 28

Clean-out and Preventing Contamination

Mechanical mixing and contamination in handling and conditioning is an ever-present risk. However, it can be completely prevented by good planning, careful management, a properly-designed and constructed facility, care before and during operations, and thorough attention to details by all managers, operators and workers.

Labeling Seed

Each bag, box or bin of seed entering the plant must be well-labeled so that its identity is constantly maintained and known to all who work with the seed. This identity must be maintained from the time the seed enters the plant until it is shipped out.

General Cleanliness

The entire conditioning plant must be kept spotlessly clean, free of trash and loose seed, inside and outside of buildings. Inside the plant, floors, corners, storages, ramps, beams, supports, access-ways, stairs, and all ledges must be kept clean and free of all trash, dust and seed. They must be cleaned up before, during and after each lot is conditioned or handled. Good sanitation practices throughout the plant not only prevent accidental contamination of seed, but also reduce fire hazards, eliminate food and cover for rodents, birds and insects, and also instill in all staff the need for and acceptance of doing everything possible to maintain high quality and purity standards.

Clean-Out Before Handling a New Kind

The entire plant, all machines and all bins must be thoroughly cleaned-out before seed of a different kind or variety is conditioned. This plant clean-out must be very thorough, to eliminate all other seed kinds and prevent mechanical contamination.

Normally, the plant staff is thoroughly trained and properly equipped to clean out the equipment and facilities. The plant's shift manager supervises the clean-out, and then the IQC inspector, or the Certification inspector, inspects the plant and identifies any further clean-out needs.

Equipment for Clean-Out

It is impossible to clean conditioning machines and the plant adequately with a brush and a broom. Every plant must have one or more industrial heavy-duty vacuum cleaners, and compressed air which delivers a high-pressure (up to 150 psi) air blast through pointed nozzles. The vacuum and air blast must be available at all points, including on top of machines, in bins, at elevator boots, etc. Some plants pipe high-pressure air throughout the plant, with quick-attaching outlets for air hoses at convenient intervals. Air hoses 30–50 feet long, with pistol-grip air nozzles, are necessary.

In addition, brooms, small brushes, counter brushes, wire brushes, and pointed probing instruments such as ice-picks must be available (Figures 28.1, 28.2). Trash containers with tight-fitting lids must be located as needed, and kept emptied regularly.

Figure 28.1 Brooms, brushes, blowers or compressed air nozzles, rubber mallets, sharp-pointed wires, etc., are essential to remove seed remaining in cracks, separator parts, and corners. Source: Kathy Reading, Seedburo catalogs.

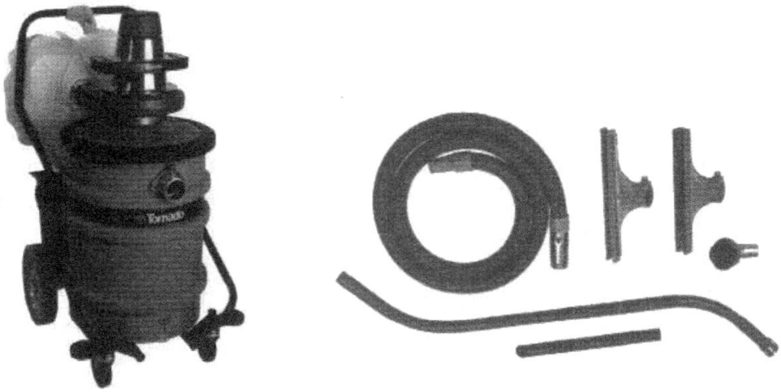

Figure 28.2 A powerful industrial vacuum cleaner, with adequate attachments for removing loose seed from floors, inside machines and bins, etc., is essential for plant clean-out. Source: Kathy Reading, Seedburo catalogs.

Clean-Out Procedure

Too often, a separator or elevator or bin is cleaned-out, then re-contaminated by undesirable leftover seed falling into it from the previous machine, which was not cleaned. The best procedure for –clean out is to begin at the point where seed enter the plant. Then, continue the clean-out through the entire plant, in the same sequence that seed would flow. Begin with the unloading ramp and dump pit, then move through the plant, first cleaning the feeding elevator, any conveyors and the bin which serve each machine in the sequence in which the seed would reach them.

Before cleaning any separator, always first clean the elevator and then the surge bin serving that separator. This prevents contaminant seed from falling into the separator after it has been cleaned.

Constructing the Plant for Easy Clean-Out

The entire plant must be constructed with one thought in mind—to clean seed to high quality standards and prevent them from being contaminated. Making the plant easy to clean out, and in some points self-cleaning, is essential. The following construction details make the plant easy to clean out.

General Plant

1. *Floors*: Should be smooth, easily swept, without cracks or rough spots that hold seed. Concrete floors should be trowelled smooth and then finished with a long-wearing and smooth floor sealer.

2. *Walls*: Should be smooth-finished, with no cracks, holes or rough spots that could hold seed or insect eggs. Do not leave exposed studs; finish the inside walls smoothly without corners which hold seed or are hard to clean.
3. *Window ledges*: Should be rounded or sloped so that seed or trash fall off rather than lodging, and birds cannot build nests on them.
4. *Structural framework*: To the extent possible, steel pipes are preferable to angle iron or I-beams, because seed and trash do not collect on pipes and birds do not build nests on them. Make structural frameworks as self-cleaning as possible, and be sure they are at least easily accessible for clean-out.

Bins

1. Inside surfaces of all bins should have no cracks or ledges where seed can lodge. Wood or cribbed bins should be lined with sheet metal, with overlapping edges of the sheet metal all pointed downward on the inside to avoid ledges, and all edges and seams brazed or soldered smooth. The ideal bin is made of heavy sheet metal, welded, with inside welds ground smooth, and reinforced with angle iron mounted on the outside, so that its entire inside surface is completely smooth and free of cracks and ledges.
2. Inverted angle iron should be welded on top edges or rims of bin walls to eliminate ledges which may hold seed. The bin discharge gate should be removable so that the slides holding the gate in place can be cleaned out. The end of the slides for the gate should be left open for easier clean-out.
3. Bins should not be mounted side-by-side with common walls, because of the danger of seed spilling over into adjoining bins. The feed spout into a bin should extend below the top of the bin wall, to avoid overflow. The feed spout should be kept in good condition so that it will not leak seed into other bins.
4. Bin covers are excellent means of preventing seed splash over the sides. However, there must be a means for air to escape as the bin is filled with seed, and there must be a hatch for a worker to enter the bin and clean it out.
5. Roof supports, or floor structure of upper floors, immediately above bins should be sealed off or covered with inverted angle iron to eliminate ledges which may hold seed.

Elevators and Conveyors

1. Where possible, self-cleaning elevators should be used. Properly operated, self-cleaning slow-moving continuous-bucket elevators do not spill seed and are completely self-cleaning.
2. Admixtures in belt-bucket elevators can be reduced by:
   A. Keeping the head and discharge spout in good condition to prevent scattering seed into bins or machines.
   B. Installing spacers or thick washers on the bolts holding buckets to the belt, between the buckets and the belt, to reduce the amount of seed which lodges behind buckets.
   C. Sealing off any open ends of head or boot pulleys, or by using elevators equipped with pulleys with closed ends.
   D. Mounting the elevator 4–6 inches above the floor, on an open frame so that boot inspection slides can be opened for clean-out. Elevators must be mounted in pits which are large enough to permit full access for clean-out. The pit should be large enough to allow a worker to get into the pit and clean out the elevator boot and the surrounding area. Floor and walls of the reinforced concrete pit should be smooth-finished, and should include a vapor barrier. If necessary, design a sump in the pit, and use a sump pump to keep it dry. Moisture and water should never be allowed to accumulate in such a pit.
   E. Augers are difficult to clean out, and impossible to clean out if their cover is not completely open-able, and should not be used in a seed plant.
   F Vibrating conveyors, properly used and installed, are self-cleaning and should be used where possible.
   G. Belt conveyors must be installed with ample space around and under the for easy access and clean-out.

Separators and Cleaners

Clean-out characteristics of separators cannot be altered greatly, but proper installation can reduce clean-out time. Separators should be mounted above the floor, on frames or posts which allow easy clean-out under the machines. Access space must be provide on all sides of machines, so that the entire machine is accessible for clean-out and maintenance. The access platform around the machine should have a floor of expanded metal grating, so that loose seed fall down to the floor for easy clean-up.

All discharge spouts, for good seed and for waste products, should be installed for easy cleaning, and should be kept free of leaks. Also, all spouts should permit easy sampling of materials flowing into them, but should not spill seed or trash onto the floor or access way during operation.

Receiving Facilities

Receiving elevators should be as self-cleaning as possible, and where possible, self-cleaning continuous-bucket types. Receiving elevator pits must be constructed without cracks or ledges, in the corners, on the edges, or on the covering grate. The pit should be of reinforced concrete, trowelled to a smooth finish, and with a built-in vapor barrier to keep out moisture.

The safety grate covering the receiving pit should be of pipes, inverted angle iron, etc., with the ends and support points welded so that they are self-cleaning.

Vibrating receiving hoppers, which receive seed and feed it into the receiving elevator, should be smooth-finished on the inside, as self-cleaning as possible, and easily accessible for clean-out.

The receiving pit must have an access hatch for a worker to enter it and clean it out, with a ladder of pipe or reinforcing rods built into the wall. There must be adequate lights and electrical outlet for vacuum, etc., in the pit.

A self-cleaning method of receiving seed is to use smooth-finished steel toteboxes which are loaded at the harvest operations, trucked to the conditioning plant, and moved by forklift directly into storage. This eliminates several handlings of the seed, and offers less opportunity to introduce contaminants.

General Plant Management

General cleanliness in the plant should be stressed, to keep plant personnel constantly aware of the necessity of preventing mixtures and contamination. Floors, corners, ramps, and similar areas should be kept well-swept. A good policy is to have one worker responsible for constantly sweeping and cleaning the conditioning, receiving and storage area. Dust or dirt should not be allowed to accumulate on ledges, machines or building structures. Unused screens, machines and components, pallets, toteboxes and other equipment should be cleaned immediately after use and before they are stored.

When harvest of different fields can be controlled to some extent, receiving and conditioning can be scheduled to reduce the number of clean-outs required in the conditioning plant.

A master schedule and plan of the entire season's conditioning plan or program should always be prepared in advance.

Only one variety or lot should be in the conditioning area at the same time. Seed which are awaiting conditioning should be kept in the raw seed storage, well-marked and completely separated from other lots and the seed which is being conditioned.

Chapter 29

Flow Management by Color Coding

Managing Flow of Materials

Efficient seed conditioning requires managing the flow of good seed and several kinds of waste materials, so that they are:

1. Immediately and completely removed from each separator and its surrounding area.
2. Not spilled onto the floor.
3. Not mixed with other types of seed or materials, but kept separate by type/end use.
4. Moved to a central, efficient collecting/handling system.
5. Not labor-intensive or time consuming.

This requires a good flow spouting and piping system, in which each kind of material can be quickly and easily identified and then managed.

Separator/Cleaner Discharge Spouts

All seed separators have one or more discharge spouts for each kind of separated material. The separator's discharge spout must:

1. Feed into a pipe which cleanly moves the material to a conveyor/elevator which moves the material to its next, or the final, destination.
2. Allow IQC to take samples.
3. Not spill materials out.

Discharge spouts and pipes should *never* be fastened to the machine. Such added weight and/or restraint may cause the machine to operate improperly, or may damage it. Spouts and pipes should always be separately supported, by connections to the machine's installation frame or if necessary to a specially-constructed support frame.

Identifying Material Flow Pipes and Machines

Most modern seed conditioning plants are installed on a single floor, so that discharge spouts are relatively short, going only to an elevator hopper or a conveyor such as a vibrating conveyor which collects and transports the discharged materials. Flow pipes are thus usually short, and fairly easy to identify. However, it still takes a little effort to identify the pipe and the kind of material flowing through it.

In multi-story plants where separators are installed in vertical series on several different flow levels, flow pipes extend to lower floors, sometimes several levels below their origin. In these plants, identity of the kind of materials flowing through the pipes requires special effort.

In both single-floor and multi-floor conditioning plants, it saves time and money, and prevents mix-ups, by having a system for quick, simple and positive identification of the kinds of materials flowing through each pipe.

In most chemical process industries, all flow pipes are color-coded to identify what they handle. A simple pipe color-coding system is also quite useful in seed conditioning.

Color-Coding System For Seed Conditioning

In a seed conditioning plant, the most important need is to identify the type of materials flowing through the pipe. In multi-story plants and in some larger plants, the separator/cleaner producing the materials may also need to be identified. A simple color-coding of pipes will identify either or both of the kind of material and the separator.

Pipes can be painted with specific colors, to identify immediately what is flowing through the pipes. Conveyors, elevators, and other seed movement and handling equipment can also be painted a specific color to identify what is being transported. The following simple system, or similar arrangement, can be used.

Fractions of Materials

The materials flowing through the pipe, elevator, or conveyor can be identified, by painting the pipe/conveyor specific colors, such as:

1. Good seed (untreated): green
2. Good seed (treated): red
3. Light waste materials: yellow

4. Heavy wastes: brown
5. Small or short waste particles: gray
6. Large or long waste particles: blue
7. Dusty air and/or air liftings: orange
8. Dusty air (treated): red
9. Middlings (mix of good seed and wastes, normally re-cleaned): black

The entire pipe/elevator/conveyor should be painted to indicate what it is carrying. However, in the case of pipes, the color coding could be used to paint only the discharge end of the pipe.

Separators/Cleaners

If the separator/cleaner producing the flowing material needs to be identified, a stripe can be painted down the pipe with a specific color which identifies the separator. If the separator color and the materials color are the same, either of two options can be used:

1. Leave the pipe as it is, with materials color and separator color the same (which can readily be seen), or
2. Paint a white stripe on both sides of the separator color stripe, to indicate that it is the stripe color identifying the separator.

A separator/cleaner color scheme such as the following can be used (see Figure 29.1):

1. Scalper: gray
2. Pre-conditioning machines (debearder, huller, etc.): black
3. Air-screen cleaner: brown
4. Length grader: green
5. Gravity separator/stoner: orange
6. Special separators (roll mill, magnetic, spiral, etc.): yellow
7. Treater: red
8. Bagger-weigher: red (treated seed)

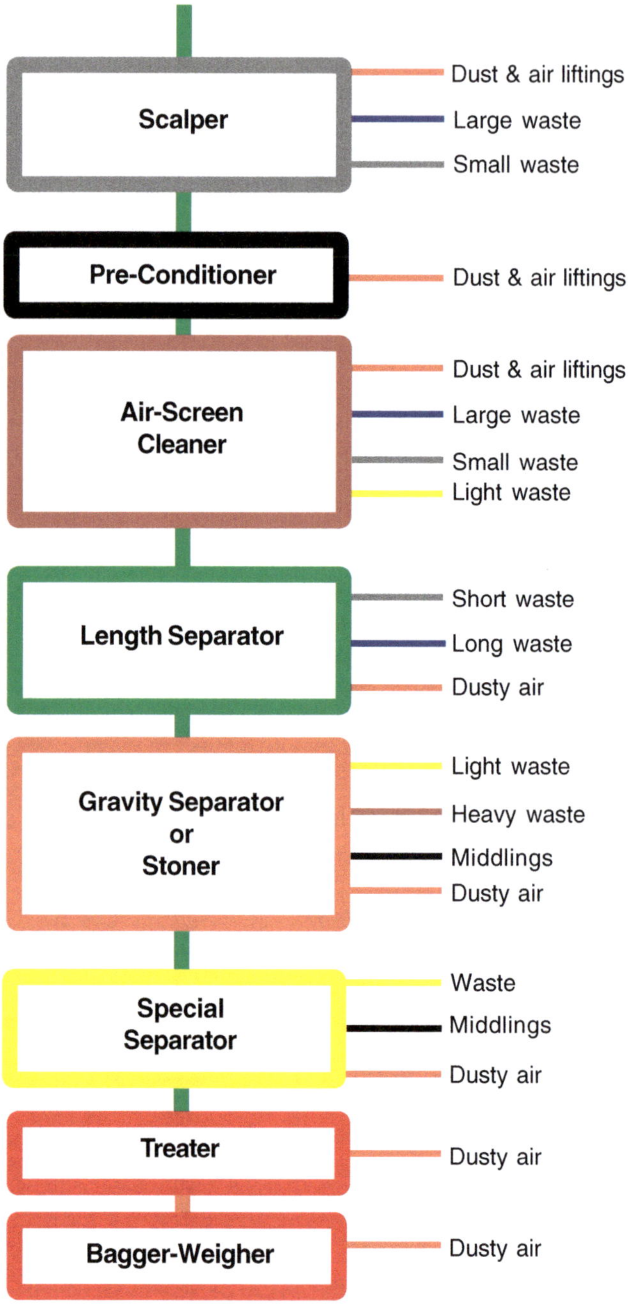

Figure 29.1 Color coding scheme for all machines and flow pipes in a seed conditioning plant.

Chapter 30

Seed Builiding Construction

Architecture

The seed enterprise is a useful economic addition to the area, in that it provides employment, creates value-added income for farmers, and helps the local community and rural development and income. The enterprise's facilities and buildings should also be attractive and add to the stature of the enterprise.

The buildings should all follow the same architectural theme, be attractive, functionally efficient, and well-built so they will have a long useful life with minimum maintenance costs.

A good facility also helps create a better impression of the seed enterprise, which is an important part of building farmer confidence and creating a market—just so long as the facility's cost does not over-burden the enterprise's finances!

Initial Design and Construction

Cost- and quality-effective operations begin with, and depend on, the initial design of the buildings of the facility. More care and effort put into the planning process can save operating costs, reduce loss of quality of stored seed, and reduce operational handling time.

It is usually more expensive, and never as efficient, to modify an existing (or not specifically designed) building. A building is just a protective shell over the operations inside it. Most efficient is to design and construct a new building to keep out as much heat from solar radiation, and pests, insects, moisture and humidity from the outside as possible, with minimum operating cost.

Design criteria should ensure an environment which is as cool and dry as possible under natural conditions, so as to keep seed quality high, improve working conditions, and protect equipment.

The design of a seed conditioning and storage building (usually including receiving and drying) must include all aspects required to ensure the highest-quality seed at the least installation/operation cost and operating loss. Even in conditioning, the seed should be in a good storage environment. Extra care and reasonable cost at the time of construction can, over the long run, be significantly profitable. This must include design and construction to achieve maximum protection of seed during all operations, lowest operating cost, and greatest efficiency:

1. Least cost in utilities (electricity, etc.).
2. Least time required to bring in raw seed, condition it, and move cleaned seed into storage.
3. Least loss of good seed in conditioning.
4. Least mechanical damage to seed during movement and conditioning.
5. Preventing loss of seed lot identity.
6. Best protection for quality during conditioning and storage.
7. Easy low-cost operations.
8. Easy access to, and identification of storage location, of specific lots.
9. Minimum absorption of solar radiation heat, to create good storage conditions with minimum mechanical equipment.
10. Protection from soil moisture and rain.
11. Protection from birds, rats, insects and other pests.
12. Ease of delivering seed into storage.
13. Ease of taking seed out of storage.
14. Maximum use of the area.
15. Safe operations such as fumigation and use of pesticides and insecticides.

Location and Access

1. Locate the plant and its buildings in a dry, well-drained, clean area to keep the environment as dry and pest-free as possible. It should not be in a crowded, polluted, or heavily-traveled industrial or residential area.
2. Storage buildings should be oriented East-to-West, to minimize solar radiation on the walls and consequent buildup of heat inside the building.
3. Well-drained, all-weather paved access to main roads should be easy, efficient, and adequate to handle the required flow of trucks. Recommended is 20 cm concrete over 25 cm well-compacted hardcore limestone.

Area Around Buildings

1. Outside area must be well-drained and clean.
2. Adequate surface drainage must be provided to drain water away from the site into adequate channels to remove water.

Office and IQC Building

The same architectural design as used for the working buildings should be used, to ensure a harmonious appearance.

The number and size of offices must be determined as the first step. A large office for each section (production, conditioning, marketing), containing desks for all the section's personnel, is often the best approach. The general manager should have a separate office; the administrative staff in one office; and the manager's secretary and other secretaries can be in another office.

There should be a reception room with sales counter to support marketing work. The reception room should be comfortable for customers, with a few simple amenities such as a tea or coffee pot.

A conference room which can be used for staff conferences, training sessions with growers, marketing meetings, etc., is very useful.

IQC is most economically located in this building. The number and size of IQC labs should be established early, and these should allow space for all operations and for growth of the enterprise.

Support Buildings and Facilities

Because they involve moisture and potential fire hazards, a separate building should be provided for the shop, with space for metal work, wood working, engine repair, electrical work, spare parts storage, etc. A shed for vehicles, trailers, spare equipment, tractors and farm equipment, etc., can be attached to the shop. This building(s) should also follow the enterprise's central architectural theme.

Drying

There are many kinds of drying systems; all dryers and the seed they contain should be protected from inclement weather. The drying system can be installed in a building, often in the seed receiving area, before seed are sent into storage or into conditioning. Alternatively, the dryer may be installed under a protective shed roof adjoining the receiving/raw seed storage/conditioning area. Because of the fire danger, drying is not usually done in the conditioning or storage building.

Sometimes, metal silo-type bin dryers are installed in the open. However, even these should be protected by a shed-type roof, as the seed handling and conveying equipment will be exposed to the weather. Also, seed cannot be safely moved in and out of the drying bins in wet weather unless the system is under a roof.

Conditioning Plant and Ambient Storage Buildings

The conditioning plant and ambient (non-conditioned) storage are the main facilities. Both should be constructed to the same technical protective specifications. Also, both are usually included in the same building, although in large-volume enterprises, additional storage buildings are often used.

Type of Building

Any building construction (concrete, bricks/blocks with concrete bearing columns, prefabricated metal with thermally-insulated panels, etc.) may be used, depending on cost and availability, so long as the technical seed quality and operating efficiency requirements are met. Because of local availability and costs, the basic construction will be of the best local type (bricks, masonry, reinforced concrete slab, insulated metal panels, etc.), on a seed-suitable reinforced concrete base.

No support columns are inside the building. Roof is supported by a clear-span truss. Support columns should not be inside the walls; they create corners which are difficult to clean, interfere with operations, and harbor insects and pests. Clear-span steel roof support truss structures are preferred. A truss constructed of round steel pipe is the best choice, to avoid ledges on which birds build nests and dust/trash collect.

The required unobstructed headroom (clear height inside the building, below the roof support structures) must be provided. The required smooth, level floor space must be provided. The walls and roof must protect seed and equipment from weather damage.

Building Style

The style of building for the conditioning section—single story, multi-story, etc.—depends on the planned installation plan for equipment. Storage, however, is a single-story rectangular layout.

One or Separate Buildings

Many conditioning facilities use a single building for receiving, raw seed storage, conditioning, and clean seed storage. However, these operations should be safely separated, to minimize risks from insects, fire, moisture and other potential damages. Note that drying, because of the fire hazard, is usually distinctly separate.

Many facilities, particularly programs, use separate buildings for different operations. This permits safe separation of operations, which reduces the scale of potential risks and often reduces insurance costs.

In either case, buildings and operations can be located so that flow of work and seed is cost/time-effective (Figures 30.1, 30.2).

Figure 30.1 A large corn (maize) seed conditioning plant with different operations in different buildings. Separate buildings are used for the office and quality control, receiving, drying, shelling, conditioning, and clean seed storage. Note the long enclosed and protected belt conveyor used to carry shelled corn from the shelling building to the intake of the conditioning building. Source: Darin Stutler, Bratney Companies.

Separating Storage

Separate storage from conditioning, so it can be closed to keep out insects, pests, unauthorized personnel, dust, etc. However, it should be close to conditioning/bagging to minimize moving seed. If in the same building, separate storage with a fireproof wall and an approved type of fire door.

Raw seed storage is preferably separated from clean seed storage, because of dust, insect, fumigation, etc., problems.

Building Shape

Most efficient for handling and movement through conditioning and into storage, is a rectangular building, wide enough (1) to include conditioning machinery and room for working, and (2) in storage, for a center operating aisle and an efficient row of storage stacks (of the appropriate size) on either side. This usually makes the building considerably longer than wide (Figures 30.3, 30.4).

Figure 30.2 A seed conditioning facility with separate buildings for different operations, but buildings are constructed adjoining each other. This allows complete separation of operations, but minimizes distances required for movement of personnel and seed. Walls between buildings are protective fire walls which minimizes fire risk and insurance costs. Source: Darin Stutler, Bratney Companies.

Layout and Design

Conditioning is laid out for efficient input of raw seed, movement through conditioning, and efficient delivery into storage of the cleaned seed. Storages are designed for maximum space utilization with minimum movement required. Each lot must be stored separately, with space for ready access to the sides. Buildings should be rectangular in shape, with doors only in the center of the ends, for efficient space use and minimum space lost to doors and operating aisles. The conditioning bagging area should be directly out from the storage door. Storage width allows the required center aisle and space for a typical-size seed lot on each side. The storage is in bays or

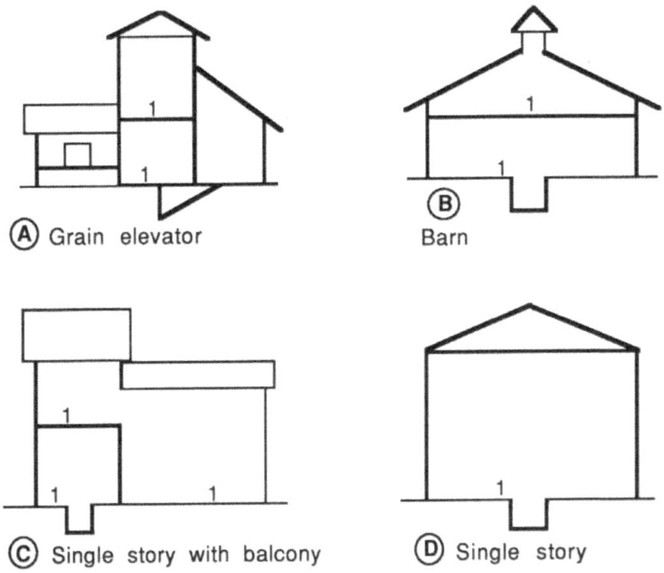

Figure 30.3 Diagrams of different shapes or styles of buildings used for seed conditioning plants. The style of building used should fit the type of equipment installation and the operating sequence. Source: USDA Handbook 354.

Figure 30.4 One single-story building housing all operations of a seed conditioning facility. The office and quality control are in a single-story building attached to the main building. Source: Darin Stutler, Bratney Companies.

sections of standard width; building length can then be uniformly and easily extended by sections to provide the required capacity. Access from storage to the conditioning bagging area should be directly out from the storage door, preferably in a straight line. Most efficient design is to have the storage width of the dimensions which allow the required center aisle and space for a typical-size seed lot on each side. The building is designed in bays or sections of standard width; building length can then be uniformly and easily extended by sections to provide the required capacity.

Fitting the Needs

The building should fit the needs of the seed conditioning plant; trying to fit the conditioning installation and operations to a building usually results in inconvenient or inefficient operations. The purpose of the building is to protect the conditioning equipment, operations and personnel. The building should be specifically designed to fit the needs of the plant (Figure 30.5).

Figure 30.5 An efficient combination of seed conditioning installation and building. The conditioning equipment was planned for efficient operations and then installed. After the installation was complete, a metal building was constructed over the equipment, to provide an exact fit to the plant's needs. Source: Crippen International descriptive literature.

Foundation and Footings

Foundation must fit local soil conditions, and carry the required load with an adequate safety margin. The foundation depends on local soil characteristics. It must be solid, deep and stable enough to prevent shifting and cracking of the floor so that moisture can enter into cracks, or cracks harbor insects. Local engineers should design the foundation according to local soil characteristics and expected bearing load.

Footings, as required, must be installed to carry the weight of machines, bins and their seed loads, loaded forklifts, etc. If loaded trucks enter the building at the receiving point, the floor must have adequate load bearing capability.

Floor Height

The floor should never be at ground level. It should be raised to the level of the bed of most trucks used, which is usually about 1.5 m. A raised floor helps control moisture and rats, and makes it easier to load trucks. A simple ramp can be used to let forklifts go from the loading platform into the truck; this way, forklifts can stack pallets of bags directly into the truck, to speed up loading/unloading and reduce labor costs.

Where trucks enter the building at the receiving station, suitable access ramps should be built at the entry and exit points. When not in use, these should be closed off with rat/bird-proof doors.

Floor

Floors must be strong, well-finished, of concrete with adequate cement content and steel reinforcement to bear the weight of stacked seed, forklifts and other loads (special footings are installed to bear machines and bins).

The floor should be at truck bed height, constructed of reinforced concrete and strong enough to carry the bearing load with a sufficient safety margin to allow for changed use.

The floor should consist of base or subfloor covered by a vapor barrier, with the finish floor applied over this. The finish floor should be of concrete without gravel, trowelled to a smooth finish.

All floors have a complete vapor barrier beneath the finish or bearing floor. After the fairly smooth subfloor is installed and is fairly well cured, the vapor barrier is installed atop it. Mop, brush or spray hot asphalt on the subfloor, and lay sheets of 10-mil thick polyethylene plastic onto this while the asphalt is still warm and sticky. Edges of all sheets are overlapped NLT 6 inches, and sealed together with hot bitumen. The vapor barrier in the floor should turn up around the walls for a few inches.

Then, pour the finish floor on top of the vapor barrier. This is usually 1.5–3 inches of cement without gravel. Finish floor is trowelled to a smooth-finish to eliminate cracks which may hold seed, and facilitate sweeping and clean-up, long-wearing, without holes which provide cover for insects (Figure 30.6).

Expansion joints should be included, in accordance with local engineering practices. Construction and expansion joints are filled with joint filler, to leave a smooth surface which is easily cleaned (Figures 30.7, 30.8).

Concrete For Floors

Concrete should have minimum compressive strength of at least 4,500 PSI in 28 days. Calcium chloride is not used in the concrete. Concrete floor final finish is hard-burnished steel trowel. After final finish and proper drying, all concrete floors are properly treated with an epoxy or other concrete sealer to prevent wear and dusting-off, and to make the floor smoother and easier to clean. Concrete is not dusted with any material to absorb surface water. Curing procedures are continued for at least 7 days.

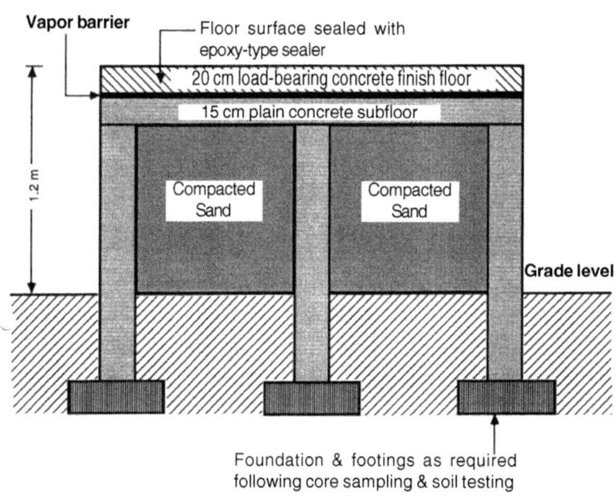

Floor to carry 300 kg/cm²
Plain concrete: Six 50-kg bags cement/m³
0.8 m³ Gravel
0.4 m³ Sand

Vapor barrier: NLT 10-Mil sheet polyethylene applied over hot-brushed bitumen

Load-bearing finish floor: 20 cm reinforced concrete; machine-troweled to smooth finish; finished with epoxy sealer.

Figure 30.6 Diagram of the construction of the floor and footings of a seed conditioning or ambient seed storage building. Source: Billups and Lasheen unpublished designs.

After the finish floor is properly cured, it should be sealed and finished a good-quality concrete floor sealer to minimize wear and dusting off. Use liquid sodium silicate or one of the new synthetic epoxy-type sealers.

Outside Rat Shield

A smooth-surfaced reinforced concrete rat shield, constructed at the same time the floor is constructed, should extend completely around the outside of the building at floor level or at least 1 m above the ground. It must be smooth, with rounded outer edges, to minimize the ability of rats to hold to it and climb around it. It should extend at least 35 cm out from the wall, or base, of the outside of the building. Walls below the rat shield are concrete, with a smooth hard finish to prevent rats from climbing up it. The ground is paved around the building; no trash, bags, equipment, etc., is where rats may use it to climb onto the rat shield (Figure 30.9)

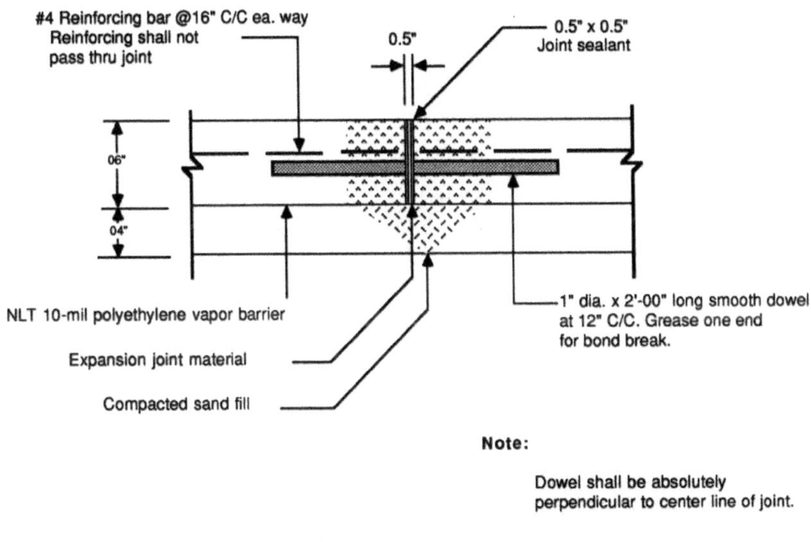

Figure 30.7 Floor slabs and joint between slabs. Floors must be smooth-finished for easy cleaning, and joints between slabs must be allow for normal expansion, but filled with sealant to make them easier to keep clean. Source: Billups and Lasheen unpublished designs.

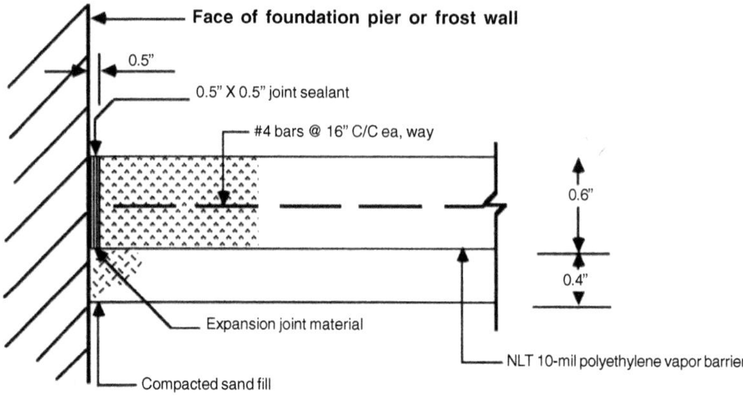

Figure 30.8 Diagram of end of floor slab, showing vapor barrier. Source: Billups and Lasheen unpublished designs.

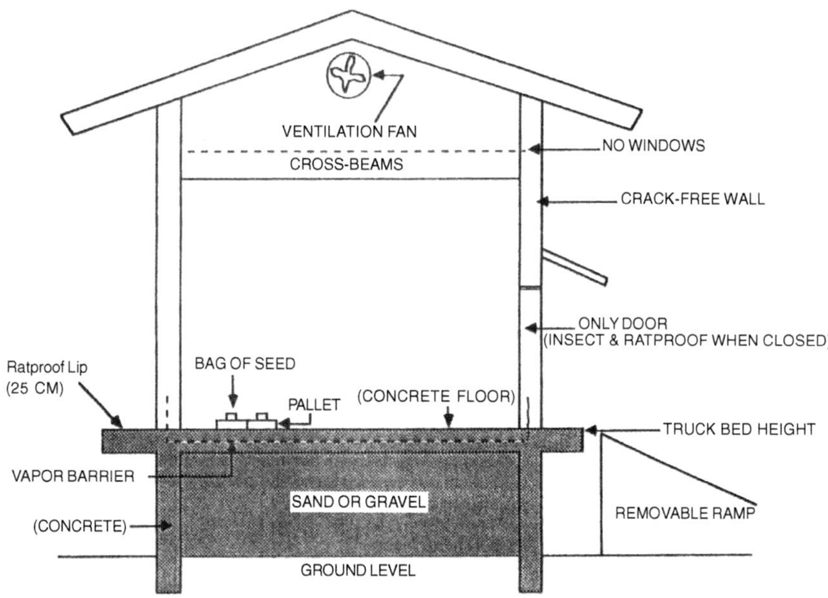

Figure 30.9 Cross-sectional illustration of the design and structure of a non-conditioned seed storage, showing the smooth-finished concrete rat shield which extends around all sides of the building. Source: Gregg *et al.*, 1970.

Walls

Walls should minimize entrance of solar and outside heat. Walls are smooth-finished to make them easy to clean, and eliminate cracks where insects lay eggs and/or dust may collect. Walls join to the roof smoothly on inside and outside, with no cracks wider than 6 mm and no ledge on which birds can build nests. Grout or fill any space between walls and roof. Inside walls are smooth and unbroken by columns.

Exterior Walls: Walls should provide exterior protection and some thermal insulation. If walls of corrugated sheet metal mounted on a steel support frame are used, some thermal insulation (and any required support frame for it) should be installed on the inside, and over this, a protective inner wall (also of corrugated sheet metal) is installed.

Lower Part of Walls: When thermally-insulated wall panels are used, the lower 1.0–1.5 m of the wall is of reinforced concrete, to prevent accidental damage. To minimize damage by the inevitable impact of forklifts, the lower 1.5 m of the walls should, if possible, be reinforced concrete. This base wall should be thick and strong enough to support the wall structure above it, and be tightly sealed to the upper part of the wall. If this lower wall construction is not possible or feasible, install a strong steel frame to about the same height, to protect the walls from accidental forklift damage.

Inside Wall: All walls which contain a frame and/or thermal insulation should have an inside facing to protect the thermal insulation inside the walls. For metal buildings, this can be a second layer of sheet metal wall panels. The inside wall surfaces should be ratproof with no openings, to prevent rat damage to the thermal insulation, and prevent rat nests inside the walls.

Surfaces of Inside Walls and Floor: All surfaces should be smooth-finished, without rough places, to (1) permit easy cleaning and (2) eliminate small cracks or indentations which will hide insects or insect eggs. Cracks along edges of sheet metal wall panels should be sealed, with a caulk type sealer or with hot asphalt.

Outside Area Around Buildings

Preferably, the entire outside area around the storage is paved with suitably reinforced concrete. If this is not possible, the area immediately around and adjoining the building, for a distance of at least 1 m out from the walls, should be paved with smooth-finished concrete, (1) for easy cleaning, (2) to prevent rats from burrowing under the edges of the building, (3) to prevent growth of weeds, trash, etc., and (4) to eliminate materials which would help rats get past the rat shield. It is slightly sloped outward to provide surface drainage. Concrete surface drains, or underground drainage, should be provided to remove rain water.

Roof

The roof is completely watertight, to prevent leaks into the building. Eave height from floor level ensures the required storage space and height. The roof should also be of the best locally-used materials, either prefabricated corrugated composition or sheet metal panels, tiles or shingles. Roof slope should be adequate to handle local rainfall without backing up water and leaking between panels. Special preparation to ensure against leaks should be installed.

Roof Support

No support columns are inside the building, to avoid interference with forklifts, working movement, etc. Roof must be supported by a clear-span truss, from columns placed in or outside the walls. Support columns must not be placed on the inside of the walls; they create corners which are difficult to clean, interfere with stacking, and harbor insects and pests. Clear-span steel roof (*e.g.*, not trusses but supports sloped along the roof line) support structures (*e.g.*, steel I-beam wall columns supporting clear-span steel roof supports which do not require a truss) are preferred. If this is not used, a truss constructed of round steel pipe is next choice, to avoid ledges on which birds build nests and dust/trash collect (Figure 30.10).

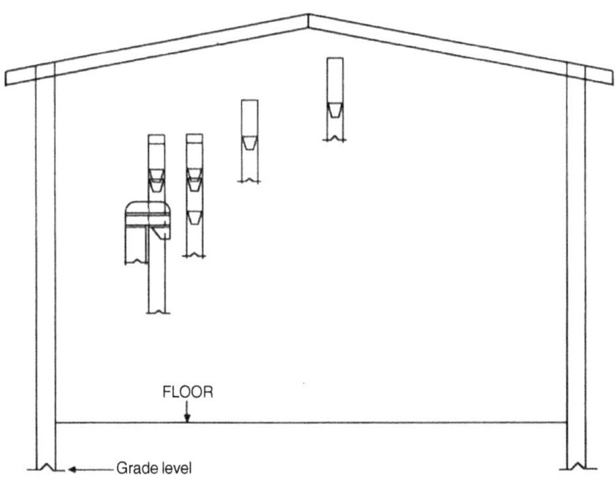

Figure 30.10 End view diagram showing roof of conditioning building, with no inside support columns to interfere with machines, and height adequate to cover elevator heads. Source: Billups and Lasheen unpublished designs.

Ventilators in Roof

Having reflective outside roof and wall surfaces, plus thermal insulation, will minimize entry and build-up of heat from solar radiation. If desired, standard available exhaust vents can be installed on the roof crest to remove warmer inside air from under the roof.

Roof Overhang

Roof should overhang (extend beyond the wall) for 1.5 meters, so that the overhang can shade the outside walls. This can significantly reduce absorption of solar radiation by the walls (Figure 30.11).

Light-Transmitting Panels in the Roof

No transparent or translucent panels should be installed in the roof. These allow entry of too much heat from solar radiation. In the long term, it is more beneficial to have the entire roof thermally insulated and provide inside lighting from fluorescent lights.

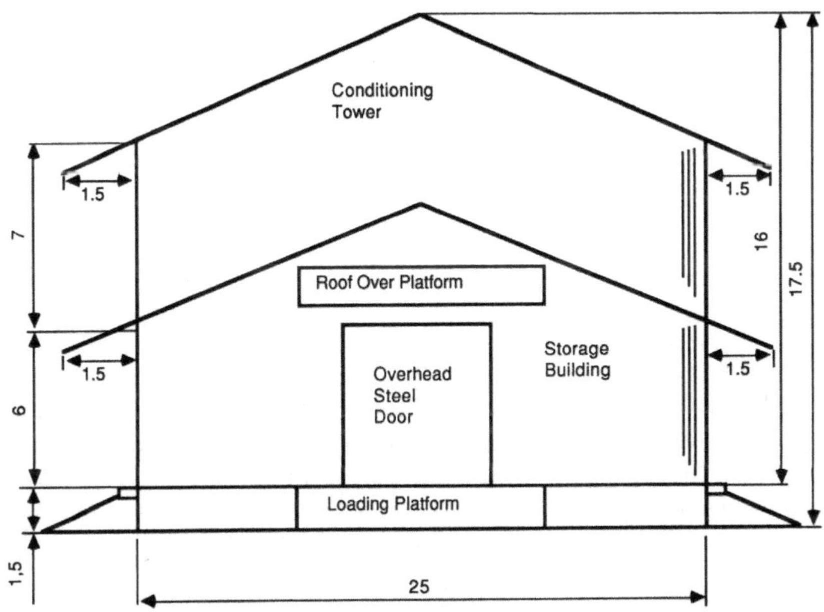

Figure 30.11 Diagram of the end of a seed conditioning and storage building, showing the roof overhang to shade the walls and protect windows installed immediately under the roof edges. Source: Billups and Lasheen unpublished designs.

Roof Support

The roof and walls are normally supported by a steel truss framework. The roof support should be a truss system which requires NO support columns or beams inside the storage. The truss system should not use up too much head room. The truss system should be above any load lifted by forklifts in stacking pallets. Stacks should not extend up between the trusses. There should be at least 0.75–1.0 m open space "head room" above the stacked seed, to allow for air circulation and stacking ease.

If the support columns are large or close together so that they would present a problem to forklift operation and storage along the walls, the support columns may be installed outside the building.

Ideally, roof support structures should be made of round pipe to eliminate ledges where birds can build nests. However, this is difficult, and birds may be kept out or their nests regularly destroyed.

Eave Troughs

A problem with eave troughs is that "standard" sizes are used to handle runoff rain water, and eave troughs and their downspouts are usually too small to handle the volume of rain water. Excess water causes backup of water under the edge of the roof. It is better not to use eave troughs and downspouts. If they are used, make sure they are specially designed to handle the volume of water during heaviest rains. If trees are close enough to the building to shed leaves onto the roof, the eave troughs should be covered with a perforated sheet metal screen, to prevent plug-up from leaves.

Painting

Outside Wall and Roof: ALL outside surfaces should be light-colored and reflective to the maximum, to minimize absorption of heat from solar radiation. The roof should be painted with aluminum paint to reflect solar heat. Outside walls should be of a light-colored (or similar process) glossy enamel paint. Aluminum paint would be best, but white or light paint is good. Paint should be durable, long-wearing, and easily cleaned.

Inside Painting: Inside walls (and under roof or false ceiling) should be painted a light color, to improve lighting and make it easier for the forklift operator to see the walls. Paint should be a durable glossy enamel, with a smooth surface.

Windows

Storages: Ideally, no windows should be used in storages, as windows allow entry of too much outside heat and humidity, and insects and pests. Especially, windows should not be installed if the building will (now or in the future) be cooled or dehumidified mechanically. During hot periods, ventilation can be achieved by using ventilators on roof crest and opening the doors. If windows are desired, they should be placed at the top of the walls, just beneath the roof, so that they are shaded by the roof overhang. Windows should be screened with a screen mesh fine enough to keep out flying weevils, grain moths, other insects and birds. The windows should have glass panels which can be opened or closed as desired.

Conditioning plant: Windows should be provided to the extent possible, to provide maximum light within the conditioning area. Windows should be screened to keep out birds and flying insects. Window ledges should be sloped, to eliminate places where birds can build nests. In areas where winter weather is severe, windows should be double-pane glass, to provide additional thermal insulation.

Electrical

Electrical connections, switches and motors should be dust-tight, explosion-proof. A central monitoring control panel may be used to control machines, and show which machines are operating. However, it is essential that each machine have a master over-ride on-off safety switch at the machine. This is to protect workers who are repairing or working on the machine from accidental turn-on by someone in the office.

Electrical work should be dust-proof, and conduits installed inside walls or in a manner which does not create risks of being hit by forklifts. A good way is to install main conduits high on the walls (or in the roof truss structure) and run conduits down the walls as required, with outlets sunk into the walls so they are flush with the wall surface.

Dust-proof electrical outlets should be placed at convenient intervals, at least close enough to enable an industrial vacuum cleaner, insect fogger-sprayers, work lights, etc., to work in all areas without excessive use of extension cords.

Lights

Floodlights or security lights on the outside walls and mounted on poles provide lights for operating at night and for security.

Inside areas are adequately lighted to permit reading tags/labels on seed bags. Fluorescent lights should be used, with switches located next to doors. Have several switches so lights can be used in only a small area, as required. There is no need to use electricity to light the entire area if you are working in only one part. Fluorescent lights give more light with less wattage, and also generate less heat than incandescent lights. Any reduction of heat in the storage, however small, helps create better storage conditions under ambient (lower cost) conditions!

Install light fixtures in, or above the bottom level of, the roof truss structure so the lights are flush with the bottom of the trusses. DO NOT have lights hanging down below the trusses—the forklift will ultimately knock them down!

In storages, light fixtures should be installed on both sides of the central operating aisle, more toward the aisle than the outside walls.

In conditioning areas, install lights over the bins so that seed level in bins can be easily seen. Lights should also be in front of and around each machine so that its operation and adjustment can be easily seen.

Doors

General: Having any door into a storage or plant wastes space, because space must then be left for movement of personnel and forklifts. Number of doors must be minimized. Doors should be placed (1) for loading and unloading storages, ONLY in the center of the ends (narrow dimensions of the rectangular building), (2) for entry and access to the receiving area to bring in raw seed, and (3) between conditioning and storage. This allows dual use of inside space for door entries and operating aisle.

For seed movement within the conditioning and storage areas, place a large (preferably overhead roll-up) door adequate for moving seed on pallets with a forklift. Close to one side of this door, install one (only) personnel door.

All outside doors should be steel, ratproof, and fit tightly into a steel frame, leaving no space in excess of ¼ inch wide. One steel personnel door is provided next to one large door. All large doors are overhead rolling steel shutter type, with a steel sill and frame which rats cannot gnaw, close-fitting with no crack exceeding 6 mm wide beneath, beside or above the door. Paint doors with adequate primer and enamel finish coat to minimize rust. A shed-type roof over the loading dock protects the door area from rain.

Never install a door in a side wall. This wastes too much of the expensive inside space for operating aisles.

Storage: A seed storage has only one large working door in the center of each end, to minimize floor space required for aisles. Doors should not be installed on sides or corners of a storage.

Conditioning plant: The conditioning area should have one entrance into the raw seed and one into the clean seed storages, with approved fire doors. One personnel door

Door Frame Protection

Each large door is protected on both sides outside and inside by heavy 5–7-inch-diameter steel-pipe door bumpers, to prevent accidental bumping/ damage to door frames. To protect door frames from damage by being struck by forklifts, on each side (both inside and outside) of each side of each door frame, install strong steel pipes filled with concrete. These should be firmly embedded in the floor structure for strength, and be strong enough to stop a forklift before it hits the door frame. These door frame protectors should be about 1.5 m high, located about 25–30 cm from the door frame. Both door frames and door frame protectors should be painted with a bright color (yellow or red, to indicate caution or danger) so they are easily noticed.

Truck Entry

Trucks enter only the receiving facility of the conditioning plant. Storage is designed so trucks cannot enter. Trucks require too much space for maneuvering, and introduce heat, moisture and dirt. They often damage seed bags and stacks. Floors close to the ground surface also increases risk of ground moisture, runoff water, etc., in the storage. Only forklifts and bag carts should enter the storage and conditioning work area.

Loading Docks and Entry Ramps

A loading/unloading dock is at the door centered in each end of the building, and at a convenient site for the conditioning plant reception. A sloping ramp for forklifts/bag carts is provided at only one dock. An angle-iron frame is inset into the concrete at all top corners of the loading dock, to prevent chipping. Docks are of adequate width and length to permit movement and loading/unloading trucks, and for forklifts to move into the trucks. A large ratproof roll-up door should open onto each dock, so a forklift can bring pallets out to load bags onto trucks. A roof must cover the loading dock to permit all-weather operations. The roof must extend far enough (usually 1.5–2 m) beyond the edges of the loading dock, to keep out blowing rain. The roof must be high enough to permit forklift operations without damage.

Air Movement and Ventilation

Exhaust fans and ventilation openings are at locations so as to remove hot air from the top of the building, and moist air from the floor level. They are screened with 6 mm wire mesh and protected by roof overhand so that rain cannot enter them. Ventilation openings in the roof top may be permanently open; ventilation openings near the floor have a tightly-sealing door or cover, so they can be closed except when conditions are suitable for ventilation.

Fire Control

Construction should be fireproof, and approved fire-doors should separate conditioning from storage, and different storage areas. Keep the conditioning plant, storages, work areas, and outside areas immaculately clean and free of dust and trash. Provide a water fire riser outside the building, near the door, in a position where it will not be damaged by trucks. Provide a fire hose inside the door. An electrical fire chemical extinguisher should be wall-mounted just inside each door.

Conditioned Storage

Conditioned storages have all the construction characteristics of non-conditioned storages, and in addition are completely vapor-sealed and thermally-insulated to minimize entry of heat and humidity.

Conditioned storages are completely vapor-sealed and thermally-insulated to create a storage space which is totally isolated and can keep out heat and humidity. Conditioned storages have the same construction features as ambient storage, with the added components to completely isolate the conditioned storage space.

A conditioned storage has no windows or ventilation openings, and only one centrally-located door. The door contains complete vapor barrier and thermal insulation, with seals or gaskets on all edges so heat and humidity cannot enter around the door. The door is reached through an antechamber consisting of an outer sealed door and a small entry room equally insulated, vaporproofed, and conditioned. The outer antechamber door is closed before the inner storage door is opened; the antechamber serves as an airlock to prevent a direct opening from the outside into the storage.

Thermal insulation keeps out outside heat, and a vapor barrier prevents the slow movement into the storage room, of outside moisture vapor through the walls, floor and ceiling. Normal construction and thermal insulation will not prevent the movement of moisture vapor into the storage room.

Floor, walls, and ceiling have a vapor barrier equivalent to at least 10 mil polyethylene film installed with bitumen. It must be continuous and uninterrupted to form a complete vapor seal around the entire storage space (Figures 30.12, 30.13).

A combination of cooling equipment and desiccant dehumidifiers is easiest to plan and operate, and is suitable for most seed storage needs.

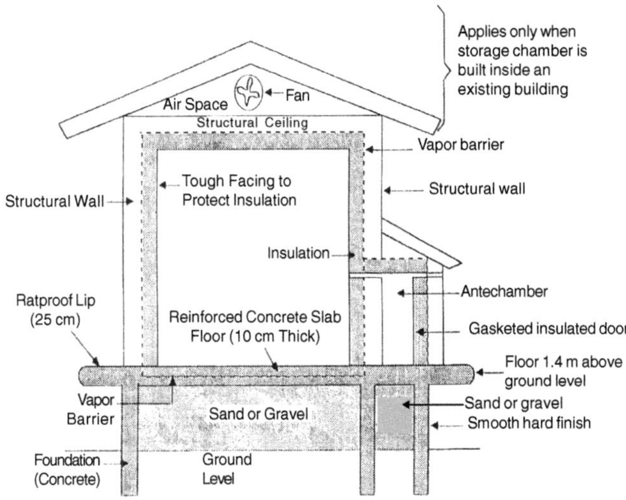

Figure 30.12 Diagram of the construction components of a safe conditioned seed storage. Source: Gregg *et al.*, Storage.

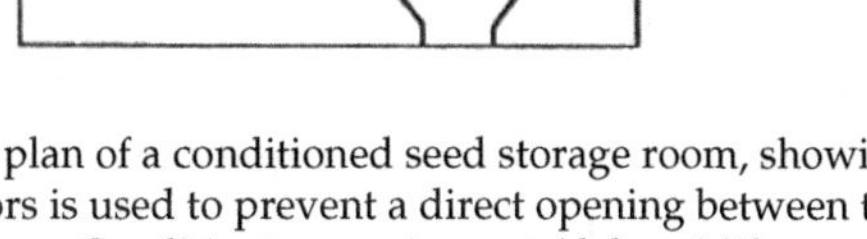

Figure 30.13 Floor plan of a conditioned seed storage room, showing how an antechamber entry with two doors is used to prevent a direct opening between the outside environment and the storage room. Conditioning equipment (dehumidifier to dry the storage air, and refrigeration to cool it) is installed outside the storage, in a special shed or room. Source: Gregg *et al.*, Storage.

Moisture Control

Seed can be kept dry by dehumidifying the air inside a storage which is completely surrounded and sealed by an adequate vapor barrier which prevents entry of outside air and moisture. This is accomplished by dehumidifying the air inside the sealed, airtight storage (Figure 30.14).

Cooled seed storages must have a complete, unbroken vapor barrier in the construction of the storage. The vapor barrier must provide a complete unbroken seal around the storage—ceiling, walls, floor and door—to keep out moisture brought in as invisible vapor in the air.

Figure 30.14 A two-drum dehumidifier, used to remove moisture (humidity) from the air in a conditioned seed storage. Because of the heat generated and the space required, such equipment is always installed outside the storage, and moist air is piped from the storage into the dehumidifier, and then dry air is piped back into the storage room. Source: Kathy Reading, Seedburo catalogs.

Seed must be kept dry as well as cool, to keep them alive in storage. The dryer seed are, the longer they will live, and the less effect that temperature has on seed viability. In a cool storage, seed survive better, and storage operating costs are less, if the storage itself is kept cool and dry, and outside heat and moisture vapor is kept out.

Typical vapor barriers are hot-brushed bitumen with 10-mil polyethylene sheets, or aluminum foil and polyethylene laminates. Sometimes, sheet metal is used. However, the vapor barrier must be completely sealed; at the joints of different sections. If sheet metal sections are not completely soldered together, moisture vapor still enters through the cracks between adjoining sections of the sheet metal.

The position of the vapor barrier in the construction of walls and ceiling is extremely important. For a "dry" seed storage, the vapor barrier MUST be placed on the warm or outside side of the thermal insulation. If the vapor barrier is on the inside or cool side of the thermal insulation, moisture vapor can and does penetrate through the thermal insulation to the vapor barrier. Installed on the inside, the vapor barrier will be cool. Moisture vapor which comes through the wall and thermal insulation will condense on the cool vapor barrier, inside the wall. This can be seen as water dripping from cracks between sections of sheet metal, moisture damage at the base of walls, etc. This causes major damage to the storage, and greatly increases humidity in the air inside the storage, which increases seed moisture content.

If the unbroken vapor barrier is installed on the outside of the thermal insulation, the vapor barrier prevents entry of moisture vapor into the thermal insulation. Because the thermal insulation prevents the outside vapor barrier from becoming cool, temperature of the vapor barrier will be above the dew point temperature of the outside air, so water will not condense on the vapor barrier.

Seed storages are always constructed with the vapor barrier on the outside (warm side) of the thermal insulation. When you build a storage for seed, be sure its construction makes it a "dry" storage as needed for seed, not a "damp" storage used for fruit and potatoes! (see Figure 30.15).

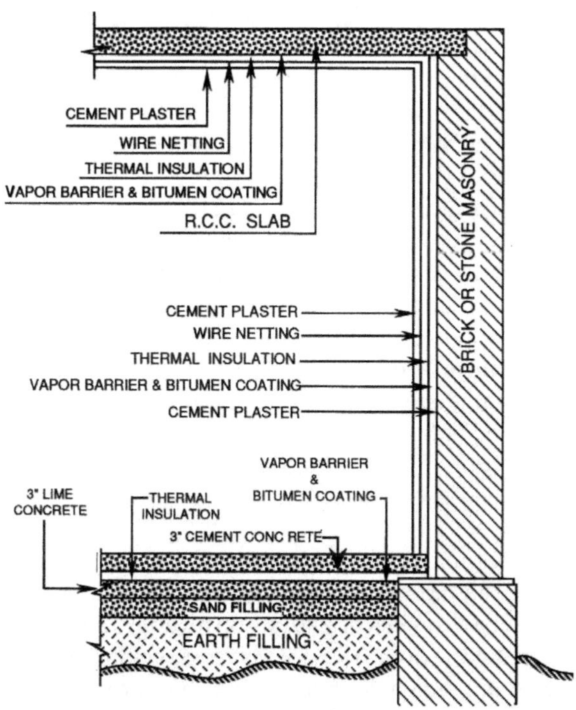

Figure 30.15 Cross-section of the construction of the ceiling, walls and floor of a conditioned seed storage, showing how vapor barriers and thermal insulation are installed. Source: Gregg *et al.*, Storage.

Temperature Control

Seed storages can be kept cooled by thermally-insulating a conditioned storage, and removing heat by air conditioning (refrigeration). Properly constructed and cooled, a conditioned storage can maintain almost any desired temperature.

To seal off the storage from outside heat, thermal insulation is installed in the ceiling and walls. Floors are usually not insulated, unless the storage is for very cold long-term germ plasm storage. However, adequate thermal insulation, without interrupted spaces, should be installed in all walls and under the roof. This should have insulating value about equal to 6 inches of Styrofoam.

Thermal insulation keeps out heat if properly installed, but does not keep out moisture vapor. Thermal insulation is installed on the *inside or dry side* of the vapor barrier, to prevent condensation inside the thermal insulation. If the vapor barrier is on the inside (cool side) of the thermal insulation, the storage *will not function properly.*

Any type of adequate thermal insulation such as styrofoam sheets, rock wool or glass wool, is suitable, so long as it has the required insulating value and is unbroken (*i.e.,* does not have thin or missed spots, or cracks between panels). Where sheets or battings of thermal insulation join/abut, they should be sealed together with warm (not too hot) asphalt or other material. Panels of polystyrene foam can be attached to walls or under roofs by warm asphalt.

A finish ceiling frame should be used to help hold the roof insulation in place and to protect it. The frame can be installed and supported between beams supporting the roof, or alternatively a false ceiling can be installed. However, the false ceiling is less desirable, as it creates a dead space where solar radiation heat will accumulate and birds may build nests.

The wall structure, on both inside and outside, must completely protect the thermal insulation to keep out rats and mice, which can seriously damage thermal insulation inside the walls.

Entry

A conditioned storage has no windows or ventilation openings, and only one centrally-located door. The door contains complete vapor barrier and thermal insulation, with seals or gaskets on all edges so heat and humidity cannot enter around the door. The door is reached through an antechamber consisting of an outer sealed door and a small entry room equally insulated, vaporproofed, and conditioned. The outer antechamber door is closed before the inner storage door is opened; the antechamber serves as an airlock to prevent a direct opening from the outside into the storage.

Cooling and Drying the Conditioned Storage Space

The storage space, completely sealed by continuous thermal insulation and vapor barrier, is mechanically cooled and dehumidified to the desired conditions. Refrigeration or air conditioning equipment removes heat to cool the storage space to the desired temperature. Heat load must be calculated for each storage, to determine the required equipment. For most moderate-sized storages in seed programs where seed are stored and/or carried over for normal periods, either standard air conditioners or refrigeration equipment may be used. Larger and/or long-term storages at low temperatures, usually require refrigeration equipment.

Air inside the storage is most efficiently dried by desiccant dehumidifiers, which can maintain lower RH at lower temperatures than can refrigeration dehumidifiers. Desiccant dehumidifiers, either dual drum or revolving bed, add to the storage heat load, but this is handled by the cooling system. Revolving bed dehumidifiers maintain more even humidity. Dehumidifiers should be installed outside, to save storage space and to minimize the added heat load.

Installing Storage Conditioning Equipment

Dehumidifiers and refrigeration/air conditioning compressors are installed outside, never inside the storage space, because (1) the inside storage space is expensive and should not be occupied by machines; (2) they give off heat, which would affect the storage; and (3) servicing the machines is easier.

Install the machinery just outside the storage, in a shed which protects it from rain and damage, but allows free air flow so heat and humidity removed from the storage can be freely discharged into the outside air.

If Much Conditioned Storage Space is Needed

If much storage space is required, *do not* construct a single large storage room. Instead, divide the required storage space into several smaller rooms. This is more cost-effective and easier to operate:

1. It is extremely difficult to maintain uniform temperature and humidity in a large open room. Usually, temperature will be higher in the top part of the room, and humidity will be higher near the floor. In a smaller room, it is easier to maintain uniform conditions.
2. "Back-up" dehumidifier and refrigeration equipment is needed, in case one breaks down or must be stopped for maintenance. If several smaller rooms (2 or more) are installed with separate dehumidifier and refrigeration for each room, it can be connected so that it serves either

one room or the other. This way, no additional back-up equipment is required.

3. Seed storage is seasonal; for short periods, the storage is full. At other times, only a small amount of storage is required. With several small storage rooms, when small amounts of seed are stored, it can all be placed in one room and the dehumidifier and refrigeration equipment in other rooms can be turned off. This is a significant saving in power costs.
4. It is dangerous to take cold seed directly from a conditioned storage into humid outside air. If several smaller conditioned storage rooms are used, seed which will be removed from storage at about the same time can be put into one room. The refrigeration is turned off, to allow the seed to warm up, while the dehumidifier continues to keep the room dry. As soon as the seed (kept dry in the room) warm up to outside temperature, the seed can be safely removed from storage.
5. If the dehumidifier or refrigeration equipment of a large storage room breaks down, all safe storage is lost. However, if there are several separate small storage rooms and equipment of one room breaks down, there is still safe storage in the other rooms.

Chapter 31

Seed Storage Principles

Storage in Seed Conditioning and Handling

The value of seed is entirely in their embryos; seed must reach the farmer with the ability to establish the desired field stand.

Seed are seldom planted immediately after harvest; they are stored for a few weeks, months, or sometimes years, before planting. Storage must protect seed from deterioration and damage in order to maintain its germination, vigor, identity, physical condition, and purity. Storage life of seed varies with condition of both seed and storage. The primary factors influencing seed storability are known, and can be largely controlled by careful planning and good management.

The conditioner/storer must store seed from harvest until marketing, and try to maintain their viability during this time span.

Good seed storage is an important phase of conditioning, and is essential to successful seed marketing. Proper storage prevents or minimizes damage to seed, and preserves seed viability, purity and identity from harvest until sale. It thus protects the investment of the producer, the conditioner, and the using farmer. To be successful, a seed program must have adequate safe storages at all places where seed will be stored.

Different kinds of seed have genetically different lifespans; for example, onion seed have short life spans, while wheat seed can be kept in good condition for longer times. In addition, the storage environment has a major influence on storability; good storage conditions (cool and dry, protected from damage) keep the seed alive and vigorous for longer periods; poor storage conditions cause seed viability to deteriorate rapidly. The conditioner/storer must know the storage requirements of each kind of seed, and maintain the best possible storage conditions so as to maintain high seed quality.

Seed Deterioration

Seed are living beings, and like all living things, have a lifespan and ultimately die. Germination and vigor are highest when seed first reach physiological maturity. Thereafter, they deteriorate or age, moving inexorably toward death, never stopping. Deterioration cannot be reversed or eliminated once it has occurred. However, it can be slowed by proper storage, so that seed life can be extended and seed kept viable over much longer periods of time.

Safe storage must be ensured for seed from the time they are harvested until they are planted by the using farmer. Storage must also be provided for seed that are carried over until the planting season in a subsequent year. There must also be long-term storage for Foundation and Breeder seed to preserve valuable germplasm. And, seed being transported from place to place is actually still "in storage" during transport. At any and all times between harvest and planting, seed is in storage and subject to storage environmental conditions which influence its quality.

All seed in a lot do not have the same life span; individual seed age and die at different rates over a period of time. Different rates of deterioration of individual seed result in three distinct phases during a seed lot's storage life (Figure 31.1):

1. A period of high viability in which relatively little deterioration occurs;
2. Then, a relatively short period in which germination declines sharply to low levels;
3. Thereafter, a relatively slower rate of decline until all seed are dead.

Good storage provides good conditions for seed, and minimizes the rate of deterioration and prolongs the first phase, in which relatively little loss of viability occurs.

Seed Quality is Not Improved by Storage

Seed never come out of storage with higher quality than they had when placed into storage. Seed, as do all living things, deteriorate with time, and deterioration is inevitable and irreversible. Low-quality seed cannot be made into high-quality seed even through the best storage conditions, nor can germination be held at high levels for many years.

Seed that have begun to deteriorate because of adverse pre-harvest environment, harvesting damage, heat damage during drying, or storage at high moisture content will lose germination rapidly in storage.

In some instances, germination increases during storage because the freshly-harvested seed had a physiological dormancy which required an after-harvest period for maturation processes to take place in the seed. This, however, is not to be mistaken for storage improving the quality of the seed.

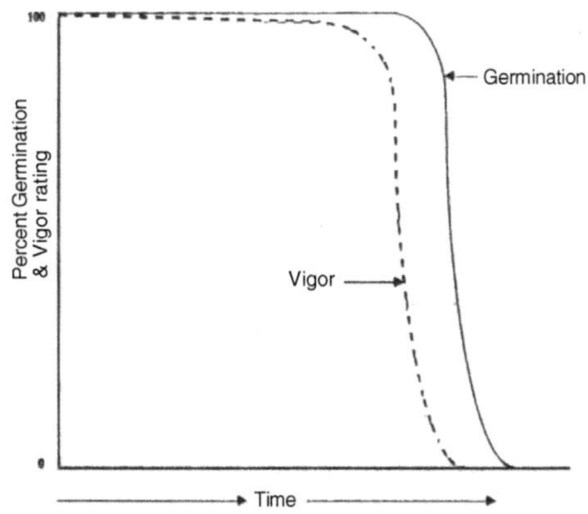

Figure 31.1 The trend of the inevitable deterioration of seed as they age over time. In general, seed maintain their viability for a certain period which depends on the kind of seed and the storage conditions. Then, viability begins to deteriorate, and falls rapidly until the seed lot has completely lost germination. Vigor begins to decline sooner that germination. Source: Gregg *et al.*, Storage.

Causes of Seed Deterioration

The basic causes of seed deterioration are unfavorable conditions, usually of high moisture, high temperature, or mechanical/insect/pest damage. Primary causes and places of damage are listed below.

Pre-harvest Field Conditions

Seed deteriorate in the field before harvest, under unfavorable weather conditions of rain, high moisture, high humidity, and high temperature. Hot dry weather with inadequate soil moisture during seed maturation also produces seed which are very susceptible to deterioration. Delay of harvest subjects seed to a longer period of deteriorative field stress and results in greater loss of quality. Seed which have begun to deteriorate in the field usually do not store well.

Immaturity at Harvest

Immature seed—smaller than normal size, wrinkled, shriveled, immature color, lightweight, etc.—do not store well.

Mechanical Injury

Mechanically-injured seed have reduced storability; they deteriorate faster, and are more susceptible to damage from storage fungi and seed treatment.

High Temperature

Storage life of seed decreases as storage temperature increases. Harrington's "thumb-rule" states that from 0o to 50°C, each 5°C increase in storage temperature halves the seed's life span. Both constant and short exposure to high temperatures cause deterioration and reduce seed storability. High-moisture seed are more susceptible to damage from high temperatures.

Temperature also plays a very important role in the storage life of seed, even though it is considered to be less important than moisture content. Within the "normal" range of seed temperature, there is increased biological activity of seed, insects, and molds as temperature increases.

In bulk-stored seed at high moisture contents, metabolic activity of seed and micro-organisms produces heat which raises the temperature of the seed mass. The increase in temperature then speeds up biological activity which causes further heat and moisture production and will completely kill the seed. This is one of the factors causing "hot spots" in seed or grain stored in metal silos under hot sunshine conditions.

Temperatures below 50°F are effective in maintaining seed quality, even with relatively high RH. Seed moisture content will increase, but the low temperature will reduce the adverse effects of somewhat higher moisture content.

High Moisture

The single most damaging influence on seed storage life is high moisture. As moisture content increases, seed deterioration increases. High-moisture seed deteriorate faster, and are more susceptible to damage from extreme temperatures, fumigation, treatment, storage fungi, insects, heating and mechanical injury. Harrington's "thumb-rule" states that for seed of 5–14% moisture content, each 1% reduction in moisture content approximately doubles seed storage life.

Loss of quality of stored seed occurs not only from water, but also from humidity in the air. Seed are hygroscopic, and their moisture content will vary according to the relative humidity (RH, expressed as a % of the air's moisture-holding capacity), but seed moisture does not vary as quickly as RH of the air may vary, because seed absorption or desorption of moisture is a slower process.

Equilibrium moisture content of seed is its moisture content which is in equilibrium (*i.e.*, does not change) with the RH of the air at a given temperature. Seed equilibrium moisture content varies with different crop seed; in general, the equilibrium moisture content of oily seed is lower than that of starch seed at the same RH and temperature (see Table 31.1). Temperature of the air and seed significantly affects RH, and moisture absorption and equilibrium moisture content of seed. No temperatures are known for all of the equilibrium seed moisture contents shown in Table 31.1. However, these are excellent guidelines for general operations.

Exposure to high atmospheric relative humidity (RH) also increases seed moisture content. Seed are hygroscopic and absorb moisture from humidity (invisible water vapor) of the air at high atmospheric RH; conversely, they lose moisture to air of low RH. Exchange of moisture with the surrounding air affects seed moisture content until it reaches equilibrium with atmospheric RH. Equilibrium moisture content varies with different crop seed; in general, it is lower for oily seed than for starchy seed.

Fungi

Storage fungi, primarily *Aspergillus* and *Penicillium* species, occur universally and may attack seed stored at RH of 65–100%. When seed moisture content and/or atmospheric RH is high, storage fungi grow rapidly, and may damage seed and cause heating. "Hot spot" damage to bulk stored seed is usually caused by insect or fungi activity in small areas where high moisture content has built up. All crop seed are not equally susceptible to fungi. While cereal seed are very susceptible, some vegetable seed seem more resistant.

Insects

About a dozen species of storage insects occur widely and may cause serious damage to seed stored at 9% or higher moisture content. Insects attack the endosperm and/or embryo, and also create high moisture which leads to heating.

Pests

Rats, mice and birds eat seed, damage bags and scatter, mix and contaminate seed.

Seed Treatment

Seed of high moisture, with mechanical damage, or stored for long periods are sometimes adversely affected by certain seed treatment chemicals. Some combination treatments affect germination; treatments safe for some crop seed may damage others; excessive treatment rate also affects storability.

Table 31.1 Seed moisture equilibrium content at various levels of relative humidity of the air. Source: B. Gregg *et al.*, Safe Seed Storage. WANA Seed network Publication no. 5/94. 55 pp. ICARDA, Aleppo, Syria. (- means no data available).

Seed Kind	Percentage Seed Moisture Content at Relative Humidity % of:															
	10	15	20	30	40	45	50	55	60	65	70	75	80	85	90	100
Field Crops																
Alfalfa (*Medicago sativa* L.)	-	-	-	-	-	-	-	-	-	-	7.8	9.3				
Barley (*Hordeum vulgare* L.)	-	6.0	-	8.4	-	10	-	-	12	-	-	14	-	-	20	26.8
@77°F	4.4	-	7	8.6	9.7	-	11	-	12	-	14	-	16	-	20	26.8
Bean, dry, Michelite	-	-	-	-	-	-	-	-	13	-	14	-	17	-	-	-
@40°F																
@50°F	-	-	-	-	-	-	-	-	14	-	15	-	18	-	-	-
@77°F	5.6	-	7.4	8.6	9.8	-	11	-	13	-	15	-	18	-	-	-
@100°F	-	-	-	-	-	-	-	-	12	-	14	-	17	-	-	-
@130°F	-	-	-	-	-	-	-	-	12	-	14	-	19	-	-	
Bean, dry, Red Mexican @77°F	6	-	7.5	8.6	9.8	-	11	-	13	-	15	-	19	-	-	-
Bean, dry, Great Northern @77°F	5.9	-	7.4	8.5	9.6	-	11	-	13	-	15	-	18	-	-	-
Bean, dry, light red kidney @77°F	6.1	-	7.5	8.7	9.9	-	11	-	13	-	15	-	19	-	-	-
Bean, dry, dark red kidney @77°F	5.4	-	7.2	8.4	9.9	-	11	-	13	-	15	-	19	-	-	-
Bean, dry, flat small white @77°F	6	-	7.1	8.3	9.6	-	11	-	13	-	15	-	18	-	-	-
Bean, dry, pinto @77°F	6.1	-	7.4	8.5	9.8	-	11	-	13	-	15	-	18	-	-	-
Bentgrass, Colonial (*Agrostis tenuis* Sibth.)	-	-	-	-	-	-	-	9.8	-	11	12	13	14	15	-	-
Bluegrass, Kentucky (*Poa pratensis* L.)	-	-	-	-	-	-	-	9.7	-	11	11	13	14	16	-	-
Bluegrass, rough (*Poa trivialis* L.)	-	-	-	-	-	-	-	9.4	-	-	12	1	14	16	-	-
Bromegrass, smooth (*Bromus inermis* Leyss.)	-	-	-	-	-	-	-	11	-	13	13	14	16	18	-	-
Buckwheat (*Fagopyrum esculentum* Moench) @77°F	-	6.7	-	9.1	-	11	-	-	13	-	-	15	-	-	19	24.5

Canary grass, Reed (*Phalaris arundinacea* L.)	-	-	-	-	-	-	-	11	-	12	13	14	15	16	-	-
Clover, alsike (*Trifolium hybridum* L.)	-	-	-	-	-	-	-	-	-	-	9.3	-	16	19	-	-
Clover, crimson (*T. incarnatum* L.)	-	-	7.5	-	-	9.5	-	-	11	-	-	-	22	-	-	33.3
Clover, ladino (*T. repens* L.)	-	-	-	-	-	-	-	-	-	-	8.7	11	15	18	-	-
Clover, red (*T. pretense* L.)	-	-	-	-	-	-	-	-	-	-	9.1	11	16	19	-	-
Clover, sweet (*Melilotus* spp)	-	-	-	-	-	-	-	-	-	-	9.3	11	13	18	-	-
Corn (maize), field (*Zea mays* L.)	-	6.5	-	8.4	-	11	-	-	13	13	-	15	15	-	19	24.2
" Yellow dent @20°F	-	-	-	-	10	-	12	-	13	-	15	-	17	-	-	-
" Yellow dent @30°F	-	-	6.7	9.2	11	-	13	-	14	-	18	-	-	-	-	-
" Yellow dent @32°F	-	-	-	-	11	-	13	-	14	-	16	-	18	-	22	-
" Yellow dent @32°F	-	-	-	-	10	-	12	-	13	-	15	-	17	-	-	-
" Yellow dent @40°F	6.3	-	8.6	9.8	11	-	12	-	14	-	16	-	18	-	22	-
" Yellow dent @50°F	6.6	-	9	9.3	11	-	12	-	14	-	15	-	18	-	22	-
" Yellow dent @60°F	7.5	-	7.8	9	10	-	11	-	12	-	14	-	16	-	20	-
" Yellow dent @70°F	-	-	-	7.1	8.3	-	9.8	-	11	-	13	-	-	-	-	-
" Yellow dent @77°F	5.1	-	7	8.3	9.8	-	11	-	13	-	14	75	16	-	20	23.8
" Yellow dent @86°F	4.4	-	7.4	8.2	9	-	10	-	11	-	13	-	15	-	17	-
" Yellow dent @90°F	4.9	-	6.6	7.7	9.3	-	11	-	12	-	14	-	16	-	19	-
" Yellow dent @100°F	4	-	6	7.3	8.7	-	9	-	11	-	13	-	14	-	17	-
" Yellow dent @120°F	-	-	-	-	8.6	-	10	-	11	-	13	-	15	-	-	-
" Yellow dent @122°F	3.6	-	5.5	6.7	8	-	9.2	-	10	-	12	-	14	-	16	-
" Yellow dent @140°F	3	-	5	6	7	-	7.9	-	8.8	-	10	-	12	-	15	-
" Yellow dent @155°F	-	-	-	-	7.4	-	8.4	-	10	-	12	-	12	-	-	-
" Yellow dent @160°F	3.9	-	6.2	7.6	9.1	-	10	-	12	-	14	-	15	-	18	-
" White dent @77°F	-	6.8	-	8.5	-	9.8	-	-	12	-	-	14	-	-	18	23
Corn, pop (*Zea mays* L.)	-	6.8	-	8.5	-	9.8	-	-	12	-	-	14	-	-	18	23
" Shelled @ 77°F	5.6	-	7.4	8.5	9.8	-	11	-	12	-	13	-	14	-	18	23
Cotton (*Gossypium hirsutum* L.)	2	-	4.5	6	-	7.5	-	8.8	9.1	-	10	-	13	-	18	-
"	-	-	-	-	6.9	-	7.8	-	9.1	-	10	-	13	-	20	-

contd. . .

. . . *Table 31.1 contd*

Seed Kind	Percentage Seed Moisture Content at Relative Humidity % of:															
	10	15	20	30	40	45	50	55	60	65	70	75	80	85	90	100
Fescue, Alta (*Festuca* spp.)	-	-	-	-	-	-	-	11	-	12	13	13	15	17	-	-
Fescue, chewings	-	-	-	-	-	-	-	10	-	11	12	-	15	17	-	-
(*F. rubra* subsp. *commutata* Gaud.)																
Fescue, creeping red (*F. rubra* L.)	-	-	-	-	-	-	-	11	-	12	13	14	15	18	-	-
Flax (*Linum usitatissimum* L.)	-	4.4	-	5.6	-	6.3	-	-	7.9	9	-	10	12	-	16	21.4
" @ 77°F	3.3	-	4.9	5.6	6.1	-	6.8	-	7.9	-	9.3	-	11	-	15	21.4
Oats (*Avena satiiva* L.)	-	6.7	-	8	-	9.6	-	-	12	-	-	14	-	-	19	24.1
" @ 77°F	4.1	-	6.6	8.1	9.1	-	10	-	12	-	13	-	15	-	19	24.1
Orchardgrass (*Dactylis glomerata* L.)	-	-	-	-	-	-	-	9.6	-	11	11	12	13	15	-	-
Peanut (*Arachis hypogaea* L.)	-	2.6	-	4.7	-	5.6	-	-	7.2	-	-	9.8	-	-	13	-
" in pod @ 50°F	-	-	-	5.5	6.1	-	7.1	-	8.6	-	9.8	-	12	-	-	
" kernels @ 50°F	-	-	-	4.8	5.5	-	6	-	6.6	-	7.3	-	9	-	-	-
Redtop (*Agrostis gigantea* Roth)	-	-	-	-	-	-	-	10	-	11	11	13	14	15	-	-
Rice (*Oryza sativa* L.)	-	6.8	-	9	-	11	-	-	13	-	-	14	16	-	18	23.6
" whole grain @ 77°F	5.9	-	8	9.5	11	-	12	-	13	-	14	-	15	5	19	-
" whole grain @ 100°F	4.9	-	7	8.4	9.8	-	11	-	12	-	13	-	15	-	19	-
" rough, @ 32°F	-	-	8.2	9.9	11	-	12	-	13	-	15	-	17	-	19	-
" rough, @ 68°F	-	-	7.5	9.1	10	-	11	-	13	-	14	-	15	-	18	-
" rough, @ 73°F	4.9	-	7.3	8.7	9.8	-	11	-	12	-	14	-	16	-	19	-
" rough, @ 77°F	4.6	-	6.5	7.9	9.4	-	11	-	12	-	13	-	15	-	17	-
" rough, @ 80°F	-	-	-	-	9.2	-	10	-	12	-	13	-	15	-	17	-
" rough, @ 86°F	-	-	7.1	8.6	10	-	11	-	12	-	13	-	15	-	17	-
" rough, @ 111°F	-	-	-	-	-	-	-	-	10	-	12	-	14	-	17	-
Rye (*Secale cereale* L.)	-	7	-	8.7	-	11	-	-	12	13	-	15	-	-	21	26.7
", @ 77°F	5.2	-	7.6	8.7	9.9	-	11	-	12	-	14	-	16	-	21	-

Ryegrass, perennial (*Lolium perenne* L.)	-	-	-	-	-	-	-	11	-	12	13	13	15	17	-	-
Ryegrass, annual (*L. multiflorum* Lam.)	-	-	-	-	-	-	-	11	-	12	13	14	16	16	-	-
Sorghum (*S. bicolor* (L.) Moench.)	-	6.4	-	8.6	-	11	-	-	12	13	-	15	-	-	19	21
", @ 30°F	-	-	8.2	10	11	-	12	-	13	-	15	-	16	-	-	-
", @ 60°F	-	-	7.5	9.5	11	-	12	-	13	-	14	-	16	-	-	-
", @ 77°F	4.4	-	7.3	8.6	9.8	-	11	-	12	-	14	-	16	-	19	21.9
", @ 90°F	-	-	7	8.7	10	-	12	-	12	-	13	-	15	-	-	-
", @ 120°F	-	-	6.6	8	9.4	-	11	-	12	-	13	-	14	-	-	-
Sorghum, kafir, @ 40°F	6.8	-	8.5	9.7	11	-	12	-	14	-	15	-	17	-	-	-
", @ 70°F	6	-	7.7	9.1	11	-	12	-	13	-	14	-	16	-	19	-
", @ 90°F	5	-	7	8.4	9.6	-	11	-	12	-	13	-	15	-	17	-
Soybean (*Glycine max* (L.) Merr.)	-	4.3	-	6.5	-	7.4	-	-	9.3	11	-	13	16	-	19	-
", @ 77°F	-	-	5.5	6.5	7.1	-	8	-	9.3	-	12	-	15	-	19	-
", @ 77°F	-	-	-	-	7	-	8	-	10	-	12	-	16	-	21	-
Sudangrass,piper (*Sor-ghum sudanense* (Piper) Stapf)	-	-	-	-	-	-	-	11	-	-	12	-	14	16	-	-
Sugarbeet seed (*Beta vulgaris* L.), @ 40°F	-	-	-	10	12	-	13	-	14	-	15	-	18	-	23	-
", @ 60°F	-	-	-	9	10	-	12	-	13	-	14	-	16	-	20	-
", @ 80°F	-	-	-	8	9.1	-	10	-	12	-	13	-	15	-	18	-
", @ 100°F	-	-	-	7	8.3	-	9.2	-	10	-	12	-	13	-	16	-
Sunflower (*Helianthus annuus* L.)	-	-	-	5.1	-	6.5	-	-	8	-	-	10	-	-	15	-
Timothy (*Phloem pretense* L.)	-	-	-	-	-	-	-	11	-	12	13	14	15	16	-	-
Trefoil, birdsfoot (*Lotus corniculatus* L.)	-	-	-	-	-	-	-	-	-	-	8.3	10	14	17	-	-
Vetch, crown (*Coronilla varia* L.)	-	-	-	-	-	-	-	-	-	-	9.9	-	14	18	-	-
Vetch, hairy (*Vicia villosa* Roth)	-	-	-	-	-	-	-	-	-	-	11	13	17	19	-	-

contd. . .

Table 31.1 contd

Seed Kind	Percentage Seed Moisture Content at Relative Humidity % of:															
	10	15	20	30	40	45	50	55	60	65	70	75	80	85	90	100
Wheat (*Triticum* spp.), @ 30°F	-	-	6.9	9.2	10	-	12	-	13	-	15	-	16	-	-	-
", @ 32°F	-	-	-	-	-	-	13	-	14	-	16	-	18	-	21	-
", @ 50°F	-	-	7.9	9.2	11	-	13	-	14	-	15	-	17	-	-	-
", @ 60°F	-	-	6.1	7.8	9.6	-	11	-	13	-	14	-	15	-	-	-
", @ 68°F	-	-	-	-	-	-	12	-	13	-	15	-	17	-	20	-
", @ 77°F	5.8	-	7.6	9.1	11	-	12	-	13	-	15	-	17	-	21	-
", @ 90°F	-	-	5.3	7	8.6	-	10	-	12	-	13	-	14	-	-	-
", @ 104°F	4.6	-	6.2	7.4	8.6	-	10	-	11	-	12	-	14	-	-	-
", @ 104°F	-	-	-	-	-	-	-	-	13	-	14	-	16	-	20	-
", @ 120°F	-	-	-	6.2	7.4	-	9.6	-	10	-	12	-	14	-	-	-
", @ 122°F	4	-	5.8	6.7	8.1	-	10	-	11	-	13	-	15	-	19	-
", @ 176°F	2.5	-	3.7	4.8	5.7	-	6.7	-	8	-	9.8	-	12	-	-	-
Wheat, white (*Triticum* spp.)	-	6.7	-	8.6	-	9.9	-	-	12	-	-	15	-	-	20	26.3
", @ 77°F	5.2	-	7.5	8.6	9.4	-	11	-	12	-	14	-	16	-	20	26.3
Wheat, durum (*T. durum* Desf.)	-	6.6	-	8.5	-	10	-	-	12	-	-	14	-	-	19	26.6
", @ 77°F	5.1	-	7.4	8.5	9.4	-	11	-	12	-	13	-	15	-	19	26.7
Wheat, soft red winter (*T. aestivum* L.)	-	6.3	-	8.6	-	11	-	-	12	-	-	15	-	-	20	25.6
", @ 20°F	-	-	-	-	11	-	13	-	14	-	15	-	17	-	-	-
", @ 32°F	-	-	-	-	11	-	12	-	14	-	15	-	16	-	-	-
", @ 50°F	-	-	-	-	10	-	12	-	13	-	14	-	16	-	-	-
", @ 70°F	-	-	-	-	9.7	-	11	-	12	-	14	-	-	-	-	-
", @ 77°F	4.3	-	7.2	8.6	9.7	-	11	-	12	-	14	-	16	-	20	25.6
Wheat, hard red winter (*T. aestivum* L.)	-	6.4	-	8.5	-	11	-	-	13	-	-	15	-	-	20	25
", @ 77°F	4.4	-	7.2	8.5	9.7	-	11	-	13	-	14	-	16	-	20	25

Wheat, hard red spring	-	6.8	-	8.5	-	10	-	-	12	-	-	15	-	-	20	25
(*T. aestivum* L.)																
", @ 77°F	4.4	-	7.2	8.5	9.8	-	11	-	13	-	14	-	16	-	20	25
Vegetables																
Bean, broad (faba) (*Vicia fabia* L.)	4.2	-	5.8	7.2	-	9.3	-	-	11	-	-	15	-	-	-	
Bean, lima (*Phas-eolus vulgaris* L.)	4.6	-	6.6	7.7	-	9.2	-	-	11	-	-	14	-	-	-	
Bean, snap (*P. vulgaris* L.)	3	-	4.8	6.8	-	9.4	-	-	12	-	-	15	16	-	-	-
Beet, garden (*Beta vulgaris* L.)	2.1	-	4	5.8	-	7.6	-	-	9.4	-	-	11	-	-	-	-
Cabbage	3.2	-	4.6	5.4	-	6.4	-	-	7.6	8	-	9.6	10	-	-	-
(*Brassica oleracea* var. *capitata* L.)																
Cabbage, Chinese	2.4	-	3.4	4.6	-	6.3	-	-	7.8	-	-	9.4	-	-	-	-
(*B. pekinensis* (Lour.) Rupr.)																
Carrot (*Daucus carota* L.)	4.5	-	5.9	6.8	-	7.9	-	-	9.2	10	-	12	13	-	-	-
Celery (*Apium grav-eolens*	5.8	-	7	7.8	-	9	-	-	10	11	-	12	14	-	-	-
var. *rapaceum* (Mill.) Gaud)																
Corn, sweet (*Zea mays* L.)	3.8	-	5.8	7	-	9	-	-	11	11	-	13	14	-	-	-
Cucumber (*Cucumis sativus* L.)	2.6	-	4.3	5.6	-	7.1	-	-	8.4	8.5	-	10	10	-	-	-
Eggplant (*Solanum melongena* L.)	3.1	-	4.9	6.3	-	8	-	-	9.8	-	-	12	-	-	-	-
Lettuce (*Lactuca sativa* L.)	2.8	-	4.2	5.1	-	5.9	-	-	7.1	8	-	9.6	1	-	-	-
Mustard, leaf (*Brassica* spp.)	1.8	-	3.2	4.6	-	6.3	-	-	7.8	-	-	9.4	-	-	-	-
Okra (*Hibiscus esculentus* L.)	3.8	-	7.2	8.3	-	10	-	-	11	12	-	13	15	-	-	-
Onion (*Allium cepa* L.)	4.6	-	6.8	8	-	9.5	-	-	11	-	-	13	14	-	-	-
Onion, Welsh (*A. fistulosum* L.)	3.4	-	5.1	6.9	-	9.4	-	-	12	-	-	14	-	-	-	-
Parsnip (*Pastinaca sativa* L.)	5	-	6.1	7	-	8.2	-	-	9.5	-	-	11	-	-	-	-
Pea, garden (*Pisum sativum* L.)	5.4	-	7.3	8.6	-	10	-	-	11	12	-	15	-	-	-	-

Fumigation

Some fumigants adversely affect viability under certain conditions. Increased seed moisture, temperature, fumigant dosage, and time of exposure increase risk of damage.

Natural Seed Lifespan

Some plant species produce seed that are naturally long-lived, while others are naturally short-lived. Seed of some crop species inherently survive or remain viable for relatively long periods; others naturally have a short life. Peanut (*Arachis hypogaea* L.), soybean (*Glycine max* (L). *Merrill*), onion (*Allium cepa* L.), and lettuce (*Lactuca sativa* L.) seed have naturally short life spans, while many legume seed (especially those with hard impermeable seed coats) survive for relatively long periods. In general, oily seed have shorter life spans.

Other Factors

Seed should be stored in a seed storage, not a storage used for fertilizer, salt, herbicides, or feeds. These may emit toxic vapors, attract insects or pests, or create moisture which damage seed quality.

In storages with concrete floors, seed bags should be stacked on pallets which maintain air space between the bags and the floor, as considerable moisture can be transmitted through concrete floors by capillary movement. Seed storages should also be well-ventilated (unless they are conditioned) and protected against rodents and birds.

Storage Operations

Figure 31.2 illustrates the sequence of operations involved in handling seed into, during and out of storage. The remainder of this section discusses management and operating techniques which help maintain seed viability, condition, purity, and identity.

Types of Storage

According to storage conditions, seed storage may be of the following types. A particular storage may also involve more than one of these characteristics.

Handling method: (1) Bulk or (2) Bagged.

Storage environment: (1) Ambient (non-conditioned) or (2) Conditioned (cooled and dehumidified).

Seed package: (1) Vaporproof bag/package or (2) Vapor-permeable bag/package.

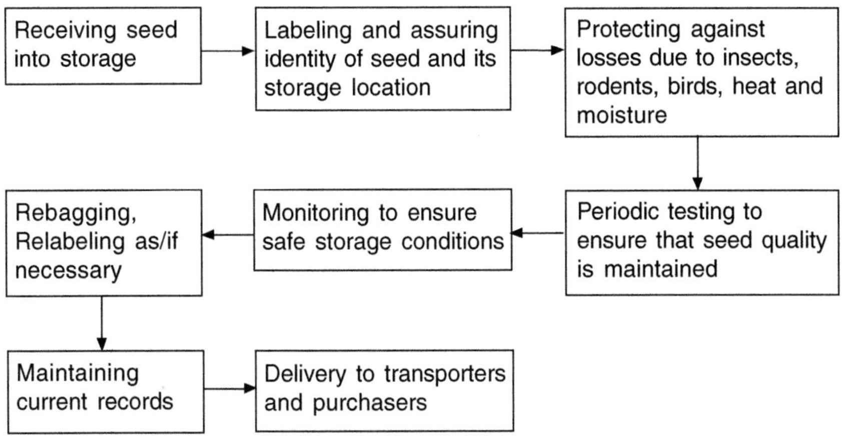

Figure 31.2 The sequence of handling operations in seed storage at the conditioning plant. Source: Gregg *et al.*, Storage.

When Seed Are Stored

Some form of storage takes place from the time seed mature in the field until they are planted by the using farmer (Figure 31.3). Seed may deteriorate from unfavorable conditions at any stage; all stages must be properly managed to create the best possible storage conditions, so seed will reach the farmer with acceptable viability. "Storage" occurs during the following periods:

On the Producing Farm

1. In the field, from seed maturation to harvest.
2. From harvest to threshing.
3. From threshing to sun-drying, to delivery to the conditioning plant.

At the Conditioning Plant

1. During receiving, drying, and pre-conditioning storage.
2. During conditioning.
3. During storage after conditioning.

During Distribution/Marketing

1. During transport from conditioning to the area of use.
2. During storage in distribution centers and at the retail outlet.
3. During transport to the using farmer.

On the Using Farm

1. While seed is being held prior to planting.

Carryover (in the Distribution System)

1. Carryover to a later planting season.

For practical operations, these may be grouped into 3 operating storage periods:

1. *Seed preparation*: All stages before seed are delivered to the conditioning plant: in the field before harvest, during harvest, threshing, and holding on the producing farm. Deterioration here determines initial storability of seed.
2. *Pre-conditioning storage*: From arrival at the seed plant until conditioning is completed: receiving, drying, awaiting conditioning, and during conditioning.
3. *Cleaned seed storage*: All storage and transport of conditioned seed at the conditioning plant, wholesale and retail distributors, etc. This is the phase usually considered as seed storage; however, the earlier two periods are also critically important to seed viability.

Where Storage is Needed

Seed must be safely stored at each point in the distribution channel, from the seed producer to the seed user. This requires a network of seed storages, each constructed to maintain safe storage conditions for the required amount of seed, and operated to maintain seed quality and serve farmers efficiently. Each storage must receive seed efficiently, kept it in good condition, and move it out efficiently when it is shipped or sold. The storage network usually includes technically proper storages of the required capacity at:

1. The conditioning plant.
2. Main distribution centers, which serve retail stores.
3. Retail seed stores.

Storage Period

Clean seed are stored for different periods of time, according to their usage. Most seed storage is short-term, only for a few weeks or months until the next planting season. However, seed storages and storage periods in a seed enterprise are of three types:

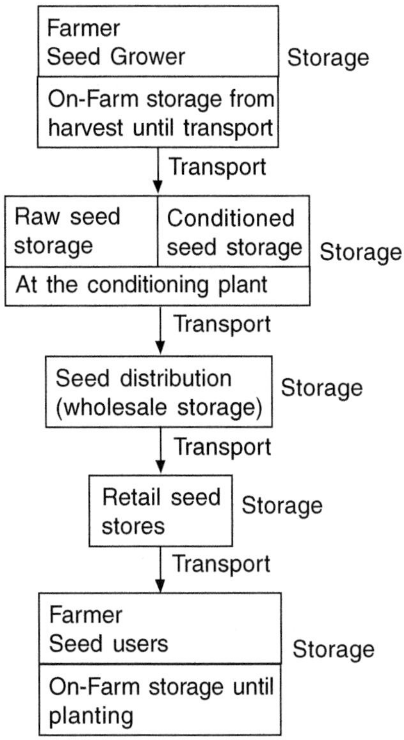

Figure 31.3 Places and periods when seed is stored. Source: Gregg *et al.*, Storage.

1. *Short-term*, until the next planting season, usually good ambient ventilated storage for seed which will be sold/planted within 6–9 months. If the Marketing Plan and purchasing/production are realistic—and operations are cost-effective—most seed storage is for this period. A successful seed enterprise should have safe short-term storages at all operating locations for all seed which will be stored there, plus an adequate reserve for growth.
2. *Medium-term*, carryover for one or two additional crop seasons. With good management, carryover storage is required only for the approximately 5% produced in excess of Marketing Plan needs, to allow for growth or unforeseen conditions. Carryover medium-term storage prevents loss of this unsold seed, has it ready for unforeseen early sales, prevents the expense of lost seed, and reduces the cost and effort of unnecessary production. A cost-effective seed enterprise will have or rent one small carryover storage, usually located at its main site or at the office in the major marketing area. This also includes good ambient

storage of moisture-vapor-proof bags for intermediate-length storage (1–3 years) of seed dried to 4–8% moisture.

3. *Long-term* storage for a period of years. Temperature-and-humidity-controlled storage for valuable Breeder, Foundation and genetic resource seed to be stored for long periods. This is normally required only for stock seed (Breeder or Foundation) or research materials. Most small-medium seed enterprises do not need long-term storage, as they are not involved in variety development, and obtain stock seed from research and Foundation seed programs.

Kind of Storage

Seed are stored either in bulk or in bags or other packages.

Bulk Seed Storage

Bulk storage is common for short-term storage of large amounts of seed awaiting conditioning, including unshelled (ear) maize seed. In temperate climates, conditioned seed is also sometimes stored in bulk for short periods. Many designs and sizes of bins, tanks, cribs, etc., are used. Seed are frequently dried in such bulk storage units. If seed is stored in them for extended periods, aeration must be provided to prevent damage due to moisture migration.

In tropical climates, wide temperature differentials which occur between outside air and the bulk seed mass, and at different points within the mass, cause moisture migration. Due to moisture movement from warmer to cooler spots in the seed mass, moisture accumulate in spots until dangerously high levels are reached. In such spots, seed deteriorate rapidly. Deterioration then spreads through the seed mass as more heat and moisture are generated. Under humid tropical conditions, bulk short-term storage may be used, but adequate mechanical aeration or drying facilities must be incorporated and properly used to keep the seed uniformly dry and cool.

Bagged Seed Storage

According to storage conditions provided, bagged seed storage is either conditioned or non-conditioned. Conditioned storages are specially constructed to create a sealed space in which lower-than-ambient temperatures are created by air conditioning refrigeration equipment, and lower RH is created by mechanical dehumidification. Temperature and RH are reduced as required for the crop seed and storage period. Non-conditioned storages do not mechanically cool or dehumidify the storage space.

Storage Conditions

Non-Conditioned

Non-conditioned storages create the best storage conditions possible under ambient conditions by using only good building design and construction. No mechanical equipment except ventilations fans is used. Non-conditioned storage is used for most short-term storage of seed for farmers. Under humid tropical conditions, it is adequate for carryover of very dry seed sealed in vaporproof packages. Under temperate climates, it is suitable for short carryover in vapor-porous bags.

Conditioned

Conditioned storage provides special conditions for longer storage. The storage is sealed to keep out ambient conditions, and is then cooled or refrigerated to lower temperature and is dehumidified to lower humidity and moisture.

Successful Seed Storage

Practical procedures for improving seed storability and keeping them in good condition for longer periods are:

1. Store, and carryover, only high-quality seed. Seed which are immature or damaged deteriorate much sooner; store only mature seed which are undamaged, dry, and cool so they will store better and maintain high viability longer.
2. Ensure that storage conditions are dry and cool, and seed are not damaged by insects, pests, chemicals, or other factors. Unfavorable storage conditions speed up deterioration, causing seed to die relatively sooner. Maintain favorable conditions which can slow the rate of seed aging, and maintain high germination longer.
3. Good sanitation should be a continuous practice, to minimize insect and pest infestations. If storage insects are a problem, use insecticides and fumigants combined with sanitation. Insect-infested lots should be fumigated before storage.
4. If conditioned or carryover storage space is limited, use it for storage of the best seed. Plan storage and marketing at the time of conditioning, and sell less storable lots immediately, and keep them in the non-conditioned or short-term storage.

Seed Deterioration in Storage

The rate of deterioration of seed during storage increases rapidly as seed moisture content increases. Both seed moisture content and temperature are critical factors in seed storage, but moisture content has a greater direct influence on seed longevity. The higher the seed moisture content, the more susceptible the seed are to damage from excessively high or low temperatures. On the other hand, very dry cereal seed will store well at temperatures up to 90°F.

Harrington's Storage Thumb Rule

According to Harrington's widely-used thumb rule, good storage is achieved when Percent relative humidity of the air plus storage temperature in degrees Fahrenheit add up to not more than 100. This thumb rule also states that:

1. A one (1) percent decrease in moisture content nearly doubles storage life potential of seed.
2. A ten (10) degree decrease in temperature nearly doubles the storage life potential of seed.

Dry Cool Conditions are Best for Seed Storage

The general conditions required for safe storage of "non-recalcitrant" seed is a dry and cool environment.

Seed of most grains will store well for one year at moisture contents of 11–13%, at 65–70°F storage temperature. For two-year storage, moisture content should be decreased to less that 10%. Storage periods longer than two years require 8% moisture content and a conditioned temperature of 60°F or less.

Oily seed such as soybeans, groundnut, cotton, etc., are difficult to store and require moisture content below 8% for 12-month storage, and even lower moisture content and cool temperatures for longer storage periods.

Conditioned storage with RH below 50% and temperature below 50°F will usually maintain seed quality for 3–8 years.

Selecting and Preparing Seed for Storage

Low-quality seed do not live long, so they do not store well. Good storage can only slow down seed aging and deterioration; it cannot remedy deterioration which has already occurred. Careful pre-storage management and handling of seed is essential, to minimize deterioration before storage begins. Only healthy, undamaged, non-deteriorated, fully mature seed of high germination and vigor should be stored. To keep storability high until seed reach the conditioning plant:

1. Plan production so seed are harvested during dry weather.
2. Harvest only when seed reach full field maturity.
3. Do not delay harvest.
4. Harvest and thresh at optimum seed moisture content to minimize mechanical damage. Use suitable equipment, adjusted and operated properly.
5. Immediately dry unthreshed plants and seed, and threshed seed, to safe moisture contents. Do not bulk or bag high-moisture seed.
6. Protect seed and unthreshed plants from rain, ground moisture, high temperatures, insects, rats, birds, chemicals, fertilizers, fuels, etc. Do not put seed heads of unthreshed plants on the ground.
7. As soon as possible, transport threshed seed to the conditioning plant for proper drying and storage. The plant should immediately check seed moisture and dry as required, and condition seed within a few days after receiving it.

Raw Seed Storage Before Conditioning

Seed may deteriorate at the conditioning center due to poor management before conditioning. While this period is usually short—seed should be conditioned and bagged within a few days after receiving—it is important to seed storability. Deterioration in this period can be minimized by:

1. Check seed on the contract grower's farm, and help him sun-dry as much as possible, so only safe seed is delivered to the conditioning plant.
2. Immediately after receiving seed at the conditioning plant, measure moisture content and dry as required, or move into safe storage. Do not keep high-moisture seed without immediate drying.
3. Immediately check for insect infestation. Fumigate and/or treat as soon as possible.
4. Store, label, and handle every bag/lot/bin of seed so as to ensure identity and purity.
5. Test germination immediately after receiving seed, preferably with tetrazolium or other quick-test method. Discard unacceptable lots BEFORE they are conditioned.
6. Condition seed as soon as possible.
7. During conditioning keep seed cool, dry, uncontaminated, and free of damage and mixtures.

It is safer and better for operations and management to have a separate storage for uncleaned ("raw") seed and another for cleaned seed. This minimizes movement of dust, insects, etc., to the cleaned seed, permits fumigation of raw seed without endangering operations of cleaning and shipping, and large operating lots can be more easily formed in a separate receiving storage, as smaller farmer/field lots come in.

If it is necessary to store both raw and cleaned seed in the same building, specific areas should be assigned for each type, and floor areas clearly marked and numbered.

Storing Conditioned Seed

After seed are processed to the desired quality, storage must maintain its identity and purity, protect it from insects and pests, and keep it cool and dry to prevent deterioration. Usual periods of storage for conditioned seed are:

1. *Short-term*: Until the following planting season, usually within 3–9 months. Most seed for farmer use goes through only short-term storage.
2. *Carryover*: until a later season. Three carryover periods are usually encountered:
   A. *Carryover for approximately 1 year*, until the next year's planting season; the usual carryover for most reserve or buffer seed stocks.
   B. *Carryover to a later planting season*, usually 2–4 years. Programs which produce seed of many varieties and have suitable storage conditions often stagger production and store seed for such periods, to reduce costs. Foundation seed is commonly carried over for such periods.
   C. *Long-term*, several years storage for Breeder seed, genetic resource collections, etc. Seldom used for farmer seed, due to cost.

Storage Management Guidelines

Practical guidelines for managing storage include:

1. At any stage in its life, seed may suffer from damage or unfavorable conditions which speed up deterioration and reduce storability.
2. Only healthy, undamaged, high-quality, non-deteriorated, fully-mature and preferably untreated seed of high germination and vigor should be stored for any period of time. Less storable lots should be planted as soon as possible.
3. Seed should be stored under dry cool conditions and protected from deteriorative influences, so seed viability is high after the required storage period. Seed deterioration, as well as insect and fungi activity, is reduced by lowering temperature and moisture content (and correspondingly, RH). How cool and how dry depends on the length of

the storage period, desired final seed quality, kind of seed, and economic considerations. The longer the storage, the lower temperature and moisture content must be. At lower temperatures, seed of higher moisture content can be safely stored; at lower moisture content, seed can be safely stored at higher temperatures. These apply to reasonable storage periods; reducing only one—moisture or temperature—cannot achieve maximum potential seed storage life.

4. For any storage period, seed quality is maintained better by providing the best storage conditions possible under existing economic constraints. It is not possible to specify exact storage conditions required, due to the many variables. However, for long-term storage of genetic resource seed FAO (1974) proposed desirable standards of –18 to –20°C at 5% seed moisture content (in equilibrium with 10–15% RH for starchy seed, and 20–25% for oily seed). Minimal standards proposed were 5°C or less at 5–7% seed moisture and/or not more than 20% RH. Delouche (1968a) considered seed moisture content of 10–13% under temperate climate temperatures safe for short-term storage (6 to 12 months) of cereal seed; 8–10% moisture content at temperate climate temperatures suitable for 1–3 years open storage; and temperatures below 50°F and seed moisture in equilibrium with less than 50% RH suitable for 3–8 years.

Provide Safe Storage Conditions

Seed of storable quality can be safely stored if the storage building or room is designed, equipped and managed to provide the required dry, cool, protective conditions. Seed can be protected from moisture and high RH by the package and/or the storage building/room. Lower temperature can be maintained only by the storage building/room.

Control Moisture

Moisture control is essential to protect dried seed from moisture and high RH. Methods include:

1. Keep short-term non-conditioned storages as dry as possible.
2. Lower storage humidity by carefully-managed and controlled ventilation of a well-constructed non-conditioned storage during periods of low ambient RH. Controlled ventilation is effective where the ambient RH is low enough for the required time periods.
3. Place very low-moisture seed in a vaporproofed storage when RH is very low, and keeping the storage sealed to prevent entry of moisture

vapor. Keeping dry seed in a closed vaporproofed but not dehumidified storage is usually most feasible in an air-conditioned storage where thermal insulation and air conditioning prevent convection buildup of high RH near the floor, and air conditioning removes some moisture.

4. Dehumidify the air inside a storage which is completely surrounded and sealed by an adequate vapor barrier which prevents entry of outside air and moisture. Air in a vaporproofed storage is most efficiently dehumidified by a chemical desiccant dehumidifer. Silica gel is a safe, easy-to-use desiccant.
5. Package very dry seed in hermetically-sealed vaporproof containers. Very dry seed may be hermetically sealed in vaporproof packages to keep seed moisture content at its original level regardless of ambient RH. This maintains seed quality at high levels for long periods.

Control Temperature

Seed storages can be cooled by:

1. Designing and constructing a non-conditioned storage to minimize entry of solar radiation and outside heat. This system can maintain the storage slightly cooler than the ambient temperature.

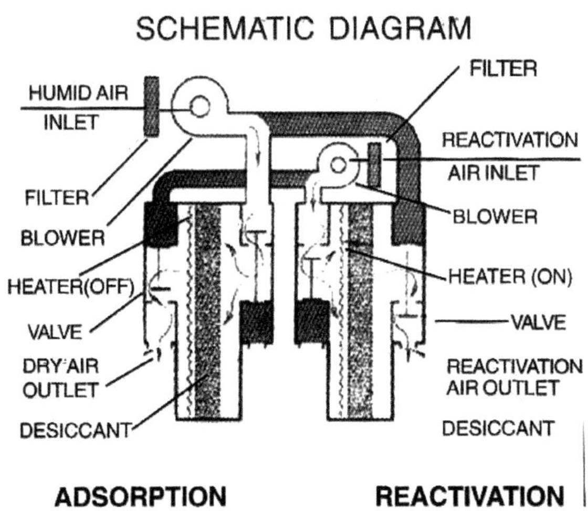

Figure 31.4 Operating diagram of a 2-tank desiccant dehumidifier to remove humidity (moisture) from air, as used in conditioned seed storage. Process air (from the sealed seed storage) enters at the "humid air inlet", is dehumidified and goes out the "dry air outlet" (both on left side of diagram). While one tank is dehumidifying the process air, the other tank is being reactivated, by removing the absorbed moisture from the desiccant (usually silica gel). After the other tank has absorbed its maximum capacity for moisture from the process air, it is reactivated and the flow of process air shifts to the second tank. Source: Ajay Sagar, Indosaw.

2. Thermally-insulating a conditioned storage, and removing heat by air conditioning (refrigeration). Properly constructed and cooled, a conditioned storage can maintain almost any desired temperature.

Mark Specific Storage Locations

Operations and management are much easier if the storage is marked off into smaller numbered areas, by painted lines on the floor. The storage can easily be marked off into numbered storage locations, to facilitate putting seed into storage, taking the proper lots out of storage, and for quality control sampling and insect control. Each smaller marked-off area is numbered. The exact storage location of each lot can then be shown on storage records, so seed is easy to find, and operations are easy to control.

With a wear-resistant paint, mark off storage space rectangles on the floor on each side of the central access aisle. The paint used should be a durable enamel, bright-colored. Lines and numbers should be wide/large enough to be easily seen by the forklift driver. The rectangles should be wide enough for loaded pallets to be stacked inside the marked rectangles. The rectangles should extend from the walls toward the center of the building; leave enough space in the building center for the operating aisle. Mark them off by painting lines on the sides and aisle ends of the storage rectangles. Paint the lines all the way to the walls. On the floor in the aisle at the end of each rectangle, paint the number of the rectangle. Also, paint the number inside the rectangle next to the wall, so inspectors can easily identify storage locations when they pass between the stacks and the walls.

Use these numbered storage rectangles in all records and storage operations, to identify where seed are to be stored, what is in storage, and where seed are to be removed from storage.

Work Aisle in Storage

To get maximum use out of the valuable storage space, have only one aisle in the center of the building, extending down the long dimension of the building. This one aisle can then be used for access to stored seed on either side. The aisle should be wide enough to permit forklift/work access (turning in, backing out) to either side without risk of damage to stored stacks.

Air Movement and Ventilation in a Non-Conditioned Storage

Exhaust fans and ventilation openings are provided at locations so as to remove hot air from the top of the building, and moist air from the floor level. They are screened with 6 mm wire mesh and protected so that rain cannot enter them. The ventilation openings in the roof top may be permanently open; ventilation openings near the floor level must have a tightly-sealing door or cover, so they can be closed except at specific times when conditions are suitable for ventilation.

If properly dry seed are stored and the storage is properly maintained, ventilation is seldom necessary. However, to permit ventilation in cases when it is essential, ventilation exhausts may be constructed on the crest of the roof. These may be left permanently open to permit exhaust of hot air building up inside the top of the storage building. They must be screened with a durable metal hardware cloth mesh with openings smaller than 6 mm. The roof or cover over the ventilators must extend far enough to prevent rain water from blowing into the ventilators and entering the building. If desired, small ventilation openings can be constructed in the side walls near the floor level. These ventilators must be properly screened with a durable metal hardware cloth mesh with openings smaller than 6 mm. They must also have a complete tight cover which can be opened when ventilation is desired, and closed at other times. All ventilators should have steel frames and covers, to prevent gnawing by rats.

Relative humidity of outside air should be checked prior to ventilation, to ensure that it is not above the level in equilibrium with the desired seed moisture content. In general, storages should be ventilated only on dry days, between mid-morning and mid-afternoon when RH is lowest.

Chapter 32

Non-Conditioned Storage

Storage Construction

Safe storage conditions are achieved by a combination of (1) construction techniques to minimize temperature and humidity in the storage, and (2) using machines (dehumidifiers, air conditioners, refrigeration) to control/reduce the storage temperature and humidity. Proper technical management can then maintain the required conditions at minimal cost.

The most cost-effective means of providing safe storage conditions is by using proper construction techniques when the storages are constructed.

Short-Term Storage

Primary management decisions in storage involve handling the seed properly, but also involve construction and maintenance of a proper seed storage. Seed storage until the following season is normally adequate when the following techniques are incorporated into the design and construction of an ambient (non-conditioned) storage (Figure 32.1).

1. Raise the storage floor about 1 meter (to local truckbed height) above ground level, and install a durable vapor barrier (not just a moisture barrier) in the floor (Figure 32.2).
2. Install an adequate rat shield around the building, at about floor level. Do not have ramps or steps which rats can use to enter the storage.
3. Keep the area around the storage well-drained and dry, to maintain a dry environment. Keep it free of weeds and trash to discourage entry of rats and birds.
4. Install at least 75–100 mm of thermal insulation in the ceiling (under the roof) to minimize entry of solar heat. If possible, also install this thermal insulation in the walls.

5. Paint the roof and outside walls with a light reflective paint such as aluminum paint, to reflect as much solar heat as possible, and thus minimize heat entering the storage.
6. Have a roof overhang of at least 1.5 meters on each side, to shade the walls and keep them cooler (Figure 32.3).
7. Use rectangular storage buildings, and orient them east-west, approximately in the direction the sun travels. This prevents the sun from heating the walls of the storage (Figure 32.4).
8. Use an absolute minimum of doors (usually one in each of the two ends), to minimize entry of rats, birds, outside heat and humidity, and unauthorized persons.
9. Construct walls of suitable materials, and strengthen the lower part of walls, to minimize damage from accidental hits by forklift (Figure 32.5).
10. Use fluorescent lights to provide the most light and the least heat at the least use of electricity.
11. Do not install "skylight" transparent or translucent panels in the roof. Windows should be only at the top of the side walls, under the roof overhang, and carefully screened to keep out insects and birds.
12. If needed, install ventilators in the roof crest (center peak) to allow heated air to escape from inside the storage (Figure 32.6).

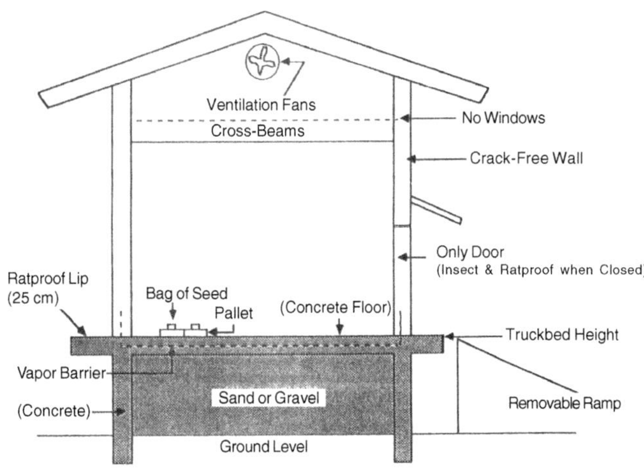

Figure 32.1 Diagram of the basic design features of a safe non-conditioned seed storage. Source: Gregg *et al.*, Storage.

When a structure such as a prefabricated steel lift-van/container is used for seed storage, it should fit the above conditions. Painting it with aluminum paint, installing thermal insulation, erecting a shading roof over it, etc., greatly improve storage conditions inside the container.

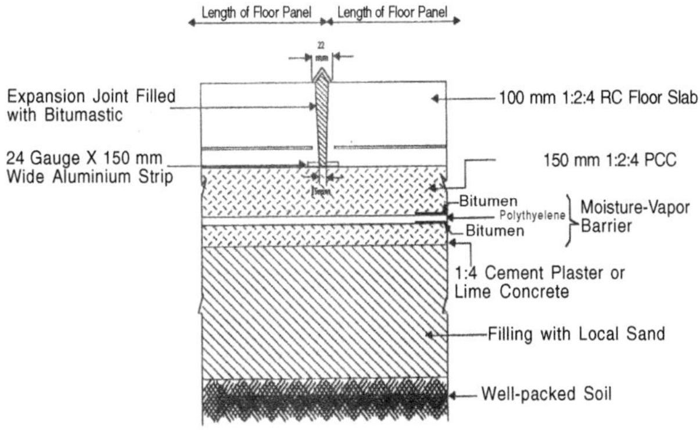

Figure 32.2 Cross-section of the construction of a seed storage floor, to prevent moisture/humidity from coming up through the floor. Source: Billups and Lasheen, unpublished designs.

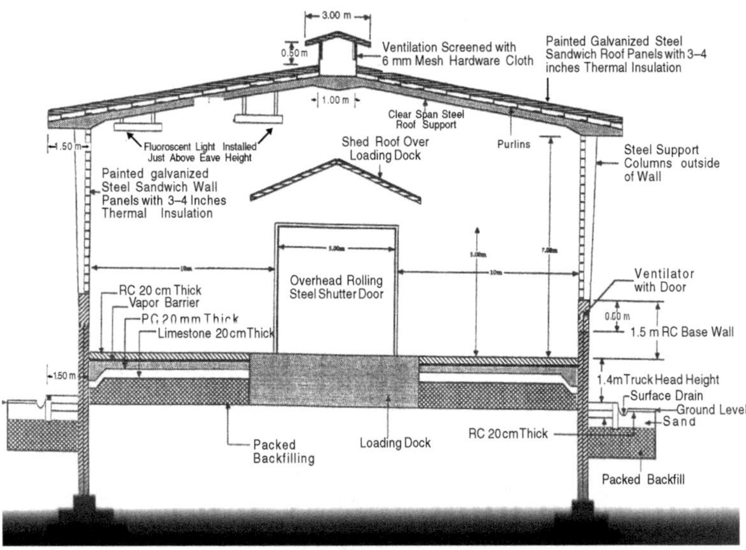

Figure 32.3 End view of design details of a safe non-conditioned seed storage. Source: Gregg *et al.*, Storage.

Medium-Term Storage

A well-managed seed enterprise normally produces or stocks some 5% (under erratic weather conditions, sometimes as much as 10%) of seed above the expected sales. Also, market conditions sometimes result in lower sales of seed; the excess seed will be a total loss unless it can be stored safely until it can be sold in the next year's planting season. This seed costs money, and

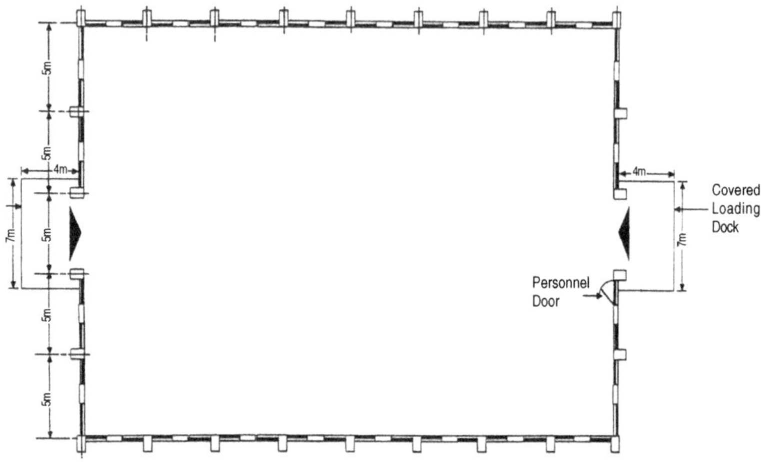

Figure 32.4 Diagram of the floor plan of an efficient layout for a safe non-conditioned seed storage, showing the support columns placed outside, and the only doors in the center of the ends. Source: Billups and Lasheen unpublished designs.

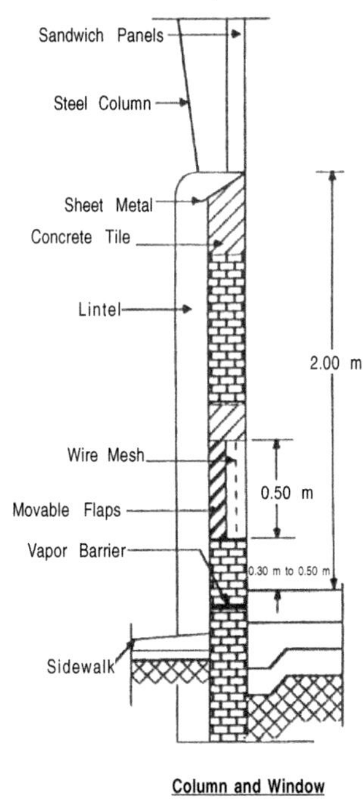

Figure 32.5 Diagram of one type of wall construction in a safe seed storage source: Billups and Lasheen unpublished designs.

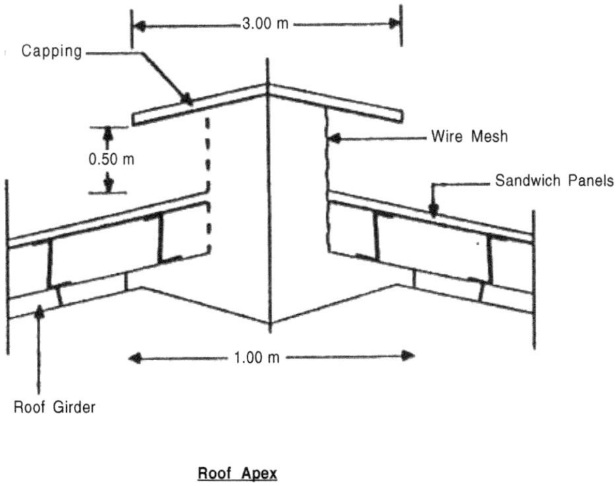

Figure 32.6 Diagram of ventilation at the crest of the roof of a safe seed storage. Source: Billups and Lasheen unpublished designs.

the enterprise cannot afford to let this seed lose quality due to poor storage. To safeguard carryover seed, a seed enterprise needs a small "carryover" storage. This is normally constructed with the conditioning plant, to reduce overall construction costs and to allow the conditioning staff to maintain and operate the storage. It may also be a smaller room constructed inside the regular storage; or, a suitable storage may be rented.

Non-Conditioned Storage

Non-conditioned storages create the best storage conditions possible under ambient climatic conditions by using only good building design and construction. No mechanical equipment except ventilations fans is used.

When Non-Conditioned Storage Can Be Used

Most important during seed storage are to keep moisture and temperature as low as possible, and protect seed from insects, fungi/molds, rats, etc. In a hot, humid tropical or sub-tropical climate, seed **cannot** be stored in non-conditioned storage (1) for long times or (2) in vaportight bags, if seed moisture is above the safe dehumidified-dried levels.

Non-conditioned storages can be used for: (1) short storage of raw seed, before processing; (2) short storage of cleaned seed, if the seed are in porous bags such as jute or cloth; (3) storing low moisture seed (dehumidified-dried to below 8% for starchy seed, below 6% for oily seed) sealed in vaportight packages.

Storage Capacity

A report from the American Materials Handling Institute noted that, on the average, only about 40% of the cubic space in a storage is actually used for product storage. The remaining 60% is used for aisles, head room, separation of lots, space around stacks, etc.

This means that the cubic capacity of the total expected seed must be multiplied by 2.5 to get the needed inside cubic capacity. Furthermore, space should be allowed for any anticipated or hoped-for expansion of handling capacity. To determine the total storage space required:

1. Determine the cubic volume (*e.g.*, cubic meters) on 1 ton of the seed in the condition in which it is stored—bags stacked on pallets, etc.
2. Determine how many tons will be stored, to determine the cubic storage space needed.
3. Always allow more space for growth in the business. How much do you think you can increase your business? If funding permits, companies often add 50–100% of the storage space they need at present.
4. Because space must be left for forklift movement, inspectors moving around lots, air circulation around/behind the stacked rows, lost space for pallets, space above the stacks for air circulation, etc., the space allowed in the storage must be at least double the space occupied by the stored seed.
5. Add these all up to get the cubic storage space needed.

Determining Storage Needs

Too often, seed enterprise managers know the value of safe storage, but try to save money by using poor storage and "accepting the loss of some seed". This is false economy, causing high operating costs, loss of inventory, extra production, and ultimately damage to the enterprise's reliability with farmers. A successful seed enterprise must have enough safe storage to keep all seed inventory alive and in good condition. Good storage is the most important aspect of a seed enterprise; the storage facilities give an indication of the quality of its management.

To determine how much of what kind of storage is needed, consider:

1. The kind of seed which will be stored, and what kind of storage conditions are needed. Starchy seed (rice, maize, etc.) store better; oily seed (groundnut, soybean, onion) deteriorate much faster, and are more difficult to store.
2. How long the seed will be stored (the longer the storage, the cooler and dryer the seed must be).

3. Where it is most efficient, in terms of marketing and operations, to store the seed.

Storage space is cubic volume; *i.e.*, the length, width, and height of the usable storage space inside the storage building is the space available for storage. Storage space is usually measured in cubic meters (or cubic feet).

A cubic meter of seed of different crops will have different weights. Variations are due to seed density and composition, seed size, air space between seed, amount of lightweight material in the seed, etc.

When seed are stacked in bags, some space is lost because the shape of the bags leaves air space. Also, a small amount of space is lost due to pallets. Seed bags are not stacked directly against the floor or wall, and space must be left between the top of the stacks and the roof supports, to ensure ventilation. Space must be left between rows of stacks to permit sampling and working, and ensure complete separation of seed lots. A central aisle must be left for forklift access to stacks of seed bags. All these use up cubic space, but are necessary parts of storage operations.

In normal use of storages which stack seed to heights normally handled by forklift, only about 40–45% of the total cubic space inside the storage building can be effectively used for seed storage. The total storage space must provide the space required to store the required amount of seed, plus floor area, space for workers/inspectors to move around stacks and for forklifts to operate, and some extra space for expansion, unforeseen needs, etc.

The dimensions of the storage building required are determined by:

1. The weight and cubic volume of seed to be stored, allowing a reasonable margin for unforeseen needs.
2. Allow additional working/operating space of 150% of the cubic volume of the seed.
3. The size and shape of the area available for the building.

Approximate weights of 1 cubic meter of bulk seed of selected crops are shown in Table 32.1.

To determine storage requirements, determine the number of tons to be stored, including the desired reserve space. Then, multiply the number of tons by the number of cubic meters per ton.

Multiply this figure by 2.5 to determine the storage and working space required, or total cubic meters of space required in the storage.

Then, determine the size and shape of building which will enclose the required cubic storage space, fit in the space available for the building, and seed pallets can be safely stacked and handled.

With a standard storage building design of specific width and height, it is easy to size the storage building. Simply adjust the storage building length by the number of sections or bays required to give the needed space.

Table 32.1 Seed/grain density, weight and volume. Source: ASAE D241. 2.

Crop Seed or Grain	% Air Space (Void in Bulk)	Specific Gravity	Bulk Density		
			Lb/ft³	Kg/m³	Lb/Bushel
Alfalfa			48.0	770	60
Barley			38.4	616	48
White Hulless	39.5	1.33			
Synasota	45.4	1.21			
Coast (6-row)	57.6	1.13			
Trebi (6-row)	47.9	1.24			
Hannchen	44.5	1.26			
Beans:					
Lima, dry			44.8	719	56
Snap			24.0	385	30
Other, dry			48.0	770	60
Bluegrass			11.2–24.0	180–385	14–30
Broomcorn			35.2–40.0	565–642	44–50
Buckwheat			38.4–41.6	616–667	48–52
Japanese	41.0	1.10			
Castorbean			36.8	590	46
Clover			48.0	770	60
Corn, shelled	40.0	1.19	44.8	719	56
Yellow dent	44.0	1.27			
Yellow dent	40.0	1.30			
Cottonseed (ginrun)			25.6	411	32
Cowpea			48.0	770	60
Flaxseed	34.6	1.10	44.8	719	56
Grain sorghum			44.8 & 40.0	719 & 642	56 & 50
Yellow milo	37.0	1.22			
Blackhull Kafir	36.8	1.26			
Hempseed			35.2	565	44
Hungarian millet			38.4 & 40.0	616 & 642	48 & 50
Kafir			44.8 & 40.0	719 & 642	56 & 50
Lentil			48.0	770	60
Millet			38.4–40.0	616–642	48–50
Siberian	36.8	1.11			
Mustard			46.4–48.0	744–770	58–60
Oats			25.6	411	32
Victory	47.6	1.05			
Red Texas	55.5	0.99			
Iowar	51.4	0.95			
Kanota	50.9	1.06			
Orchardgrass			11.2	180	14
Peanuts, unshelled					
Virginia type			13.6	218	17
Runner, SE			16.8	269	21
Spanish, SE			19.7	316	25

Crop Seed or Grain	% Air Space (Void in Bulk)	Specific Gravity	Bulk Density		
			Lb/ft³	Kg/m³	Lb/Bushel
Spanish, SW			19.8	318	25
Popcorn, shelled			44.8	719	56
Poppy			36.8	590	46
Rapeseed			40.0 & 48.0	642 & 770	50 & 60
Redtop			40.0 & 48.0	642 & 770	50 & 60
Rice (paddy)			36.0	577	45
Honduras	50.4	1.11			
Wataribune	46.5	1.12			
Rye			44.8	719	56
Common	41.2	1.23			
Sesame			36.8	590	46
Sorgo			40.0	642	50
Soybean			48.0	770	60
Manchu	36.1	1.18			
Wilson	33.8	1.13			
Spelt			32.0	513	40
Sudangrass			32.0	513	40
Sunflower			19.2 & 25.6	308 & 411	24 & 32
Timothy		36.0	577	45	
Velvetbean, hulled		48.0	770	60	
Vetch		48.0	770	60	
Wheat		48.0	770	60	
Hard					
Turkey winter	42.6	1.30			
Turkey, yellow	40.1	1.29			
Soft					
Harvest Queen	39.6	1.32			

Based on: 1 m = 3.28 ft; 1 m^3 = 35.29 ft^3; 1 kg = 2.2 lb.

To plan seed storage requirements, the amount and kind of seed to be stored, allowing a reasonable amount for extra unforeseen or temporary requirements, is determined. Weight per cubic meter (m^3) and volume per ton are then determined. When several kinds of seed will be handled, it is best to use the weight/volume of the lightest seed.

Using the planning base of 750 kg/m^3 of bulk wheat, 1 mt of bulk wheat fills 1.33 m^3 of space; *i.e.*, 100 mt fills 133 m^3. Thus:

1. 1,000 mt of bulk wheat seed would occupy 1,330 m^3.
2. 2,000 mt of bulk wheat seed would occupy 2,660 m^3.
3. 5,000 mt of bulk wheat seed would occupy 6,650 m^3.

Considering the extra space required for seed in bags, operations, handling, etc., the volume of these amounts of bulk seed is multiplied by 2.5 to give the total effective storage/working space required in the storage building. Including working space, 1 mt of wheat seed (1.33 m^3 of wheat) requires 3.33 m^3. Thus:

1. Storage for 1,000 mt of seed would require 3,330 m^3 of storage and working space.
2. Storage for 2,000 mt of seed would require 6,660 m^3 of storage and working space.
3. Storage for 5,000 mt of seed would require 16,650 m^3 of storage and working space.

Using a standard storage design whose effective inside width is 25 m and height is 7 m, each 1 m of building length provides 175 m^3 of storage/working space; *i.e.*, space for 52.5 mt. Assuming that the standard design can be built most economically in bays (sections) of 6 m length, each 6 m bay provides 1,050 m^3 of storage/working space, which handles approximately 315 mt of seed. Using the above standard storage design, storage size for specific amounts of seed would be as follows.

1. A storage building of the above standard design for 1,000 mt of seed would require 19 m of length to provide the required 3,330 m^3 of storage space. Three bays of 6 m each would not provide the required storage; it is better to build 4 bays and provide some extra space.
2. A storage building of the above standard design for 2,000 mt of seed would require 38 m of length to provide the required 6,660 m^3 of storage space. Seven 6 m bays could be built, providing a total length of 42 m.
3. A storage building of the above standard design for 5,000 mt of seed would require 95 m of length to provide the required 16,625 m^3 of storage space. Using 6-m bays, 16 bays could be built, providing a total length of 96 m. If the space available makes it more suitable, this could be divided into 2 buildings of 8 bays, or 48 m length, each.

Chapter 33

Conditioned Storage

Conditioned Storage

Conditioned seed storage is a sealed, insulated storage space in which machines are used to create artificially cooled and dried conditions.

Good storage minimizes the rate of seed deterioration, and keeps the seed alive, with high germination, longer. This requires dry cool conditions and protection from damage (by heat, moisture, insects, fungi, rats, birds, etc.), so seed viability is high after the required storage period.

Seed deterioration, and insect and fungi activity, are reduced by lowering temperature and moisture content (and correspondingly, RH of the air). How cool and how dry depends on the length of the storage period, desired final seed quality, kind of seed, and cost. The longer the storage, the lower temperature and moisture content must be. At lower temperatures, seed of higher moisture content can be safely stored; at lower moisture content, seed can be safely stored at higher temperatures. These apply to reasonable storage periods; reducing only one—moisture or temperature—cannot achieve maximum potential seed storage life.

Safe Conditioned Storage Conditions

Seed can be protected from moisture and high RH by the package and/or the storage building/room. If seed are dried to very low moisture content (8% or lower for starchy seed, 6% or lower for oily seed) and kept sealed in airtight vaporproof packages, they can be safely stored.

If this is not possible, the seed must be kept in a conditioned storage which provides temperatures low enough to keep the seed alive. Lower temperature can be maintained only by the storage building/room.

Seed of storable quality can be safely stored if the storage building or room is designed, equipped and managed to provide the required dry, cool, protective conditions.

Removing Seed From Conditioned Storage

If seed in vapor-porous packages are taken from conditioned storage into warm humid outside air, seed temperature may be below the dewpoint of the outside air. When this occurs, moisture from the air condenses on the seed, seed moisture increases, and the seed lose viability rapidly.

This can be prevented by: (1) Packaging low-moisture seed in vaporproof packages for cold storage; or, (2) Quickly moving seed in vapor-porous packages from cold storage into a dehumidified room at ambient temperature, and storing them there until they warm up to outside temperatures; or, (3) If all seed will be removed from a cold dehumidified storage at about the same time, stop the cooling equipment and operate only the dehumidifier until seed warm up to outside temperatures.

Chapter 34

Managing Seed Storage

Managing Storage

Seed are in some kind of storage from the time they are produced until they are planted by farmer-users. Every place that seed stay for any time is a part of this storage process; no matter what is done or how short each period is, it affects the final quality of the seed either positively or negatively. Careful management is essential.

Technically, storage is a special science in itself. From the management standpoint, storage is a technical operation essential to support marketing and delivering high-quality seed to farmers. The seed enterprise manager must install, manage and use storage as a cost-effective part of his operating and managing plan and system.

Managing to Prevent Problems

Storage must keep seed alive and in good condition until it is planted by farmers. This means that it must provide conditions—low temperature, low humidity and moisture, protection from damage—which help keep the seed alive and in good condition.

Problems cost money and time, and often lose seed. Good management practices prevent problems, and include the following.

Pre-Storage

Before seed storage begins, pre-storage management requires careful and complete implementation of two aspects:

1. Have enough good storage, at the right places.
2. Be sure that only high-quality seed go into storage.

During Storage

Then, during storage, management and preventing problems requires:

3. Operate and manage storages so as to prevent loss of seed quality/viability, identity of seed bags, damage to seed or seed bags, or loss of seed.
4. Be sure:

   A. The storage is clean, dry, and cool.
   B. Seed bags/packages are in good condition, not torn or damaged.
   C. Proper labels are on each bag and up-to-date stack cards are on each stack.
   D. Rats and birds are properly controlled and excluded.
   E. Regular germination testing of all stored seed is up-to-date.
   F. Records are kept current, and management receives daily reports (during the sales period) of sales and remaining stocks.

Removing Seed From Storage

Management guidelines for removing seed from storage include:

1. Keep seed in good storage as long as possible. Remove them as close as possible to the time when the using farmer takes actual delivery of the seed.
2. When seed are removed from storage, immediately record the amount/kind removed, on the stack card in the storage and on the permanent record of the lot.
3. Handle bags carefully. Do not drop or throw them.
4. The first seed removed/sold should be the oldest seed and/or the seed with the lowest (yet still above the required standard) germination. Leave the newest seed, and the highest-germination seed, until last.

Removing Seed From Conditioned Storage

If seed in vapor-porous packages are taken from conditioned storage into warm humid outside air, seed temperature may be below the dewpoint of the outside air. When this occurs, moisture from the air condenses on the cold seed. The seed absorb this condensed moisture quickly, so that seed moisture content increases rapidly to unsafe high levels. And, as the seed warm with this high moisture, seed respiration rate increases rapidly, molds become active, and germination may be lost within a few days.

This can be prevented by:

1. Packaging low-moisture seed in vaporproof packages for cold storage; or,

2. When seed is removed from cold storage, keep it in a dry (dehumidified) ambient-temperature storage for several days before going into warm humid outside conditions. If cold seed is moved directly into the ambient (outside) environment, moisture will condense on the seed, and they will rapidly lose viability. Quickly moving seed in vapor-porous packages from cold storage into a dehumidified room at ambient temperature, and storing them there until they warm up to outside temperatures; or, if all seed will be removed from a cold dehumidified storage at about the same time, stop the cooling equipment and operate only the dehumidifier until seed warm up to outside temperatures.

As a safety factor, cold storage for seed should not exceed 60% in RH. Most commercial cold storage facilities are designed for succulent or moist materials such as potato or fruits, and RH is maintained above 80% to prevent desiccation. Avoid such storages for seed, which are dry materials.

Storage Problems and Their Causes

Loss of Viability

The primary concern in storage is to prevent seed from deteriorating or being damaged while in storage, awaiting sale. This requires keeping the seed as cool and dry as possible during storage. And, this applies to all the time between seed harvest and when it is planted by the using farmer. The time that seed are in pre-conditioning, during conditioning, at the conditioning plant, in transport, and at the retail sales outlet are all parts of the storage period. Storage problems arise because:

1. Low-quality seed are placed in storage.
2. Inadequately-dried seed, with too-high moisture content, are stored.
3. The storage structure is not properly constructed or proper storage conditions are not maintained.
4. Storage temperature is too high.
5. Seed are stored for too long a period for the seed kind or the storage conditions

Loss of Identity

Seed have no value if their identity—crop, variety, lot number—is lost. This requires positive identification of each bag or package at all times. A descriptive tag (label) is normally attached to each bag. As the tag may be torn off in handling, a lot number is also indelibly printed/written on the bag. Seed bags and packets must also be handled carefully, to prevent losing or smudging their labels. To prevent mistaken identity, a "stack card" with detailed information should be kept up-to-date on each stack of seed. Only one kind (crop, variety, lot) of seed should be stored in one stack.

Loss/Damage of Seed and Packages

Fire, damage, insects, birds and rats can damage or complete loss of seed and seed bags. Management control to prevent loss and damage include:

1. The seed storage must be kept dry and cool to prevent deterioration of seed and damage to bags and packages. Keep the storage clean, to avoid contaminating seed, discourage rats and birds, and prevent fire. Maintain good rat and bird control, to prevent damage to bags and loss of seed.
2. In moving or stacking seed packages, handle them carefully; never throw or drop seed bags from stacks. Especially with low-moisture seed (in vaporproof bags), the impact suffered when bags are dropped will cause severe mechanical damage—cracking, breaking—to seed.
3. Provide a water fire riser outside the building, near each forklift door, in a position where it will not be damaged by trucks. An adequate fire hose should be provided inside the door, in a suitable protective container. An electrical fire chemical extinguisher should be wall-mounted just inside each forklift door.
4. Do no allow fire risks in or around the building. This includes smoking, leaving flammable trash, using flammable materials such as gasoline, having fires or flammable operations (except for welding, under carefully controlled conditions), have all electrical work of fire/explosion-proof quality, etc.

Follow-up and Supervision

Procedures for putting seed into storage, handling during storage, and removal from storage, should be carefully planned to ensure high seed quality, minimum loss, and least-cost operations. The manager should regularly check to see that:

1. The storage is clean, dry, and cool.
2. Seed bags/packages are in good condition, not torn or damaged.
3. Proper labels are on each bag and up-to-date stack cards are attached to each stack.
4. Rats and birds are properly controlled and excluded.
5. A regular schedule of germination testing of all stored seed is kept up-to-date.
6. Records are kept up-to-the-hour, and management receives daily reports (during the busy sales period) of seed sales and remaining inventory.

Records and Reports

Seed storages maintain the seedstocks or inventory which are the business of the seed firm. The storage manager must keep accurate records, and keep them up-to-date on a daily basis. The seed firm manager and the marketing unit must receive regular reports which show the current status of all seed which is available for sale.

Records

Complete records should be kept up-to-date on a daily basis, with entries made immediately when a sale/delivery is made. Records should be in the form of a lot record card, with space for individual entries for sales/ deliveries from the lot. These can be maintained on a computer, for easy handling.

The storage manager should maintain records for each seed lot which shows at least the following data on each lot.

1. Lot number
2. Crop and variety (or hybrid)
3. Year produced
4. Initial total lot weight and total number of bags of seed
5. Weight of each seed bag
6. Date the lot was tested, and complete quality information
7. Date the lot must be retested
8. References (*e.g.*, lot number) which can trace the lot to its conditioning and production records
9. Exact storage location (*i.e.*, number of the small marked-off storage block where the seed is stored)
10. Date and amount (weight and number of bags) of each delivery/sale from the lot, with name and address of the receiver
11. Amount (total weight and number of bags) of seed remaining after each sale

Storage records are of three types:

1. *Labels on bags/packages*: usually specified by the Seed Law, data on the label includes the crop, variety, and lot number; test results (germination, purity, etc.) and date of test; year (and sometimes place) of production; name and address of the supplier/producer; etc. A Certification tag includes appropriate Certification information.
2. *Stack cards*: includes the crop, variety, and lot number of the seed in the stack; location in storage; bag weight; number of bags; retesting (for germination) schedule and dates, along with latest germination; year produced; date and number of bags removed, and number of bags remaining.

3. *Permanent records*: usually combined into the permanent lot record along with complete information of production and conditioning. Storage aspects included in the permanent record include storage location; bag weight and number; date and amount of each sale, with name and address of purchaser.

Reports

Reports of (1) seed inventory in storage and (2) seed removed from storage, are made to management at regular intervals. During busy sales seasons, these reports are made daily; in off-seasons, they are usually made monthly.

Seed inventory reports on seed still in storage include crop / variety / lot number; bag weight and number of bags; year produced; latest germination test results and period of legal validity of the germination test; and storage location.

Reports on seed removed from storage include data on crop / variety / lot number; amount and number of bags; date removed; name and address of purchaser / receiver of the seed.

Up-to-date reports of seed sales / deliveries, and of seed stocks remaining in storage, must be made to the program manager and the marketing unit regularly. During off-seasons when seed are not being sold, once a month is usually sufficient. However, during planting and sales seasons, reports are often made daily. This permits the marketing unit to move excess seed from one storage to a different location which needs that seed kind. Reports can be maintained on a computer, for easy handling; with proper modem connections, reports can be transmitted from the storage to the manager and marketing unit by telephone computer connections.

Reports should include at least the following:

Sales/Delivery Report:

1. Storage name and location
2. Lot number
3. Crop and variety (or hybrid)
4. Amount and dates sold / delivered
5. District(s) to which the seed went

Seedstock Status Report:

1. Storage name and location
2. Lot number
3. Crop and variety (or hybrid)
4. Amount (total weight of seed, bag weight and number of bags) of seed remaining in storage
5. Seed quality information, especially germination and date of test
6. Year produced

Records and Reports as Management Tools

The owner/manager needs up-to-the-hour records and reports from storage to:

1. Know how much seed has been removed from storage, and where/to whom it was delivered.
2. Know how much of what kind of seed remains in storage and is available for sale.
3. Know which kinds of seed are selling well, and which kinds are not.
4. Determine if some seed is losing viability in storage, and should be immediately marketed.
5. Identify problems (such as poor storage conditions) with storage, in a manner which permits identifying remedial actions.
6. Determine the efficiency and cost-effectiveness of storage facilities and operations, and modifications which could improve them.

Records and reports can give details on operations and current status, if they are complete and up-to-date. Records and reports should include at least the following information (see the chapters on Records and Reports):

1. How much seed of which kind and lot was removed, and to whom it was delivered.
2. How much of what seed is in storage and available for sale.
3. Which kinds of seed sell well, and which kinds do not.
4. Determine if some seed is losing viability in storage, and should be immediately marketed.
5. Identify problems with storage, to identify remedial actions.
6. Determine efficiency and cost-effectiveness of storage operations, and modifications which could improve them.

Records for planning storage include (see following tables for sample formats):

Storage needs for planned operations: Summarizes expected commodities to be stored, storage period, and amount/kind of storage required. Helps plan use of storage, and obtaining additional storage (Table 34.1).

Planned shipping/sales delivery schedule: Gives details of planned shipping or sales of stored commodities; helps plan storage use, storage location of different lots, and arranging transport and labor/handling (Table 34.2).

Records used in storage operations include (see following tables for sample formats):

Receiving receipt: Acknowledges receipt of seed into storage, and provides basis for following documentation and management of storage (Table 34.3).

Planning Record Formats

Table 34.1 Storage needs for planned operations. Source: Gregg, Records and Reports. Unpublished.

Page No. ______________________ For the Period ______________________________

Crop/Input	Variety/Kind	Amount	Expected Date Received	Storage Needed	Expected Date Out	Storage Period	Kind of Storage Needed	Storage Available	Remarks

Table 34.2 Planned shipping/sales delivery schedule. Source: Gregg, Records and Reports. Unpublished.

Page No. ______________________ For the Period ______________________________

Crop/Input	Variety/Kind	Amount	Expected Date Start Ship/Sell	Expected End Ship/Sell	Peak Ship/Sell Period	Ship Method	Buyer Pick-Up	Remarks

Storage Operating Record Formats

Table 34.3 Storage receiving receipt. Source: Gregg, Records and Reports. Unpublished.

Receipt No. ______________________ Date: ______________________________

Received From ____

Type of Containers ____

No.	Crop/Variety or Other Goods	Field/LotNo.	Bag Size	No. Bags	Total Net Wt.	Disposition	Remarks

Receiver ______________________ Supplier/Deliverer ______________________

Table 34.4 Storage shipping/delivery receipt. Source: Gregg, Records and Reports. Unpublished.
(sales invoice may be used instead of this, to avoid a second form)

Date ________ Delivery Receipt No. ____________ Sale/Ship Order No. ________
Sold/Delivered to ____

No.	Crop/Variety or Other goods	Lot No.	Bag Size	No.Bags/ Units	Total Net Wt.	Remarks

Receiver ____________________ Storage Operator ____________________

Table 34.5 Storage stack/bin card. Source: Gregg, Records and Reports. Unpublished.

Storage Location ____________ Crop ____________ Variety ____________

Date	Amount In	Amount Out	Balance on Hand	Remarks

Shipping/delivery receipt: Documents removal of seed from storage; used to manage and verify disposal of stocks and inventory of stored materials (Table 34.4).

Stack/bin card: Shows current status of kind and amount of seed/other sales products in a stack or bin (Table 34.5).

Receiving log/register: Records details of commodities received into storage; helps manage storage and receiving operations, and provides data on what/when commodities come into storage (Table 34.6).

Storage inventory: Summarizes amount of sales commodities in storage on a given date, their location and condition/readiness for sale (Table 34.7).

Delivery/shipping log/register: Records date, kind and amounts of sales commodities removed from storage (Table 34.8).

Internal quality control testing schedule for stored seed: Provides a reminder of dates each lot must be retested, and a log/register of quality of each lot on the required testing date (Table 34.9).

Forms used to report on storage inventory and operations include the following:

Goods delivered and goods remaining in storage: Details kinds and amounts of sales commodities sold, and remaining in storage. Helps manage re-supply and purchase/sales (Table 34.10).

Quality of seed in storage and recommended disposition: Helps monitor performance of the existing storage facilities in terms of seed deterioration in storage, and helps prevent accidental sale of below-standard seed, or using storage space for unsalable lots (Table 34.11).

Table 34.6 Receiving log/register. Source: Gregg, Records and Reports. Unpublished.

Register Page No. ____________ For the Period ____________

Crop/Input	Variety/Kind	No. Bags Or Units	Total Wt/ Details	Date Received	Received From	Expected Ship Date/Period	Storage Location	Remarks

Table 34.7 Storage inventory (use same form for daily, weekly, monthly inventory report). Source: Gregg, Records and Reports. Unpublished.

Date ____________ Inventory Record Page No.____________

No.	Lot No.	Crop/Variety orInput/Kind	No. Bags/ Units	Bag Net Wt.	Total Wt.	Germination		Storage Location	Condition/ Remarks
						%	Test Date		

Table 34.8 Delivery/shipping log/register. Source: Gregg, Records and Reports. Unpublished.

Register Page No. ____________ For the Period ____________

No.	Lot No.	Crop & Variety	Shipped/Sold to	Date	Amount			Ship. Receipt No.	No. Bags Remaining
					No. Bags	Bag Wt.	Total Wt.		

Table 34.9 Internal Quality Control testing schedule for stored seed. Source: Gregg, Records and Reports. Unpublished.

Schedule Page No. ________ For the Period________

No.	Lot No.	Crop Variety	Year Grown	No. of Bags	Scheduled Re-Testing (date; X when tested; plus 0 if germination below standard)											
					J	F	M	A	M	J	J	A	S	O	N	D

Storage Report Formats

Table 34.10 Seed delivered and seed remaining in storage. Source: Gregg, Records and Reports. Unpublished.

Date ________ Report for the Period________

Date	Crop/Input	Variety/Kind	Lot No./ Details	Delivered to	Amount	Amount Remaining	Remarks

Table 34.11 Quality of seed in storage and recommended disposition. Source: Gregg, Records and Reports. Unpublished.

Date ________

Date Tested	Crop	Variety	Lot No.	Amount	Quality	Recommendations (Discard, Keep for Sale, etc.)	Remarks

Management Procedures

Even with good storages, seed deteriorate and costs are high if management and planning are inadequate to maintain seed in good condition. The following practical management procedures should be followed.

Maintaining Lot Identity

If identity of seed is lost, its value is lost. Good storage management maintains seed identity by:

1. Identify every bag adequately with secure, complete labels and stenciled lot numbers.
2. Mark off the storage into smaller numbered areas, to permit easy identification of exact lot location. Use painted lines on the floor to mark off the storage.
3. Store or stack seed lots separately. Never stack a lot on top of a different lot.
4. Use stack cards to identify the seed in all stacks.
5. Keep adequate, up-to-date records.

Maintaining Purity

To prevent contamination of seed:

1. Keep bags in good condition.
2. Prevent scattering of seed by birds, etc.
3. Keep bags in their correct lots and well-identified.
4. Keep the storage spotlessly clean.

Protecting Seed From Insects, Pests, Fungi, and Damage

Management protective procedures include:

1. Keep seed cool and dry to minimize damage by storage fungi and insects.
2. Prevent insect and pest entry into the storage.
3. Keep the storage and surrounding outside area clean, to reduce attraction, food and cover for insects and pests.
4. Control insects and pests by a complete program of fumigation, seed treatment, space sprays, residual surface sprays, poison baits, etc., before damage occurs.
5. Do not store chemicals, feeds, fuels, fertilizers, etc., in the seed storage.

Stacking

Dry bagged seed may be stacked to any efficient storage height without weight damage to seed at the bottom. Damage to stacked seed occurs from:

1. If bags fall or are dropped off the stack, mechanical seed injury results. Remove bags from stacks by conveyor, forklift pallet, or hand-carrying, never by dropping.
2. With poor ventilation (especially if the floor has no vapor barrier) air convection currents due to temperature differences move air and RH vapor-moisture within the storage. RH of cooler air near the floor increases, causing increased moisture content and rapid deterioration of seed at the bottom of stacks.

Good stacking management includes:

1. Ensure proper ventilation around seed, as well as providing space for workers, inspectors, and samplers. Leave 75 cm open space between stacks and walls, and 175 cm above stacks (Figure 34.1).
2. Provide adequate aisle space so operations do not damage or misplace bags.
3. Never stack a lot on top of a different lot; to ensure that workers don't move the wrong seed. Also, if a seed bag is torn, seed of a different lot will fall down onto bags of a different lot, causing potential mixtures in the field.

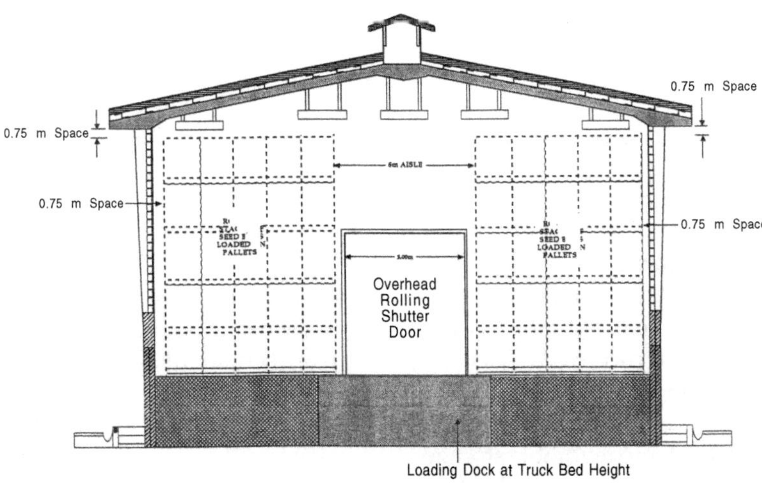

Figure 34.1 Diagram of the end view of a safe non-conditioned seed storage, showing how seed on pallets are stacked in the storage. Note that a door is in the center of each end of the building, and the operating aisle is in the center Source: Gregg *et al.*, Storage.

4. Do not stack bags directly on the floor; stack stack them on pallets, with at least 10 cm open space for air circulation between the floor and the bags.
5. Do not stack bags in piles. Stack loaded pallets; do not remove bags from the pallets.
6. Stack bags on pallets in a pattern which "ties the bags together", so they do not fall or slide.
7. Remove bags from stacks carefully and gently. Never drop bags off of stacks.

General Storage Management

Some good management techniques to keep storage efficient and safe include:

1. Keep buildings in good repair.
2. Maintain cooling and dehumidification equipment in good condition.
3. Keep adequate spare parts.
4. Practice preventive maintenance.
5. Regularly monitor and record storage conditions; be sure they are suitable.
6. Regularly check seed germination; improve storage conditions as required to keep seed viable. Chart germination against storage time and conditions to obtain storage management data.
7. Sell stored seed so they are planted before they deteriorate. Follow the "first in, first out" principle of selling seed.

Carryover

Carryover must be planned and managed to minimize seed loss.

1. Provide some good storage suitable for the required carryover, in addition to short-term storage.
2. Test seed at harvest. Immediately sell lots with lowest acceptable viability. Carryover only lots of highest viability and vigor, to ensure high viability after storage.
3. Estimate anticipated carryover at harvest, and immediately condition and move possible carryover seed into the better carryover storage.
4. Possible carryover "less storable" crop seed should be conditioned first and put immediately into carryover storage. For example, leave maize or rice seed for later processing, while soybean is immediately moved into favorable storage.

Keep Power Costs Low

Use the following measures to save energy and reduce costs:

1. Safeguard seed quality so no seed is lost. Loss of seed wastes all energy and expense invested in it.
2. Storage construction—thermal insulation, vapor barriers, door construction and sealing, etc.—should be over-designed (*i.e.*, significantly more then the minimum required) to minimize operation of and dependence on, mechanical equipment to maintain storage conditions.
3. Paint building roof and walls with a light-colored heat-reflecting paint to reduce absorption of solar heat.
4. Keep the area around the storage clean and dry to minimize dehumidifier operation.
5. Use sealed, vaporproofed and insulated antechamber entries for conditioned storerooms.
6. Plan storage layout and stacking to minimize distance and number of times seed are moved.
7. Proper air vents in non-conditioned storage remove hot air and reduce convection-current moisture.
8. Plan seed production for harvest in dry seasons; at full field maturity; dry by sun or shaded solar heating.
9. Plan removal of seed from storage and transport, to reduce number of times storage rooms are opened, and truck trips required.
10. Eliminate re-drying of seed, by proper humidity and moisture control.
11. Pack properly-dried seed in vaporproof packages; dehumidifiers are not required, but higher storage temperatures are safe, re-drying is eliminated, non-conditioned storage is useful, seed is protected during transport and storage without special equipment, and carryover seed is still salable.
12. Construct all storages with vapor-barriered floors raised to truckbed level, to reduce storage moisture and facilitate loading trucks.
13. Permit only essential workers into conditioned storerooms for the shortest time possible, to reduce heat and moisture load.
14. Install spring-loaded closers on doors of conditioned storages, to reduce entry of heat and moisture.
15. Use fluorescent lights to reduce electricity use.
16. Locate main storages as near as possible to conditioning, to minimize seed movement.
17. When large conditioned storage capacity is required, build several smaller rooms, not one large room. When seed volume is reduced during off-seasons, turn off equipment in empty rooms. Smaller rooms with

separate cooling and dehumidification equipment also ensure some conditioned storage during equipment failures.

Managing Retail Sales Storage

The same good storage conditions must be maintained at retail seed outlets. Construct a small good storage at the retail outlet. At the least, have a cool, dry room where seed can be stored. Do not store seed in the sunshine, on sidewalks, on floors, etc.

Retail seed storages must keep seed in good condition until it is sold to using farmers. A retail sales storage usually handles small amounts of seed and stores them for short periods. Small technically-proper seed storages are needed. Management procedures, in addition to those for all seed storages, include:

1. Carefully plan the kinds and amounts of seed needed, and the time they are needed. Order the proper seed from the producer/ distributor in time to have it available when farmers need it.
2. Do not provide credit for farmers; help them obtain credit from regular credit sources. A seed sales/storage is not in the credit business.
3. Maintain close contact with farmers in the sales area, to determine in advance what seed they will buy. Educate farmers on benefits of improved seed and varieties. Make follow-up contact to be sure farmers are happy with their seed.
4. Keep good up-to-date records to show how much of what kind of seed is sold, and maintain close contact with the seed producer/distributors. Have them deliver additional stocks in time to supply farmers.
5. Regularly sample and test all seed lots, to be sure seed sold to farmers is of high quality.
6. Maintain a clean, technically-proper storage. Keep seed in it. Let farmers see the storage, and see their seed come out of it. This has considerable sales promotional value.
7. Do not keep seed on the sidewalk, in the sun, in glass windows, in hot or damp places, etc. Keep seed in the special seed storage; put out posters, signs, pamphlets, etc., to show what seed is available.

Storage Network Management

Establishing a Seed Storage Network

Best service to seed suppliers and farmer seed users is maintained by establishing a network of seed storages through the following procedures:

1. Providing free designs and planning assistance to all interested firms and individuals, to help ensure that their storages are technically proper.
2. Providing special concessional credit to construct technically proper seed storages.
3. Providing assistance and guidelines on operating and managing storages, and helping design record/report systems.
4. Requiring specific storage structures and/or conditions as part of the requirements for obtaining a permit or license to supply and sell seed.
5. Permitting seed firms and programs to maintain and operate their own storages, under normal regulatory guidance of the Seed Law, Certification and quality control requirements.

Operating the Storage Network

A network of safe seed storages, serving the entire area where seed are produced and used, is necessary. However, like other seed industry operations, storage is solely a tool to help deliver high-quality seed to farmers. In a developing or developed seed industry, storage is not a separate business; it is an integral part of the overall operation.

A centrally-owned and operated network of seed storages, storing seed for different agencies, is seldom if ever economically viable. Most efficient operation is each firm, program or component of the seed supply industry operating its own storages. This may require several different storages in the same area; however, they will be of different sizes, so there is no wasted investment. Operating efficiency is usually improved.

Normally, the seed producing firm owns its own conditioning and distribution storages. It supplies seed to retail dealers, each with its own small storages. Each seed producing firm maintains his own storages at the conditioning plant. Large distribution storages are owned by the seed producing firm or special seed wholesaler firms. Local retail sales stores are owned by local seed merchants.

The primary requirement is technical: each storage must provide the conditions which keep seed in good condition, and be located where seed must be stored.

When a centrally-owned network of storages is established, it usually restricts or limits the development of seed supply enterprises, especially in the private sector. When a central storage system tries to store seed for different programs/suppliers/ firms/dealers, the many unavoidable problems associated with seed quality, viability and longevity create problems which override any potential benefits from the system. Also, such storages are seldom constructed in the exact site which best serves the needs of the users.

Managing Main Distribution Storage

Main distribution storages are centrally-located in an area, region, governorate, province, etc. Relatively large amounts of seed are sent from conditioning plants to these storages, by relatively large-volume, cost-efficient transport. They then supply smaller amounts of seed to retail sales stores throughout the area it serves. Distribution storages are located closer to the retail seed sales stores, so they can quickly deliver seed when needed during the planting season.

Distribution storages must be properly constructed to provide the required safe storage conditions. They must follow good storage management practices, to keep costs low, while seed quality and service are high.

A distribution storage must be able to:

1. Receive relatively large amounts of seed quickly and efficiently.
2. Handle seed quickly, at low cost, and with minimum loss.
3. Keep stored seed in good condition, with minimum damage and deterioration.
4. Maintain contact with the retail stores it serves, to determine their needs.
5. Quickly and safely ship out and deliver the right kinds and amounts of seed to retail seed stores.
6. Hold carryover seed in storage, and protect its quality.
7. Maintain a regular program of testing seed in storage, to keep seed properly labeled and be sure that low-quality seed are discarded rather than sold to retail stores and farmers.

Distribution storages also should have either an internal quality control laboratory or ready access to quick testing at a service laboratory. All stored seed lots should be regularly tested for both germination and vigor. Lots which show decline but are still acceptable should be sold first for immediate planting. Because they can survive longer in storage, more vigorous lots should be held for later sale.

Distribution storages should establish a system of daily contact by telephone with all their retail store customers, to determine their needs for seed. A delivery truck can then be sent on a regular route, to carry seed to several retail stores in one trip. This system can keep all retail stores properly stocked with seed needed by farmers, and can avoid the problems of poor storage in consignment sales.

Consignment Sales

In the consignment system the seed producing firm or distributor delivers to retail sales storages a certain amount of seed. After the planting season, the retail dealer pays the seed producer for the seed which was sold to farmers, and returns the unsold balance to the seed producer.

This system is inefficient and costly; it is strongly recommended that this system not be used. Among the reasons it is inefficient are:

1. Since the retail dealer has no personal investment in his seed stocks, he does not work hard to promote farmer use of the seed. Thus, fewer farmers use the improved seed sold through the organized seed industry.
2. The seed producer's money is tied up in a system which essentially provides free credit to retail dealers. However, the seed producer must pay the cost of using these funds.
3. Returned seed have seldom been properly stored, so they deteriorate or are weakened so they cannot be safely carried over until the next planting season. Thus, the investment in this seed is lost to the seed producer. He must either absorb a loss, or increase seed prices so that the price of seed sold covers the cost of the estimated amount lost through the consignment system.
4. It is very difficult for the seed producer to plan production to meet needs. Excessive, wasted seed production is common.
5. Handling and storage costs are excessive. Record-keeping is more complicated.

The seed distributor should help retail stores with advertising and promotion, but when seed are sold to retail stores, sales should be final, with no seed to be returned. This not only improves production planning and marketing, but helps in storage management.

Chapter 35

Handling Certified Seed

Seed Certification

Seed certification is a special external quality control program designed to help producers/conditioners supply seed of higher quality, especially genetic purity. Certification helps seedsmen produce seed of high and known quality, and helps farmers identify and buy high-quality seed of superior crop varieties.

This requires that seed be handled, conditioned and stored so as to (1) prevent mixtures, (2) maintain high genetic varietal purity, (3) maintain high seed quality, and (4) preserve the identity of the seed. This requires special procedures in handling, conditioning and storage, which are usually specified by the Seed Certification Agency. Details vary, but the following illustrate the more common conditions for handling and conditioning Certified seed. The Certification Agency usually has specific forms used for recording and reporting each activity.

Requirements for conditioning Certified seed may seem intensive and tedious, but most Certification requirements should be standard procedures for any good seed conditioning plant.

Certified Seed Conditioning

Certification is a limited-generation quality control system which requires specified genetically-pure stock seed to plant seed-producing fields, fields free of specified problems, growing seed crops under conditions which keep them pure, isolating seed fields, then harvesting/drying/conditioning/storing so as to ensure high seed quality and purity, with known identity of every bag.

All generations of seed under Certification must be handled and conditioned only on plants approved for handling Certified seed.

Applying to Handle Certified Seed

To be approved to condition Certified seed, the conditioning plant must apply to the Certification Agency and request approval. The Agency then inspects the plant, all its facilities and storages, and may make recommendations for improvement. After the plant meets all requirements, the Agency approves the plant for handling Certified seed, and issues a permit or license.

Fees

The Certification Agency normally charges specified fees for each activity (inspections, tags, sampling, testing, etc.) which it conducts. The fees are usually nominal, and provide income to maintain the Certification services.

Conditioning Facility Requirements

All Certification requirements and activities are designed to prevent problems and ensure maintaining seed identity and traceability, as well as specified levels of genetic, physiological and physical seed quality. Requirements for facilities in a Certified seed conditioning plant vary, but generally include:

1. Handling systems which can be efficiently operated and managed so as to maintain seed purity and identity.
2. Drying facilities, as required by local conditions.
3. Conditioning equipment required to condition the specified crop seed up to required standards of purity and germination.
4. Buildings, bins, elevators, conveyors, equipment installations, etc., which are built and designed so as to permit complete clean-out between lots and to maintain seed purity.
5. Clean-out equipment such as industrial vacuum cleaner, compressed air, hand tools, etc.
6. Well-trained personnel.
7. Storages which can maintain seed quality, condition, purity and identity, and permit easy access for samplers and inspectors.

Clean-Out of Equipment

Before handling a lot of Certified seed, all equipment must be thoroughly cleaned-out to remove all seed remaining from the previous operation. This applies to all buildings and equipment used for planting, harvesting, drying, transporting, handling, conveying, conditioning, and storage.

Clean-out is required before handling a new lot of Certified seed, especially of a different variety or crop. Usually, with advance approval of the Certification Agency, different lots of Certified seed of the same variety and crop, can be conditioned without re-cleaning the plant between the different lots. If different Certified generations of seed are conditioned, the higher class should be conditioned first; *e.g.,* first condition Foundation seed, then Registered seed, and then Certified seed. If a different crop or variety, or non-Certified seed, was previously conditioned, the plant must be completely cleaned out before handling seed under any class of Certification.

In most cases, the Certification Agency must inspect and approve the clean-out before the new kind of Certified seed is handled.

Inspections

A trained inspector from the Certification Agency must inspect and approve all operations which have the potential for affecting seed quality. Inspections are made at the stages where the quality of the seed can best be protected.

If the facilities and their cleanliness do not meet all requirements in the inspection, the seed or the conditioning plant may be rejected. In the case of inspecting clean-out of the conditioning plant and some other instances (such as re-conditioning a seed lot to bring it up to standards), the conditioner may carry out further work and request re-inspection.

If a field is harvested, or a seed lot conditioned, before the required inspection is conducted, the field or seed is automatically rejected. If the required inspection cannot be conducted at the proper time, for any reason, the field or seed is rejected.

Plant Representative

At its discretion, the Certification Agency may appoint a trained, responsible staff member of the conditioning plant as the Authorized Plant Representative for this plant. This is normally done only by plants which have shown careful performance and high seed quality for several years. The plant's bond also covers any liability due to negligence of the Plant Representative.

Although the Plant Representative is an employee of the conditioning plant, he/she may act for the Certification Agency in activities delegated to him/her, such as affixing tags and seals to bags, drawing seed samples, or inspecting the plant for clean-out before conditioning different lots.

The Plant Representative must be a well-trained, experienced seed specialist whose integrity and dedication to quality is without question.

Records

A conditioning plant must always maintain complete, up-to-date, detailed records of all seed and all operations on it. Certification requirements for records usually involve normal management records, plus a special tabular record/report form to send to the Certification Agency with the following information on each lot of seed:

1. Name and address of the seed grower.
2. Previous crop variety conditioned on the plant's facilities.
3. The Certified variety conditioned.
4. Total weight of raw seed received.
5. Date raw seed received.
6. Date(s) conditioned.
7. Lot number of the conditioned seed.
8. Number and weight of bags of conditioned seed in the lot.
9. Number of Certification tags received for the lot.
10. Number of Certification tags used on the lot.
11. Number (if any) of "advance-issued" tags received for the lot.
12. Date plant inspected before conditioning the lot.
13. Other specified information.
14. Name and signature of the plant's authorized representative.

The conditioner normally keeps a copy of the record, and sends the original to the Certification Agency within a specified time after the lot is conditioned and bagged.

The Certification Agency may, at any time, inspect any part of the plant's record system.

Forming Lots

Seed from different growers, fields, varieties, or crops must be maintained in different lots which are completely identified. Lot size should not exceed sizes specified in ISTA or other rules, and the Certification Agency may limit lot size.

A final lot is normally formed from the specified number of bags which are consecutively-filled as the seed is conditioned and then bagged.

Sampling

The conditioner must advise the Certification Agency when the seed are ready, and request that the seed be officially sampled for Certification testing

Certification tests only the final seed quality, to ensure that seed meet standards and to provide information for labeling the seed. Seed must be completely conditioned and bagged in the final labeled new bags, before the Certification agency is requested to draw samples. Under some conditions, this procedure may be varied.

Samples for official testing for labeling Certified seed are drawn only by the authorized representative of the Certification Agency. This is normally an inspector from the Certification Agency, but at the Agency's discretion, it may appoint a competent officer of the conditioning plant to act for the Certification Agency in drawing samples and/or affixing Certification tags (especially pre-issued tags) and seals to seed bags.

Samples are drawn in accordance with officially-established procedures of the ISTA or other similar organization (*e.g.*, AOSA). Size of the sample shall be as specified by ISTA or the Certification Agency.

The seed lot must be stacked and stored in such a manner that the lot is readily accessible for sampling in the approved manner, and the sampling inspector can accurately count the number of bags in the lot. If the lot is not so stacked or stored, the inspector may refuse to sample it and either require it to be re-stacked or reject it for Certification.

Testing Seed Samples

Official samples drawn to determine the quality of seed lots submitted for Certification are tested and analyzed in the seed testing laboratories approved by the Certification Agency.

Seed Identity

The identity of every bag of seed must be maintained at all times. This must include, or provide unmistakable reference to, the crop, variety, lot number and grower. All bags/containers shall be labeled with tag and/or stencil to show the lot identity, and stacked/stored in a manner which prevents any possible mis-identity. Lot number or identity can be changed only with permission of the Certification Agency.

Bags and Bagging

As a safeguard against possible contamination, Certified seed can be bagged only in new, unused bags. In certain cases, the Certification Agency may inspect and accept used bags, but this procedure is not common.

Tags/Labels

Each bag or container of Certified seed must be labeled with the information required by the Seed Law. In addition, it shall carry a special tag showing its

Certification status. Both these requirements may be combined into one tag, or may be on separate tags.

Certification tags must be ordered from the Certification Agency, which is the sole supplier of such tags. Normally, the exact number of tags as required by the number of bags, is provided. This information is supplied by the conditioning plant, and is also shown on the official seed testing report.

Tags and seals are normally provided only after the seed lot is conditioned, bagged, sampled, tested, and the test results as shown on an official test report indicate that the seed lot meets all requirements of Certification.

Seals

To prevent adulteration of seed in the bag, Certified seed bags carry a special Certification seal which prevents opening and then re-closing the bag. Normally, the seal is broken if the bag is opened, and cannot be re-attached to the bag.

Advance Ordering of Tags and Seals

Certified conditioners which have established a reputation for careful and reliable quality in handling Certified seed may, by special arrangement with the Certification Agency, "pre-order" tags and seals, so the tags and seals can be attached to the bags when the seed are bagged. This eliminates a separate tagging operation, and is a significant saving in time and cost.

Tags pre-issued normally show seed quality as the minimum quality standards required by Certification, even if final test results indicate higher quality.

Lots carrying pre-issued tags and seals must be kept in the conditioner's storage until the official test results indicate that the lot meets or exceeds the labeled minimum quality standards. Then, the seed lot may be distributed and sold. If the lot does not meet Certification minimum standards in any factor, the Certification tags must be removed from the seed bags and returned to the Certification Agency.

Storage

Certified seed must be stored in pre-approved dry well-ventilated or controlled-condition, weather/pest-proof storages. If seed are stored in bulk, they must be separated from any other seed of different quality or variety, in a manner which prevents any possibility of mixture. Identity of each bag/ container of each lot as to crop, variety and lot number must be maintained at all times.

CERTIFIED SEED CONDITIONING PLANT BOND

Bond No. ____________

____________________________________ Seed Conditioning Plant
____(address)
____ (representative)

KNOW ALL MEN BY THESE PRESENTS: That we, ________(name of the conditioning plant), as Principal, and ____________________(name and address of the bonding company/bank), as SURETY, are held and firmly bound unto the ________________(name and address of the Certifying Agency), as Obligee, in the full and just some of _______________(amount of the bond) for the payment of which well and truly to be made, we jointly, and each of us individually, do hereby bind ourselves, our heirs, executors, administrators, successors and assigns, firmly, by these presents.

WHEREAS, the said Principal operates a seed conditioning plant and desires to qualify with said Obligee to perform in accord with said Obligee's established and published rules and regulations governing the conditioning of seed which have been, or are to be, certified by it.

The terms "Condition" and "Conditioning" referred to in connection with bonded seed conditioning plants is defined as any handling procedure used in connection with seed which may or may not improve the quality of such seed. If bonded conditioning plants bag, tag, seal and/or store Certified seed or seed which may be Certified, any one or more of such operations shall be performed entirely in accordance with the regulations of the ________________Seed Certifying Agency.

NOW, THEREFORE, THE CONDITION OF THE ABOVE OBLIGATION is such that if the said Principal shall, from the ____________day of ____________to the ____________day of ____________well and truly indemnify and save harmless the said Obligee from any and all liability it may incur by reason of the failure of said Principal to comply with and abide by the established and published rules and regulations of said Obligee as respect any one or more of the functions mentioned above, then this obligation to be void, otherwise to remain in full force and effect.

PROVIDED, HOWEVER, that no new liability shall accrue under this bond fifteen days after receipt by the Obligee or written notice from the Surety of its desire to cancel this bond.

PROVIDED, FURTHER, that this obligation may be continued for any subsequent year by continuation certificate, signed and sealed by the said Surety.

SIGNED, SEALED, AND DELIVERED this the _________day of ____________

______________________	______________________
Bonding Firm	Principal
______________________	______________________
Attorney in Fact	Conditioning Plant Owner

Figure 35.1 Form of the bond for Certified seed conditioning plants used by the Mississippi Crop (Seed) Improvement Association. Source: Bennie Keith, MC(S)IA.

Storages must be inspected and approved by the Certification Agency, and may be inspected at any time(s) which the Agency considers important to ensure seed quality.

Stacking Bags in Storage and Handling

Bagged Certified seed must be stacked in such a manner that (1) inspectors have access to the entire lot, so that a representative sample can be readily drawn; (2) number of bags can be accurately counted by the inspector; (3) each lot is separately stacked, so there is no chance for mis-identifying any bag and/or mixing seed from different lots; and (4) stacking protects seed quality, for example, places bags on pallets rather than directly on the floor or against walls.

Seed in the Conditioning Area

As a normal management procedure, no seed except the lot(s) of the crop variety being conditioned should be brought into the conditioning area. Any other lots, varieties, and kinds should be kept in the storage/handling/drying areas, properly labeled and handled so as to avoid any chance of mixture.

Conditioning Non-Certified Seed

Any non-Certified seed (as well as all Certified seed) brought into the plant's storage or operating areas must be properly bagged, tagged, labeled, and stored so as to avoid mixture. The Certifying Agency may inspect it to ensure this.

Non-Compliance With Requirements

If inspections show that any of the requirements of Certification are not met and cannot be rectified so that the seed can be certified, or if the plant manager declines to correct any correctible condition, the plant will lose its approval to handle Certified seed, and any involved seed lots may not be Certified.

Bulk Seed Carried Over

Certified seed which is bagged and labeled may be carried over and sold in a later year, if it still meets Certification quality standards. However, bulk seed carried over to later years is normally not eligible for Certification unless the conditioner applies to the Certification Agency for such carryover in the year the seed is produced, before the normal season when the seed would be planted.

Blending Certified Seed

Normally, the Certification Agency will permit, upon prior application, the blending of Certified seed lots of the same variety and Certification class. Blending may be done to ensure uniform acceptable seed quality, or to form large lots for conditioning. If different classes are blended, *e.g.,* Registered and Certified, the blend carries the lower Certification class. The conditioning plant must have adequate blending facilities and procedures, and these must be observed in operation by the Certification Agency, to ensure that the resulting lot is uniform in quality.

Performance Bond

A legal bond, cash or security, in a specified amount is often required of conditioning plants which handle Certified seed for several growers, or handle both Certified and non-Certified seed. The bond is made payable to the Certification Agency, and may be forfeited if the plant or its staff violate any Certification procedure and this may cause financial loss to farmers, due to negligence by the plant in handling the seed.

The Mississippi Crop (Seed) Improvement (Certification) Association uses the following format for a performance bond for a Certified Seed Conditioning Plant.

Chapter 36

Training Operators and Staff

Technical Competence Essential

All the staff and equipment operators in a conditioning plant have essential roles in ensuring cost/time-efficient operations and high seed quality. They must know and understand their equipment and operations, and how to handle them efficiently and effectively.

In-depth training, and constant renewal and upgrading, are essential to ensure high personnel competence and performance. In seed conditioning, this requires not only personal intelligence and dedication, but also in-depth training in the use, operation and maintenance of a range of equipment and activities.

The seed enterprise must take the initiative and provide or secure training, and to encourage staff to take the training.

Rewarding Employees for Training and Improving Competence

When employees improve their competence through training, self-study or other means, they should be rewarded with either promotions or salary increases. For example, one seed program with several conditioning plants holds annual 1–3-day training courses in the off-season, with in-depth training in specific subjects (fumigation, clean-out, gravity separator, etc.). All concerned personnel are expected to attend; attendance and demonstration of their added value to the program are required for promotion or salary increases.

Training Formats

Although constant improvement and upgrading of work skills is essential, it is difficult to release personnel for long-term training and still maintain an efficient work schedule. To accommodate this, training format is usually short (1–5 days, in off seasons) training courses and/or self-study provisions and on-the-job training.

Publications and Manuals

Few "practical operations" publications are available on general seed conditioning and storage. Those that are available should be obtained, and made available to staff. A "lending library" can be established within the enterprise, to make publications available to staff. Close monitoring (possibly by some of the administrative staff) of the library is essential to maintain its stock of publications.

This book seeks to fill the gap of available management and operational technology information. Every person in the seed enterprise with management or supervisory responsibility, or responsible for operation and maintenance of equipment, should have a personal copy.

The enterprise should obtain copies, or photocopy, of operating manuals for all equipment. These should be provided individually to concerned staff for their personal use. They should also be available in the plant near the machines, in the break room, and in the "lending library".

Guidelines can be prepared by the IQC and/or the conditioning supervisors, for specific operations. These can be printed and given to staff and posted in the plant during the concerned operations.

Training Programs

Some agencies have the capability to arrange short training programs, either in-country or in other countries. These courses may provide free tuition from the sponsor, or require the seed enterprise to pay the costs of its participants. The seed enterprise should contact and work with these agencies, and send key staff to them.

Training courses of short duration, from one day to several days, can also be arranged in-country, or within the seed enterprise and especially for seed enterprise personnel and their training needs. Instructors in these courses can be seed enterprise personnel, specialists provided by the equipment manufacturers, and/or specialists in government agencies, development assistance programs, international specialists, etc.

On-The-Job Training

Technical and supervisory improvement for enterprise staff can also be provided in the form of on-the-job training. This can be in the form of instruction provided while staff is actually working on seed lots. Trainers can be seed enterprise staff or "outside" specialists obtained for this special guidance.

Regular Upgrading

The focus of training before each season should be to "brush up" and sharpen the knowledge of the concerned staff, so they are familiarized with operating details before the conditioning season begins. And, new and additional knowledge should be imparted each year.

Evaluating Personnel Competence

A regular annual review of each employee's performance and professional growth should be conducted. This should be done by fully competent persons, in a manner which is entirely helpful to the rated employee. The results of such ratings must be kept completely confidential, but must be discussed privately with each employee. The employee must be given a chance to respond to any point in the rating, and to request changes, special training, etc. At the time of the confidential rating review, the employee should also be asked for his viewpoints and suggestions for improving the enterprise and its operations.

Important: it is critically essential that ratings and evaluations be completely friendly and supportive, and entirely without any semblance of bias or personal animosity whatsoever. The easiest way to destroy a "good working team" is to allow feelings that personal biases enter into the management picture.

Training Information Content

Training should include in-depth information presented so that the trainee can understand it and see how it can be applied to his work and responsibilities. Information covered should include all aspects of the trainee's responsibilities. Additionally, trainees should receive information on closely-related activities; this "cross-training" has proven to be quite helpful in coordinating work activities, improving staff understanding, and improving overall operating efficiency. And, when staff are trained in other jobs, it is easier to find competent temporary or permanent replacements in case of sickness or other absences.

A third kind of training should also be included: management principles and organization and management of the seed enterprise. Developing and broadening staff competence increases operating efficiency and reduces losses. It also helps generate a pool of competent persons which can be drawn upon to fill staff openings. This promotes the concept of "promotions from within the organization", which helps retain competent staff and develop employee loyalty.

Chapter 37

Procuring Equipment for a New Conditioning Plant

After a market survey shows there is adequate and accessible market potential, and the business plan and financial analysis show that the business can be done at a profit, the conditioning facility must be planned, and equipment procured and installed.

It is rarely possible, or most economically feasible, to purchase an existing conditioning plant, especially if it is old. It is generally more efficient and more economic to build a new plant. Since plants operate for many years, the latest available technology should be incorporated.

The objective is to get the plant that you need, a plant which can fit your needs for a long time into the future. A well-planned new plant should fill the operational needs at minimal investment, minimum operating time and cost, and produce the best-quality seed with the least loss of good seed. It should be easy to monitor, operate, adjust and manage, with efficient seed movement into, through, and out of the plant.

Information Required

To initiate the process of deciding what seed conditioning is needed, careful and accurate consideration must be given to the details of what really must/should/can be done. The chapter on "Planning a New Plant" in "Technology") gives detailed guidelines for the planning process, and getting the necessary information. This includes:

1. Crop seed to be handled.
2. Number of varieties, and number of times the plant must be cleaned out between varieties.

3. Size of seed lots, and need/no need to clean out the plant between lots of the same crop variety (important if Certified seed is conditioned).
4. Impurities (weed seed, crop seed, trash, etc.) which must be removed, and amounts.
5. Need/no need for drying.
6. Total amount of seed to be conditioned of each crop variety.
7. Number of days between harvest and moving the conditioned seed to market.
8. Availability, and reliability, of electric power. Is power off often? For how long?
9. How seed are received (bulk, bags, toteboxes).
10. How seed are handled/stored when they are received.
11. How long raw (uncleaned) seed must be stored before conditioning.
12. Storage and field insect problems.
13. Labor availability, quality, reliability, and cost.
14. Trained personnel available, training needed, and sources of training.
15. Quality testing and control before, during and after conditioning.

Based on all the information above, the most efficient and effective seed plant must be designed. This includes the separations which must be made, the sequence of separations, and operating capacity required.

A complete plan and layout must be prepared, to ensure that nothing is overlooked and the layout planned is complete and most efficient.

Technical Assistance

Planning an efficient, cost-effective conditioning plant is a complex task, requiring in-depth knowledge of conditioning equipment, separations, seed and seed quality. High-quality technical assistance is essential. Engineering firms, often supplying equipment, technical assistance and servicing, are available.

Many equipment supply/engineering firms have significant experience and knowledge, and can provide good guidance. However, one must be certain that the chosen supplier handles a complete range of equipment and will provide the most effective and integrated system. The supplier should not limit himself to "what he sells" instead of to "what the buyer needs".

A seed conditioning plant should never be planned or erected without getting the best possible technical assistance. Regardless of cost, good technical assistance is the best insurance for getting an efficient operation that will work cost-effectively for many years.

Procurement Documentation

Several different manufacturers provide different models of the same kinds of machines. The initial planning should be quite complete in terms of separations, machines and capacity. However, the initial plant design prepared for procurement should be sufficiently general to allow different suppliers to offer a bid. This should help ensure getting the best possible price quotation.

The offer document, and the responding bid document, should be sufficiently detailed to ensure that the bidder offers the desired equipment. There have been cases when different equipment was supplied, and did not fit the requirements.

Selecting and Evaluating Bidders

If potential suppliers are known, the request for bids can be sent directly to them. Bidders should be evaluated, and the winning offer selected, on the basis of overall quality. Consideration should include:

1. Fit to the desired conditions and specifications.
2. Experience of the bidder, and knowledge of prior performances.
3. Completeness of the offer.
4. "Fit" of the offered equipment, in matched capacity, separations, etc.
5. Price, although price is a minor consideration as compared to quality of the offer.
6. Timeliness and completeness of the supplier's service in supplying spare parts, technical assistance, etc.
7. Conditions and timing of delivery.
8. Installation and service.

General Purchase Requirements

When equipment is purchased, the documents for requesting bids or offers must be legally valid and binding, and must specify the conditions desired and under which the purchase will be made. Generally, these documents are based on the situation, and should be prepared with the input of an experience legal or procurement specialist.

Table 37.1 gives an example of general requirements and procedures in the document for requesting bids for supplying a new seed conditioning plant, under which the supply and purchase will be made.

Table 37.1 General specifications and requirements for purchasing equipment for a seed conditioning plant.
(Source: Gregg unpublished data)

Equipment General Specifications and Requirements

1. *Application*: These apply to all equipment, goods, services and materials offered and/or supplied.
2. *Offer*: The bidder will offer all required goods, equipment and materials, at a price CIF the port of Port Sudan (on the Red Sea) and valid for ninety (90) days after receipt by AMAD. Shipping and insurance costs may be listed separately in the price breakdown. The buyer will clear the goods from port customs and transport the goods to the buyer's plant site, and notify the bidder when this is done.
3. *Payment*: For equipment and materials, the supplier will be paid 100% of the total price upon receipt of the inspection certificate issued by a licensed and bonded independent International Shipment Inspector showing that the goods are complete, in good condition, and EX works in the producing country, when the international shipping insurance now covers the goods. For services such as installation, 50% of the cost will be paid upon beginning the work, and 50% upon completion of the work.
4. *Installation Information and Guidance*: The bidder will provide complete installation stands, bins, pipes, and detailed installation plans for the entire plant, for all required installation stands, supports, pipes, bins, etc., required for mounting, support, operation and flow of seed and waste spouting and flow required for operation of each and all machines.
5. *Installation*: The bidder will provide at least one trained and experienced supervisor for installing the equipment at the buyer's site with the buyer's labor, at the time advised by the buyer when equipment is delivered and the site and installation facilities are ready for installation. Detailed installation instructions, plans, specifications and drawings will also be provided by the supplier at the time of purchase.
6. *Plant and Equipment Desired*: Attached hereto are plans and/or drawings, illustrating the separators and flow layout desired.
7. *Crop Seed Handled and Varieties Per Crop*:
8. *Operating Capacity Required*: ___ mt per hour output of cleaned seed of ___.
9. *Balancing/Matching Machine Capacities*: Each machine's capacity will match that of other machines (3–5 mt/hour) in the conditioning line; if necessary, two or more similar machines will be mounted side-by-side in a parallel-flow installation.
10. *Receiving*: Seed will be received into the conditioning line by a vibrating receiving pit installed beneath the floor level in a concrete waterproof pit in the floor. Seed can be fed into this receiving pit by dumping bulk seed from trucks driven over the grate-covered pit, from forklift toteboxes, from bags, or from conveyors or pipes bringing seed from drying or storage bins.
11. *Bins*: Each machine will have an overhead steel feed bin of NLT 3 mt useful capacity (not counting waste space in bin top due to angle of repose of wheat seed), EXCEPT

contd. . . .

Table 37.1 contd. . . .

	the air-screen cleaner bin with NLT 5 mt and bagging bin which has a 10 mt capacity. Each bin will have an adequate support frame from the floor to handle the bin and seed load weight; a separate shutoff at the bin discharge; bin discharge will be shaped to fit unusual shape of machine intake hopper (such as the air-screen cleaner); bin will be smooth-welded to eliminate interior cracks which may hold seed; safe personnel ladder from floor to top of bin with ladder side rails extending NLT 1 m above bin top level; bin cover with inspection hatch adequate for cleaning bin inside, and inside personnel ladder of smooth-welded round steel rods; screen-covered air exhaust in top; feed inlet for feed pipe in exact center of bin; safely-mounted rubber-padded seed stepdown ladder mounted inside bin center so that no seed freefall more than 50–60 cm; each bin will have load switches to indicate near empty or full.
12.	*Drives, Controls and Switches*: Each machine will have complete and adequate drives, controls and switches; main switch will be located on the central control panel, with an over-ride safety off-on switch at the machine.
13.	*Central Control Panel*: The plant will have a central control panel, with on/off switch for each machine (but this switch is over-ridden by the machine safety switch); control panel will have lights to indicate operation of each machine which is running, and in pipes to show where seed flows. Status of each bin filling is to be indicated by lights and alarm.
14.	*Elevators*: Elevators will be slow-speed "seed" belt-and-bucket elevators (as opposed to high-speed grain elevators); with attached component feed hopper with shutoff gate; elevators mounted on raised angle-iron stands to facilitate opening bottom for clean-out; adequate inspection doors, easily removable, for inspection and clean-out of boot, legs and head; spacers/washers installed on bucket (cup) bolts to hold buckets out from belt for easier clean-out. Elevator will be installed in a floor pit to allow safe clean feed from the central vibrating conveyor and/or from bags or totebox; pit will have adequate space to facilitate clean-out. Elevator height will be adequate to permit 45° slope of discharge spouts which feed into exact center (NOT edge) of bin, into the seed letdown ladder installed inside the bin.
15.	*Elevator Height Specification*: Bidder/supplier will specify for each elevator (1) overall height, plus 6 inches for mounting base frame to allow boot cleanout; (2) base height (plus 6 inches) to top of feed intake hopper (so as to specify depth of floor pit for mounting); and (3) total height to discharge height. This information is essential for building and floor design and construction.
16.	*Central Vibrating Conveyor*: This conveyor will receive clean seed (and separately, waste materials) from the air-screen cleaner (whose mounting base height, elevator, etc., must allow for sufficient clearance height) and convey them the length of the installation. Leak-proof adjustable (use or not use) side discharge gates must be provided for feeding (or not feeding) seed to length separator, gravity separator, and treater (see bypass for each separator below).
17.	*Bypass for Each Separator*: For each separator and/or machine between the air-screen cleaner and the bagging bin, the central vibrating conveyor shall have an adjustable (use or not use, as desired) bypass arrangement so that the machine can be used or

not used, according to the needs of each seed lot. The bypass arrangement will be mechanically simple, quick and easy to adjust, and will not allow seed to leak through the discharge when it is closed, or leak through the main conveyor channel when the side discharge is used.

18. *Waste Products Handling and Removal*: Waste products from each machine (except the treater) will be spouted into the single "waste product channel" of the main plant vibrating conveyor. All waste products will be conveyed and elevated into a single waste product bin of NLT 5 mt, from which the waste products can be emptied into toteboxes or bags.
19. *Dust Control and Removal*: Machines with included air and dust removal systems (*e.g.*, air-screen cleaner) must be supplied with complete dusty air ducting and collecting systems. Other machines, in their manufacture, should have access points where in-plant dust collection facilities can be connected. A central plant dust exhaust system (with adequate ducting, fans and collectors) is to be provided by the seller, to remove dust from elevator heads, bin tops, dusty machines, etc.
20. *Operator Safety*: In addition to the master over-ride off/on switch at each machine, all belts, chains, pulleys, and other drives will have adequate safety shields. Adequate warnings shall be attached where useful.
21. *Spare Parts*: Each machine and/or elevator/conveyor will be supplied with the manufacturer's recommended spare parts for 3 years normal operation.
22. *New Goods*: All equipment and components supplied will be new, used only for normal post-assembly testing to ensure quality.
23. *Detailed Specifications*: In his offer, the Bidder/Supplier must include complete detailed specifications of his offered goods, including brand name and model number.
24. *Electricity*: In the plant area, single-phase electricity is ____ volts, and three-phase electricity is ____ volts.

Detailed Equipment Requirements

After an intensive and detailed study of the needs of the specific conditioning operations to be conducted, a planned flow pattern, separation sequence, and plant design are prepared. Then, based on this, separators and seed handling equipment are selected, and a layout is planned. Then, the detailed specifications for each item are prepared, and these are compiled into the Procurement Specifications Document.

These specifications must be specific for the planned installation. However, they are usually technical in nature in such a way that different suppliers can meet the requirements and offer a bid.

Table 37.2 lists specific specifications for machines require in this particular modern layout and handling design; other layouts and designs are possible, as required for specific situations.

Figures 37.1 through 37.4 show additional information which should accompany the procurement or bid request documentation.

Table 37.2 Detailed procurement specifications for equipment for one modern design of a seed conditioning plant.
(Source: Gregg unpublished data)

Item No.	Number Required	Description
		Equipment Specifications NOTE: Bidder/Supplier shall provide complete detailed specifications of his offered equipment, along with detailed plant layout installation plan and equipment and plans for individual machine installations.
		Receiving
1	1	*Receiving Pit*: Capacity NLT 3 mt, vibratory type; installed beneath floor so trucks can drive over it for dumping; floor level grate cover of pipes to eliminate ledges which holdseed; shutoff gate control above floor; emptying seed into hopper of receiving elevator; constructed of steel, painted to resist corrosion; with all required motors, drives and controls; WITH: (1) vibrating conveyor with dimensions and specifications as required to safely and cleanly move seed from receiving pit into receiving elevator; (2)complete detailed installation plan and equipment/materials; (3) complete detailed erection plans for required support and spouting; (4) detailed construction plan for concrete pit in building floor to contain the receiving pit.
2	1	*Receiving Elevator*: Belt-and-bucket elevator, capacity NLT mt/hour to allow rapid filling of air-screen bin; with all required motors and controls; with receiving hopper; installed in pit to receive seed from vibrating receiving pit; height adequate to spout seed at 45° slope to center of air-screen bin and/or to conveyor carrying seed to bulk storage bins; with 3-way discharge spout control to allow sending seed to air-screen cleaner or to either of two (2) conveyors carrying seed to drying or storage locations; all required motors, drives and controls; with adequate easily-opened inspection doors for clean-out of head, leg and boot; elevator shall be mounted on a small angle-iron frame which permits opening boot for clean-out. Opening (with cover) in head for attaching dust suction pipe. Slow-speed seed elevator (as opposed to high-speed grain elevator); spacers installed on bolts mounting cups (buckets) to belt. WITH: (1) complete detailed installation plan and equipment/ materials; (2) complete detailed erection plans for required installation and support (including below elevator for clean-out of boot) and spouting.
		Basic Cleaning
3	1	*Air-Screen Cleaner*: Capacity NLT # mt/hour on wheat; with NLT 2 air-columns (initial "dustless" and final clean seed) and 3 OR 4 screens (scalp and bottom grading) in series/sequence flow to handle all seed.

Item No.	Number Required	Description
		Complete with all required motors, drives and controls; dust ducting and collectors as required; spouting for all waste products; WITH: (1) complete detailed installation plan and equipment/ materials; (2) complete detailed erection plans for required bin and bin support, machine support, and spouting required. SCREENS WITH PERFORATION SIZES TO BE INCLUDED FOR: 1. (crop): 2. (by crop): PLUS the next size smaller AND larger perforation for each of the supplied recommended screens. (above screen perforation sizes are in English measurements)
		Conveying, Separating And Cleaning
4	1	*Main Conveying Vibrating Conveyor*: Capacity NLT # mt/hour of cleaned seed, plus separate channel to receive/commingle/handle waste products; installed so as to receive clean seed (without splash loss) from air-screen cleaner and convey them the entire length of the cleaning line, with open-or-close outlets along side to allow discharging seed into any of the following machines (*i.e.*, into the elevator feeding such machine), with outlets not allowing leak of seed when they are closed, and when outlets are open, not allowing seed to pass on down the conveyor; designed so as to receive seed from any machine and carry it on to the desired later outlet feeding a selected machine; total side outlets 4, to feed length separator, color sorter, treater, and/or bagging bin. Conveyor must also be able to receive seed from any of these units (except bagging bin) and convey it to the next desired machine; seed feed into the conveyor must be designed so as to avoid splash loss of seed. With all required motors, drives and controls, WITH: (1) complete detailed installation plan and equipment/ materials; (2) complete detailed erection plans for required bin and bin support, machine support, and spouting required.
5	1	*Elevator Feeding Length Separator*: Belt-and-bucket elevator, capacity NLT # mt/hour; with all required motors, drives and controls; fed with attached receiving hopper with shutoff gate and hand lever control; height adequate to spout seed at 45° slope to center of its fed bin into installed seed letdown ladder; with adequate easily-opened inspection doors for clean-out of head, leg and boot. Opening (with cover) in head support (including below elevator for clean-out) and spouting.
6	1 (set)	*Length Separator*: Set of parallel-feed units as required to ensure clean wheat seed output of NLT # mt/hour, to match capacity of air-screen cleaner; incorporating indented cylinders to make 3-way separation of both long and short materials from wheat seed; with recommended cylinders installed; preferably of indent sizes which could also be used (with proper adjustments) to clean barley seed;

contd. . . .

Table 37.2 contd. . . .

Item No.	Number Required	Description
		all required discharge spouting; all required motors, drives and controls; vent(s) for installing dust exhaust suction pipe(s). WITH: (1) complete detailed installation plan and equipment/ materials; (2) complete detailed erection plans for required bin and bin support, machine support, and spouting required.
7	1	*Elevator Feeding Gravity Separator*: Belt-and-bucket elevator, capacity NLT # mt/hour; with all required motors, drives and controls; fed with attached receiving hopper with shutoff gate and hand lever control; height adequate to spout seed at 45° slope to center of its fed bin into installed seed letdown ladder; with adequate easily-opened inspection doors for clean-out of head, leg and boot. Opening (with cover) in head for attaching dust suction pipe. Slow-speed seed elevator (as opposed to high-speed grain elevator); spacers installed on bolts mounting cups (buckets) to belt. WITH: (1) complete detailed installation plan and equipment/ materials; (2) complete detailed erection plans for required support (including below elevator for clean-out) and spouting.
7A	(1)	*Elevator Returning Middlings into Gravity Feed Bin*: To salvage the good seed from the middlings fraction discharged from the gravity, a belt-and-bucket elevator, capacity NLT # mt/hour; with all required motors, drives and controls; fed with attached receiving hopper with shutoff gate and hand lever control; height adequate to spout seed at 45° slope to center of the gravity feed bin into installed seed letdown ladder; with adequate easily-opened inspection doors for clean-out of head, leg and boot. Opening (with cover) in head for attaching dust suction pipe. Slow-speed seed elevator (as opposed to high-speed grain elevator); spacers installed on bolts mounting cups (buckets) to belt. NOTE: This elevator is not required IF it is decided not to salvage good seed from the middlings fraction. However, this recirculation of the middlings fraction is recommended. WITH: (1) complete detailed installation plan and equipment/ materials; (2) complete detailed erection plans for required bin and bin support, machine support, and spouting required.
8		*Gravity Separator*: Rectangular deck, multiple fan; capacity # mt/ hour on beans; with wire mesh deck for medium-large seed (beans); with all required motors, drive and controls; with built-in adjustable feed hopper; discharge system designed to permit unlimited selection of separating different discharge fractions. WITH: (1) complete detailed installation plan and equipment/ materials; (2) complete detailed erection plans for required bin and bin support, machine support, and spouting required.
9	1	*Elevator Feeding Treator*: Belt-and-bucket elevator, capacity NLT # mt/hour; with all required motors, drives and controls; fed with attached receiving hopper with shutoff gate and hand lever control;

Item No.	Number Required	Description
		height adequate to spout seed at 45° slope to center of its fed bin into installed seed letdown ladder; with adequate easily-opened inspection doors for clean-out of head, leg and boot. Opening (with cover) in head for attaching dust suction pipe. Slow-speed seed elevator (as opposed to high-speed grain elevator); spacers installed on bolts mounting cups (buckets) to belt. WITH: (1) complete detailed installation plan and equipment/ materials; (2) complete detailed erection plans for required support (including below elevator for clean-out) and spouting.
10	1	*Treater*: Capacity NLT # mt/hour; suitable for applying liquid and/ or slurry treatment solution at precise rates; with adequate mixing/ blending provisions to ensure uniform coverage of all seed; with range of easily-changed chemical metering system/components; relatively simple treater, easily operated and/or adjusted; all required chemical supply tank, hoses, stirrer/mixer, and other operating components; all required motors, drives and controls; dust/fume control system with separate ducting and collector. WITH: (1) complete detailed installation plan and equipment/ materials; (2) complete detailed erection plans for required bin and bin support, machine support, and spouting required.
11	1	*Elevator Feeding Bagging Bin*: Belt-and-bucket elevator, capacity NLT # mt/hour; with all required motors, drives and controls; fed with attached receiving hopper with shutoff gate and hand lever control; height adequate to spout seed at 45° slope to center of its fed bin into installed seed letdown ladder; with adequate easily-opened inspection doors for clean-out of head, leg and boot. Opening (with cover) in head for attaching dust suction pipe. Slow-speed seed elevator (as opposed to high-speed grain elevator); spacers installed on bolts mounting cups (buckets) to belt. WITH: (1) complete detailed installation plan and equipment/ materials; (2) complete detailed erection plans for required support (including below elevator for clean-out) and spouting.
12	1	*Bagger-Weigher*: Semi-automatic; accurate, simple to operate and adjust; suitable for open-mouth bags of cloth, jute, or paper; net weight range from 10–40 kg or more; for mounting beneath the bagging bin; WITH: (1) complete detailed installation plan and equipment/ materials; (2) complete detailed erection plans for required bin and bin support, machine support, and spouting required.
13	1	*Bag Closing System*: Heavy-duty bag sewing machine, mounted on bagging bag conveyor; for installation beneath the bagger-weigher; suitable for bags of cloth, jute, or paper; complete system including sewing machine, conveyor, all required motors, drives and controls. WITH: (1) complete detailed installation plan and equipment/

contd. . . .

Table 37.2 contd. . . .

Item No.	Number Required	Description
		materials; (2) complete detailed erection plans for required bin and bin support, machine support, and spouting required.
14	1	*Elevator Feeding All Waste Products into Waste Product Bin AND Waste Product Bin*: Belt-and-bucket elevator, capacity NLT 3 mt/hour; with all required motors, drives and controls; fed with attached receiving hopper with shutoff gate and hand lever control, to receive combined waste products from waste product channel of central plant vibrating conveyor; height adequate to spout waste products at 45° slope to center of the waste product bin; with adequate easily-opened inspection doors for clean-out of head, leg and boot. Opening (with cover) in head for attaching dust suction pipe. Slow-speed seed elevator (as opposed to high-speed grain elevator); spacers installed on bolts mounting cups (buckets) to belt.Waste Product Bin, capacity NLT 5 mt, with all required mounting and access, and sliding shutoff gate at bottom center discharge. Bin design, access and components shall otherwise be the same as seed bins. WITH: (1) complete detailed erection plans and equipment/ materials for elevator, bin and bin support, and spouting required.
		Support
15	5	*Rubber-padded seed "letdown ladder"*: for installation in each bin. Length adequate to reach within 50–75 cm of the bin bottom. With hardware and parts for mounting in the center of the designated bin, designed and constructed of pipes to eliminate ledges which could lodge seed.
16	8	*Pipe Elbows*: Rubber-lined 45° pipe elbows, same diameter as seed spouting pipe; to connect pipe from elevators to either seed letdown ladder in bin or to conveyor from receiving elevator; flanged, with two flange clamps each.
17	1	*In-Plant Dust Control System*: complete system recommended by supplier for removing dust from equipment and general plant. To include all pipes, fans, supports, collectors, etc., required.
		Monitor And Controls
18	1	*Central Control Panel*: Panel with visual representation of plant's machines and flow operations; with on/off switches, but these cannot over-ride the safety switch at each machine.
19	6 (sets)	*In-Bin Level Indicators*: To signal to central control panel and light mounted outside the bin, when seed level is the bin is (1) nearly full or (2) nearly empty; one for each bin in the conditioning line. WITH: All required components, including wires and lights and connections to central control panel.

Item No.	Number Required	Description
		Bins Required
1	1	Bin feeding air-screen cleaner, capacity NLT 5 mt
	1	Bin feeding length separator, capacity NLT 3 mt
	1	Bin feeding gravity separator, capacity NLT 3 mt
	1	Bin feeding treater, capacity NLT 3 mt
	1	Bin feeding bagger-weigher, capacity NLT 10 mt
	1	Bin collecting all waste products, capacity NLT 5 mt

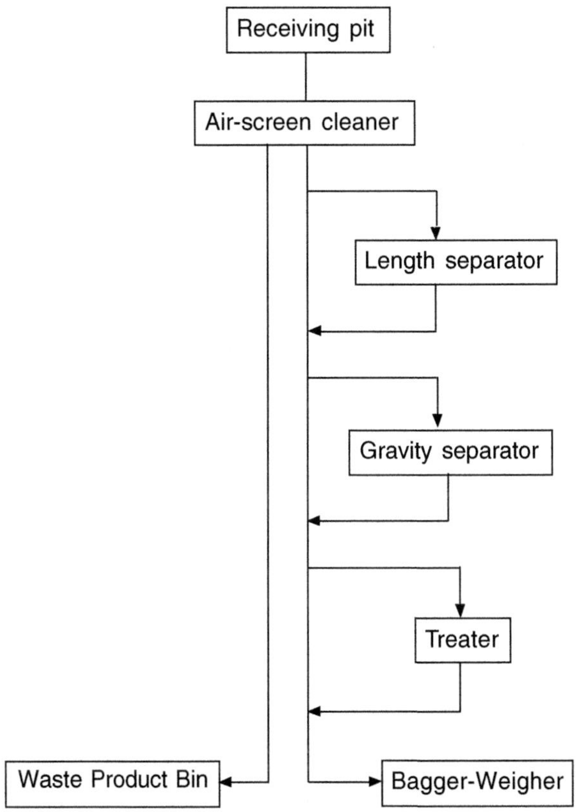

Figure 37.1 Diagram of desired layout, flow and machines for the example plant equipment listed in Table 36.2. This should be a part of the request for bids or procurement documents. The double line is the central vibrating conveyor which moves (with separations in the good seed channel) good seed from machine to machine, and moves all waste products (in a single separate waste product channel) to the elevator feeding the waste product bin (Source: Gregg unpublished data)

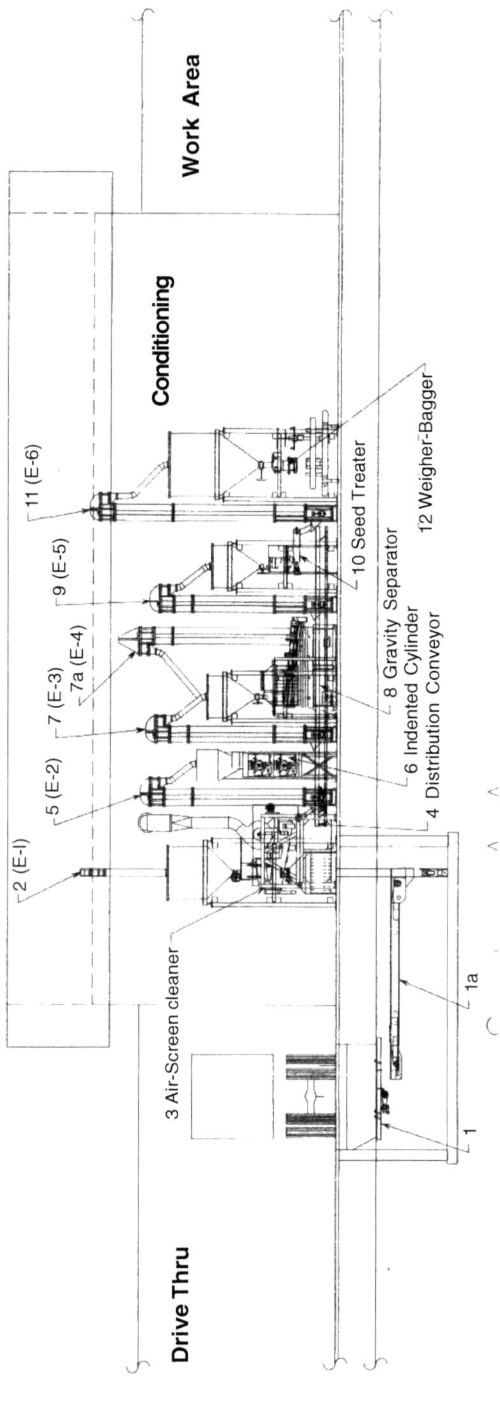

Figure 37.2 Elevation side view of proposed installation detailed in the example equipment specifications of Table 36.2. This should be a part of the request for bids or procurement documents (Source: Gregg and Billupps unpublished data).

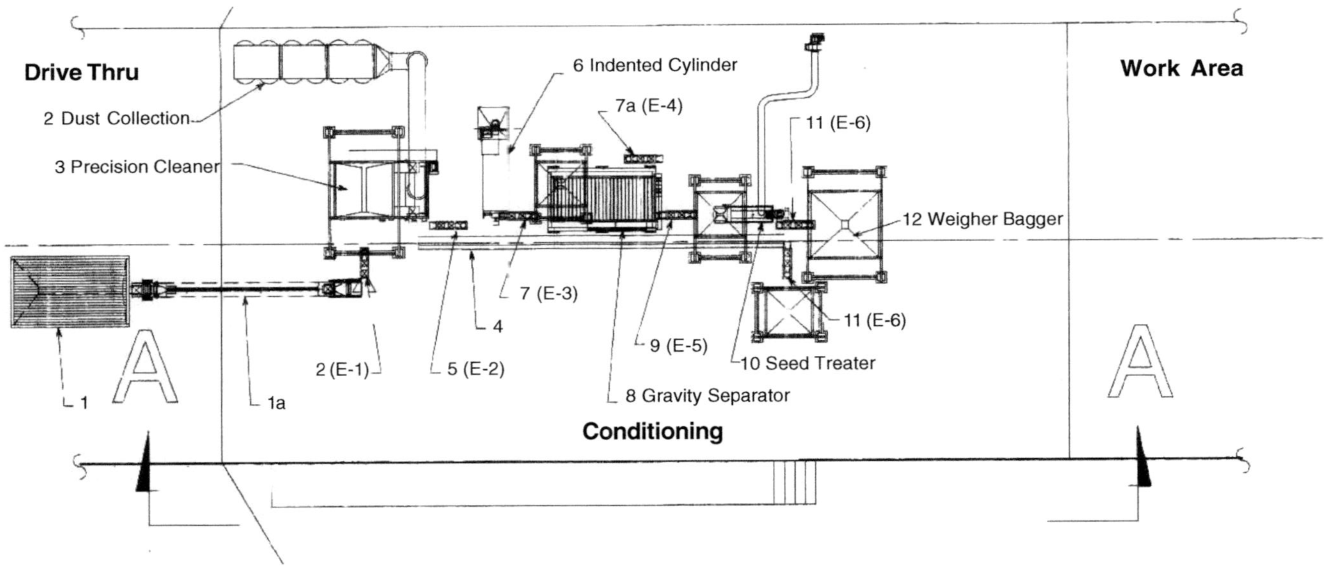

Figure 37.3 Floor plan layout of proposed installation described in the example specifications of Table 36.2. This should be a part of the request for bids or procurement documents (Source: Gregg and Billups unpublished data).

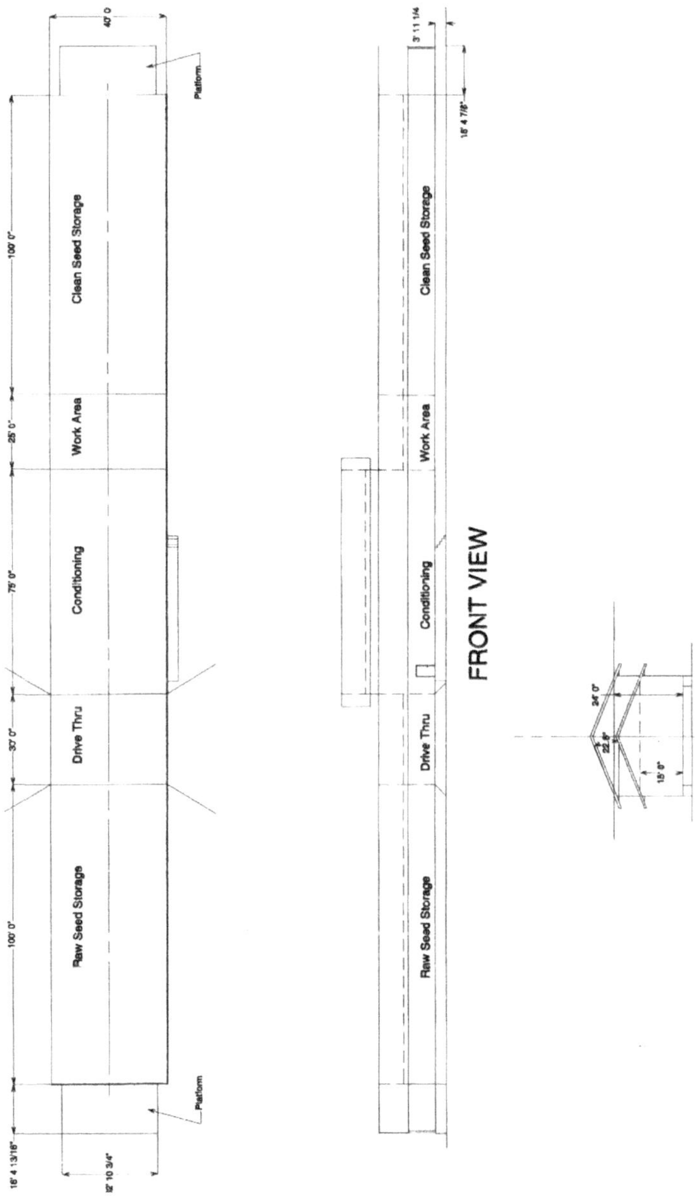

Figure 37.4 Building layout for equipment installation specified in the example shown in Table 36.2. This should be a part of the request for bids or procurement documents (Source: Gregg and Billups unpublished data).

GLOSSARY

Definitions herein are those commonly meant in "seed conditioning industry parlance". ISTA definitions are shown as in ISTA publications. This does not preclude other usages of the words or terms, in different situations.

MANAGEMENT

A

Accounting System: The established method or system used by a business enterprise to maintain its financial records and reports.

Audit: A regular detailed examination and evaluation of the accounting records of a business enterprise, conducted by a properly-qualified external agency.

B

Business Laws: Laws which guide or control the establishment, operation, management, reporting, etc., of a business establishment.

Business Operating License: A legal permit, issued at regular intervals by the appropriate government agency, to operate a specified kind of business in a specified area. A license, required by many governments, allows the enterprise to conduct certain kinds of business operations and facilities, attesting that the enterprise has fulfilled all government requirements therefore.

Business Plan: The overall plan of all aspects of the operations of an enterprise. It is a special plan prepared to show all planned operations, inputs and results of conducting business (such as a seed enterprise or program) for a specified period, usually one year. A detailed Business Plan is usually a standard requirement before a loan can be obtained.

C

Cash Flow: The actual flow of cash into the enterprise from sales, and flow of cash out of the enterprise as expenses are paid.

Cash Flow Plan: An annual (or even 2–3 years) plan showing the expected cash inflow and outflow in each month of the year. This is an essential financial management tool, to plan cash supply and needs.

Company: Legal structure or form of an organization which conducts business activities.

Company (or Corporation), Limited Liability: Legal structure of a business enterprise in which the liabilities of the company are limited to the company's assets, and do not include the personal assets of those who own stock or shares in the company.

Cooperative: A specialized form of limited liability corporation, in that the owners are usually the staff (employees cooperative) or customers (*e.g.*, farmers cooperative) of the business.

Costs: The expenses which must be paid in conducting the business and technical operations of a seed enterprise. The accounting system categorizes costs/expenses.

Costs, Capital Investment: The costs of purchasing land, buildings, machines, vehicles, etc., which are capital assets of the enterprise.

Costs, Operating: Costs of activities which are directly involved in handling seed, such as seed purchases, operating expenses, supplies, etc.

Costs, Overhead: The enterprise's costs for things which are required to support the operations; *e.g.*, salaries, interest paid, wages, utilities, etc.

Cross-Training: Training conditioning staff not only in their own duties, but also in the duties of related positions, so that when needed, a person can be efficiently used in other positions or work.

Custom: Said of the operations of an enterprise when only the services (*e.g.*, drying, cleaning, etc.) are provided to other persons/enterprises who actually own the seed.

D

Distributor (or Distribution): Wholesale seed movement or marketing; the agency (or operations or facilities) of moving seed from the conditioning plant to supply retail dealers who sell to farmers. Used to describe the operations of moving seed from the conditioning plant to the using farmers. This often involves facilities of vertically-integrated enterprises, as well as wholesale and retail sales units.

Distributor (in Seed Conditioning/Movement): A multi-spout outlet device which permits the conditioner/storer to select any of different outlets which direct seed into pipes which go to different places. For example, a distributor may receive seed from an elevator, then allow selecting the outlet pipe which goes to the air-screen cleaner, dryer, storage bin, or other destination.

Distributor (in Marketing): A wholesaler or seed enterprise facility which supplies seed stocks to local retailers in an area.

E

Enterprise: An organized (usually legally) private-sector operation; may be sole proprietorship, partnership, or limited liability company. In the seed industry, may produce, condition, store, or market seed, may do any or all of these activities.

Environment, Environmental Conditions, Environmental Quality, etc.: In terms of an enterprise, the conditions or "environment" in which an enterprise's employees work. These conditions are usually evaluated in terms of their effects, positive or negative, on the health and wellbeing of the employees.

F

Financial Analysis: The evaluation and analysis of the enterprise' financial income and outgo which forms the basis for the Financial Plan. It also can indicate whether or not the intended operations can be profitable, or will lose money.

Financial Plan: The planning process of estimating income and expenses, so as to manage the enterprise's financial aspects. Usually annual, the Plan shows and analyzes all cost items involved in the year's operations, how/ when the money will be spent, and where the necessary funding will be obtained.

Forklift: The vehicle used in a seed conditioning plant to move pallets of bagged seed, or toteboxes of loose or bulk seed.

G

H

I

Income: Money received by the enterprise. This is usually from sale of seed, sale of waste products, services provided, etc.

Informal Seed Sector: The unorganized sale/exchange of seed between farmers. Usually not controlled by the Seed Law, and quality of the seed is unknown, in terms of the measured seed quality factors. Refers to the supply of seed from farming and operations which are not specifically organized to produce high-quality seed. For example, "farmer-saved" seed are exchanged from farmer to farmer in this sector.

Inputs: Anything the farmer uses to produce a crop: seed, fertilizers, fuels, herbicides, pesticides, etc. In a seed enterprise, "other inputs" refers to things other than seed which are sold to and/or used by farmers.

Input Supply: Many enterprises which market seed also supply other inputs required by farmers. This provides more complete services to farmers, which increases the enterprise's income and spreads out cash flow over the year.

Interest: The direct cost/charge paid for obtaining a loan, usually a specified percentage of the amount of the loan.

Internal Quality Control: Within a seed enterprise or program, the facilities, highly-trained staff and work of evaluating every detail of every operation, to ensure high seed quality, minimum seed loss, and cost/time-efficient operations.

IQC: Internal Quality Control.

J

K

L

Law, Seed: Law regulating the sale and quality of seed offered for sale.

Liability: Legal responsibility for debts and/or obligations.

Limited Liability Company or Corporation: The form of ownership of a company legally registered as a company or corporation under the appropriate laws. The owners, whose ownership is usually denoted by a number of shares of stock, are not personally liable for debts or obligations accrued by the company. Liabilities are limited to the assets of the company or corporation itself.

Line Responsibilities: Said of those positions in a seed enterprise which are directly responsible for work on the seed. For example, manager or operator in conditioning, seed production, storage, sales, etc.

M

Management (Manager): The operations (responsible person) involved in ensuring that operations are efficient in time/labor/cost/seed cleaning, the desired seed quality is achieved, and schedules are met so that the seed can be sold when needed, at a profit.

Marketing: The operations involved in taking seed to farmers, having the right seed available at the right place/right price/right time, and convincing farmers to buy it.

Marketing Plan: The advance plan of what seed are expected to be sold, how/when/ where/price/etc. of the marketing and selling operations. The plan is detailed, showing what the seed enterprise plans to sell, where and how it will sell the seed, at what prices, and the income expected from the operations.

Market Survey: A detailed survey of potential customers, to learn what seed they will buy, the quality they seek, how much seed they will buy when, where they plan to buy and what prices they are willing to pay. Based on this realistic survey, the enterprise can plan its seed production and marketing.

Marketing: The operations involved in reaching and convincing farmers to buy the seed, and the operations involved in delivering the seed so it is available to farmers when, where, and in what packages they want.

Mechanic: The specialist in a seed conditioning plant who can weld, install machines, do major and minor repairs to machines, and keep the machines in good operating condition. May also handle electrical repairs and installations.

Monitor (or Monitoring): The process of constant supervision and checking to ensure that a machine operates properly at all times.

N

O

Operating Plans: The detailed plans covering planning and preparation for all aspects of seed production, conditioning, storage, transport and distribution. Operating plans are made so as to be certain of meeting the requirements of the Marketing Plan in the areas of Production, Conditioning, Transport and Distribution, to ensure having the required seed and getting them to farmer customers cost-effectively and efficiently.

Operator: The trained and experienced person who can operate the machines concerned, and adjust them accurately so they make a close and efficient separation and/or operation.

Organization: Of an enterprise or operation, the structure, sequence and management system.

P

Partnership: The legal ownership of a business by more than one individual, usually only a few persons. Some countries limit the number of partners in a business.

Permanent-Hire Staff: Persons who are employed continuously, year-round, by the seed enterprise. Highly recommended for efficient handling of the technical work of a seed enterprise.

Personnel: The sum of the employees of an enterprise.

Planning: The process(es) of advance identification, organization, and preparation for operations. Advance planning is a major need for business success, and a major tool of efficient management.

Prevent Problems: The primary objective of good management and IQC; preventing problems from arising saves the cost of correcting them.

Q

R

Raw Seed: Seed which has not been cleaned or conditioned.

Record System: The system and/or the information maintained on a seed lot and each bag or container of the lot. Specific systems and information are usually required by law.

Record: Recording of operations, seed quality, seed sales, etc., which document each seed lot and actions involving it. Records include specified types of information on the seed lot, and are maintained for specified periods of time.

Regulations (Under the Seed Law): In general, a Law expresses the policy, concept and general terms of a government requirement. Detailed regulations are then issued under authority of the law, to specify details of how the law will be applied and implemented. A law is usually difficult to modify, whereas a regulation can be modified sooner and easier as or when changing conditions require a change in the law's implementation.

Reports: In a seed operation, a report summarizing operations, seed available and/or sold, and operational status, and delivered to the responsible person(s). Reports are delivered at regular intervals, so that management is always current in knowing the operational status.

S

Seasonal Staff: Persons who are hired temporarily, to handle the work load during busy seasons. Usually laborers, and not recommended for work which requires skill in operating or adjusting machines or handling seed efficiently.

Seed Identity: The exact definition of the kind, quality, origin, etc., of a seed package and lot. This includes the information on the seed, and the methods (labels, stencils, records, signs, etc.) used to ensure seed identity.

Seed Law(s): A law or laws which control the quality, handling, labeling, records, sale and/or trade in seed for planting purposes.

Sole Proprietorship: Said of the legal form of ownership of a business establishment which is owned by a single individual, and is not legally incorporated.

Seedstocks: Used in different situations with either of two different meanings, to indicate the seed used to plant a crop. More commonly, it denotes the seed used to plant a crop which will produce planting seed.

Staff Responsibilities: Operational responsibilities which are not directly involved in the sequence or flow of seed, but support the work or responsibilities of others. For example, quality control, finance, personnel, maintenance, etc., are staff responsibilities.

Stocking: Having on hand, or supplying to dealers, the seed required for sales to farmers.

T

Tax: A fee assessed by government, on a business, individual, sales, etc.

Transport and Distribution Plan: The planning done to be ready to deliver and distribute seed when and where they are needed for sale to farmer-users.

U

V

Vertical Integration: Said of a seed enterprise which conducts all (or several) of the operations from producing seed to selling seed to farmer-users. The "vertical sequence of operations" are integrated into a single enterprise.

W

X

Y

Z

Zone or Zoning: A designation or limitation established by local government, on the type(s) of business, buildings, activities, etc., which may be established or conducted in a specified area.

TECHNOLOGY

A

Accelerated Aging (AA) Test: A test procedure to estimate the potential storability of a seed lot. It exposes the seed sample to high humidity and temperature (which is fatal to weak seed with reduced life span) for a specified period, and then germinates the seed. The seed which germinate are stronger and more likely to survive a period of storage.

Accepts: Good seed which have been "accepted" by a separator such as a color sorter.

Access: Buildings, machines, machine parts, etc., placed so that they are safely and easily reached for necessary operations. Access is essential in seed conditioning.

Acid Delinting: Using acid (usually sulfuric and/or hydrochloric) to burn the linters (short fibers) off cottonseed, to make them free-flowing to enable close separation and upgrading in conditioning.

Air Blast Aspiration: The separation system in a cleaner or separator which uses air flow to remove light-weight dust and undesirable materials.

Air Liftings: The lightweight particles and dust lifted and separated by an air blast or aspiration separation.

Airlift Elevator: A seed-moving and lifting elevator which moves seed by introducing them into a pipe containing a fast-flowing air stream which sweeps the seed along to the discharge outlet.

Air-Screen Cleaner: The basic seed cleaner which uses screens with perforations of specified sizes to remove particles larger and smaller than the good seed, and also uses airblasts to remove particles lighter than the good seed. This is usually the first "cleaner" used on all seed lots; sometimes, no further cleaning is required.

Ambient: Used to describe existing or natural conditions such as air temperature and humidity, which are not artificially controlled or changed by any process other that the characteristics of the building construction.

Angle of Repose: When seed is poured or fed onto a surface or into a bin or hopper, it will pile up in a mound like an inverted cone. The angle, from the horizontal, of the side of this pile is called the "angle of repose", the angle at which the seed mass will change from a stable pile to a flowing or sliding mass. In a bin, space above the angle of repose will not be filled with seed; in a seed flow pipe, seed will not flow if the pipe's angle from the horizontal is less that the angle of repose.

Antechamber: A small sealed room at the door of a conditioned seed storage. Workers open the outside door and enter the antechamber, then close the outside door. Then, they open the door to the conditioned storage and enter it. The antechamber thus prevents a direct opening to the outside atmosphere and prevents outside air of high humidity and temperature from entering the storage.

Aspiration: The process of pulling a flowing air stream through a thin layer of seed, and adjusting the air flow precisely to blow out and remove specific lighter seed and particles.

Aspirator: A seed separator which uses aspiration to remove undesirable light seed or particles from the seed lot.

Auger Conveyor: A bulk seed and grain conveyor which consists of a screw-like auger inside a close-fitting housing. As the screw turns, the "threads" on the auger move the seed mass forward.

B

Backleg: Seed falling back down the "leg" of a belt-bucket elevator; an undesirable seed-damaging condition caused by poor condition or operation of the elevator.

Bag Closer (or Bag Sewing Machine): A special sewing machine which stitches shut the open mouth of a seed bag after it has been filled with the proper amount of seed.

Bag/Package Sealer: A device used to seal the open mouth of a vaprproof bag or packet after it has been filled with seed. Usually, a heat-sealing is done of the entire mouth, to ensure that outside moist air cannot enter the bag/package.

Bagger-Weigher: A machine, ranging from simple hand-operated to complex automated, to measure a weighed amount of seed into a bag or other container.

Bagging (Packaging): Bagging (or packaging) the desired quantity of seed into the desired bag/package, and then closing it. The kind and size of bag/package is generally selected to safeguard and contain the amount of seed required (or an equal part of it) by most customers-users.

Beater (or Beater Arm): One of the rods extending from the revolving shaft or from the inside of the housing of a debearder. The beaters create the rubbing and force which removes beards, husks, and extraneous materials from the seed during debearding.

Belt-and-Bucket Elevator: An elevator which lifts seed to a higher level, by using a series of cups or "buckets" attached to a vertically-moving belt. As the belt moves, the buckets are filled with seed at the bottom, and the filled buckets are moved to the top of the elevator, where the seed are discharged into a discharge spout.

Bin: A structure which holds bulk seed (seed not in bags). Many sizes of bins of different shapes are used to hold uncleaned or raw seed, cleaned seed, and seed being fed to various separators in the conditioning process.

Bin, Bagging: The bin which receives seed after they are completely cleaned (and treated, if required). Seed are accumulated in this bin, and then bagged off for removal from conditioning and transfer to storage. Sometimes called hopper or hopper bin.

Bin, Drying: A bin fitted with a system which forces drying air (often heated) through the seed mass contained in the bin. This air flow absorbs moisture from the seed and moves the moisture out of the seed mass, thus drying the seed.

Bin, Feed (or Surge or Holding): A bin installed over a seed cleaner or separator. It receives seed from the previous machine or elevator, holds the seed mass, and feeds it uniformly into the separator it serves.

Bin, Storage: A bin used to store seed for a desired time period. During this period of storage, the seed are kept in good condition, but no operations (except sometimes drying and/or fumigation) are conducted on the seed during this storage period.

Blending: The process of (1) blending together different lots/components to produce one final lot of uniform quality throughout, or (2) blending within a non-uniform lot so that all its parts are uniform in quality.

Boot (of a Belt-and-Bucket Elevator): The bottom part of a belt-and-bucket elevator. Seed are fed into the boot, where the buckets on the moving belt pick up the seed and carry them to the top of the elevator.

Brine Flotation Separation: A method of separating heavy from light seed (used primarily on pea seed) by their sinking or floating in a brine solution which was adjusted in density until the light seed floated.

Brush Machine: A pre-conditioning machine which uses different kinds of brushes to brush off extraneous material (husks, linters, etc.) from seed, to make them more flowable and easier to condition.

Bucket Elevator (Belt-and-Bucket Elevator): An elevator which uses buckets or cups mounted on a moving belt to elevate seed to the top of bins or machines.

Bucket (Cup): One of the series of containers mounted on the vertically-moving belt of a belt-and-bucket elevator. In the bottom or boot section, the buckets scoop up a load of seed and are then moved to the top (head) of the elevator where the seed is dumped into a discharge spout.

Buckhorn Separator: A machine which did not actually separate, but prepared seed of buckhorn plantain so they could then be separated from clover seed by air-screen cleaner or gravity separator.

Buffel Grass Seed Cleaner: One of several machines designed to clean seed of buffel grass, which are not free-flowing.

Bulk: Refers to seed which are still in a "loose, flowing" condition, and not in the final bag or package.

Bypassing (a Separator): Skipping or not using a separator in the conditioning "line", by (1) having a two (or more)-way valve or distributor on the elevator discharge, or (2) using a vibrating conveyor to transport seed through the entire "line's" length, so that seed can be fed into any machine, or can be moved to the next machine without using the by-passed machine.

C

Capacity (of a Seed Cleaner or Separator): Expressed either as (1) amount of raw uncleaned seed fed into the machine per hour, or (2) amount of clean seed discharged from the machine per hour. Because of variation in cleaning requirements, capacity usually expresses the amount of raw seed fed to the machine.

Carryover: Holding seed beyond the planting season following the seed harvest, for planting in a later season or year.

Certification: An external quality control limited-generation system which ensures both high quality and high genetic (varietal) purity of seed.

Certified Seed: Seed produced under the supervision of, and according to the rules of, the Seed Certification System. Certified seed is commonly the end generation in the limited-generation system, and is intended for planting by general farmers. Previous generations are, in order, Breeder, Foundation and Registered.

Chain-Bucket Elevator: An elevator which uses a system of chains (as opposed to a belt) to move the buckets which carry seed.

Class (of Seed): One of the generations or "classes" of seed under Certification. In descending order, the commonly-used system includes the classes of Breeder seed, Basic or Foundation seed, Registered seed, and Certified seed.

Clean(ed) Seed: Seed which has been conditioned and cleaned up to the desired quality standard, and may be bagged and sold to farmers.

Clean-Up: The procedures by which leftover seed are cleaned out and removed from elevators, machines, bins, etc., in a seed conditioning plant. The entire plant is cleaned-up before handling a different kind of seed, in order to prevent contaminating the new seed lot with leftover seed from the previous lot.

Cleaner or Seed Cleaner: A machine used to remove undesirable materials from a seed lot. In the seed industry, this term has come to mean specifically the air-screen cleaner, which is the basic seed cleaning machine.

Cleaning: The process or processes of separating undesirable materials from a seed lot, to purify the seed lot up to specified quality standards. This term may be used for (1) cleaning seed only on the basic or air-screen cleaner, or (2) to all the machines and separators used to condition seed.

Clean or Cleaned Seed: Seed which has been completely conditioned or cleaned up to equal or exceed the desired quality specifications.

Clean-Out: The process of thoroughly removing all seed from a cleaner or separator before handling a new seed kind, so that absolutely no seed from the previous lot remain in the separator to adulterate the new lot.

Clean-up: The process of cleaning up the conditioning plant and its surrounding area, to remove trash, loose seed, dust, etc., so as to maximize cleanliness and sanitation, and to minimize fire hazard, cover and attraction for insects and pests, and unsightly signs of poor management.

Coating (Seed): Coating seed with a thin layer of material which may include several different substances. Coating applies more volume of material than common seed treating, but not as much as the processes of encrusting or pelletizing.

Color Sorter: An electronic machine which views each seed individually and separates them according to pre-set differences in seed color.

Components (of a Seed Lot): The different categories of materials which are contained in the seed lot. These may include pure crop seed, other crop variety seed, weed seed, and/or inert materials.

Condition Quality: Expresses the condition of seed as relates to factors such as moisture content, seed size and uniformity, mechanical damage, etc.

Conditioned Seed: Cleaned seed.

Conditioner (Seed): A person, program or enterprise which conducts seed conditioning.

Conditioning: The processes and operations involved in cleaning or conditioning seed to remove undesirable materials until the seed meet desired quality standards.

Contaminant: An undesirable particle or material in a seed lot which reduces its purity and quality.

Continuous-Bucket Elevator: A type of elevator in which a series of specially-designed buckets are connected together with swiveling joints, rather than using chains or a belt to move the buckets.

Conveyor: A machine used to move bulk seed horizontally (or on a slight incline) during the operations of seed conditioning, handling and storage.

Conveyor, Bag: A flat-belt conveyor used to convey bags to/from the top of a stack, or horizontally as from bag closing to stacking on pallets.

Conveyor, Bulk Seed: Any of several designs of conveyors which move seed horizontally or up a slight incline.

Conveyor, Ear Corn: A conveyor used to convey unshelled ear corn (maize). This conveyor is also often used as a picking conveyor, to allow removal of undesirable ears as the ears are conveyed to the drying bins. Also used to convey ears from the drying bins to the sheller.

Conveyor, Flat-Belt: A bulk seed conveyor which uses a flat belt as the conveying method. To keep the seed on the belt, the belt usually moves through a sheet-metal trough whose vertical sides keep seed on the belt.

Conveyor, Mass-Flow (Drag): A conveyor, often using a belt, which uses baffles or similar devices to force seed to move as the conveyor system moves along through its housing.

Conveyor, Pneumatic: An airlift type of conveyor which uses the pressure of a flowing air stream to move seed through the delivery pipes. A pneumatic conveyor may use positive or negative (vacuum) air pressure to move seed. Or, positive pressure may be used in part of the system, with negative pressure used in the second part of the system.

Conveyor, Squeeze-Belt: A conveyor which uses two belts. After the bulk seed is fed onto the lower belt, an upper belt presses against the load on the lower belt, and holds it in place until it is transported to the discharge point.

Conveyor, Trough-Belt: A bulk seed conveyor which uses a trough-shaped belt as the conveying method.

Conveyor, Vibrating: A trough-like conveyor which uses a vibrating, very slightly inclined, trough to cause seed to move down the length of the conveying trough.

Cooperative: The legal form of a business enterprise which is owned by the people who work in it or are served by it.

Counter-Balances: Weights attached to a shaft driving eccentric bearings. The eccentrics generate considerable thrust force which causes the machine to shake or vibrate; counter-balances offset this and eliminate the shake or vibration.

Cylinder Separator (Trieur): A length separator which separates long from short particles by the use of indentations (of specific size for each separation) in the wall or body of a revolving cylinder. Short particles lie in the indentations and are lifted out and discharged into an upper discharge conveyor; long particles are not lifted by the indents in the revolving cylinder, so they discharge from the end of the cylinder.

Cull (Culls): Another term used to denote waste or undesirable materials which are, or will be, separated out of the seed lot.

D

Date of Testing: The month and year in which the germination (and/or other quality factors) of a seed lot were tested. Seed laws generally require seed to be tested within a specified time period prior to their sale.

Debearding: Removing the beards or awns from seed such as grasses and barley. Usually accomplished by a controlled brushing.

Debearder: A machine which rubs seed to remove awns (beards) and other appendages, and to break seed clusters into individual seed. It does not separate different particles, but makes them more free-flowing so they are easier to separate with later machines.

Deck: The separating surface of the gravity separator, stoner, and some other separators. The seed mass flows across the deck, where the desired conditions are created to separate the seed mass into the desired fractions.

Decorticator: A machine used to remove excess corky material from seed, usually by abrasion. The decorticator was used on multi-germ sugarbeet seed.

Dehumidified (Air): Air which has had its moisture (relative humidity, RH) artificially removed so that it is dryer than ambient air. In a closed-circuit dehumidified dryer, the drying air is artificially dehumidified so that seed can be dried to low moisture contents.

Dehumidifier: A machine used in conditioned storage and low-moisture drying. It takes in air from the storage or seed, removes moisture humidity from the air, and then blows the dried air back into the storage or seed.

Delinting: Removing the short fibers or linters from cottonseed.

Delinting, Acid: Removing linters from cottonseed by burning them off with acid.

Delinting, Flame: Passing cottonseed through an open flame to burn off linters.

Delinting, Machine (Mechanical): Abrading or rubbing cottonseed with a mechanical system to remove linters.

Desiccant: A substance used to absorb moisture (RH) from air in a closed space. In seed drying to low moisture contents, silica gel is a common desiccant.

Design (a Conditioning Plant): To determine the conditioning requirements, and plan the design of the equipment installation, sequence and layout, and flow and handling of seed and waste products.

Deterioration (of Seed): Loss of seed quality, usually physiological quality of vigor and germination, of seed due to poor storage conditions or handling.

Discharge Spout: A spout on an elevator, cleaner, or separator through which seed or separated waste products discharge.

Disk Separator: A separator which uses revolving disks with pockets or recesses in their side walls to separate seed by differences in their length.

Distributor (in Seed Conditioning): A device installed in a seed flow pipe which allows the incoming seed to be sent to either of several discharge outlets. A properly-positioned distributor provides great flexibility in seed movement, so that different machines can either be used or bypassed.

Dockins Grader: A separator which used dimensional sizing to separate red rice seed from cultivated variety rice seed.

Dodder Mill: Another name for the roll mill, which separates smooth legume seed from rough-surfaced dodder and other weed seed.

Draper Belt Separator: A separator which separates round seed from flat or rough seed by their relative abilities to roll when fed onto a slanted, moving belt.

Dryer, Bag: A seed dryer which dries seed while they are in partially-filled air-pervious bags. Drying air is forced through the bags and the seed, to remove moisture from the seed.

Dryer, Belt: A continuous-flow dryer in which the seed are carried on a slow-moving conveyor-type belt. Drying air flows up through the porous belt and passes through the seed.

Dryer, Drum (or Cylinder): A continuous-flow type of seed dryer in which seed pass through a long cylinder slightly tilted from the horizontal. The cylinder turns, and baffles installed on its walls lift seed up near the top, and then drop the seed down through the drying air flowing through the cylinder.

Dryer, Portable: A dryer mounted on wheels (or is otherwise movable) so that it can be moved from place to place and used where needed.

Dryer, Tray: A dryer in which the seed are held in a thin layer in a tray with a screen mesh bottom. Drying air is forced up through the mesh and passes through the seed layer.

Drying: The facilities, operations and/or processes of removing moisture (water) from seed.

Drying Air: In seed drying, seed lose moisture to the surrounding air. The drying air must be of lower RH than the level in equilibrium with seed moisture content, so that moisture will move from the seed into the surrounding air. Such air is called the drying air.

Drying, Batch: Drying seed in a system which holds a given quantity of air in place until drying is completed.

Drying Bin: The bin (of different shape or form or capacity) which holds bulk seed while drying air is forced through the seed mass to remove moisture from the seed.

Drying, Dehumidified: Drying seed by forcing very dry, artificially-dehumidified air through the mass of drying seed. Dehumidified drying is used to dry seed to moisture contents below equilibrium with ambient air, for sealed vaporproof packaging.

Drying, Continuous-Flow: Drying seed as it slowly flows through the drying system. In some cases, drying is completed in one pass through the drying system; in other cases, seed are partially dried, then tempered in a holding bin, and then returned to the dryer. This process may be repeated several times to achieve the desired drying.

Drying, Sun (Natural Air): Drying seed by spreading thin layers in the open air under direct sunlight. This is effective only in dry, low-humidity weather.

Dust Collector: A device or system used to collect dust generated in the seed plant, and allow its safe disposal.

Dustless Air Blast System: Said of the airblast or aspiration separation which removes dust and light particles as the seed mass first falls from the feed hopper. This removes dust before it has a chance to enter the main parts of the machine, where it would be agitated and would contaminate the surrounding air.

E

Eccentric: A bearing (and its housing) which is "off-center" in its opening which fits onto the drive shaft, so that it can generate a back-and-forth motion to the connected component. It is used, for example, in the air-screen cleaner and gravity separator to impart a back-and-forth motion to the screen shoe or the gravity deck.

Electrode: The component of an electrostatic seed separator which created an electrical field or "sprayed" a charge on the seed, to enable separating seed of different electrical potential.

Electrostatic Seed Separator: A specialized seed separator which uses differences in seeds' abilities to conduct an electric charge.

Elevator: A machine which lifts seed vertically in order to place the seed into a bin or the hopper over a separator or cleaner.

Equilibrium (Seed Moisture Content): The seed moisture content which is in balance or equilibrium with the relative humidity (RH) of the surrounding air. At equilibrium moisture content, moisture will move into and out of the seed, but there will be no net gain or loss of seed moisture content.

Extraction (of Vegetable Seed): Removing vegetable seed from the fruits, which are "wet" or "dry". Seed are extracted from such fruits as tomato, squash, melon, pepper, etc.

F

False Vibration: An undesirable shaking or vibrating of a machine, created by its mechanical action. Unless it is offset by a heavy foundation and/or counter-balancing, false vibrations can destroy the separating action of a separator.

Feed Hopper (Hopper): The device or system by which seed is fed into a machine at a controlled rate.

Fingers, Separating: When seed discharge in a flowing curtain over a sloping surface, separating fingers are movable dividers which can separate the curtain of seed at desired points, to direct different seed to different discharge spouts. Separating fingers are used in a number of separators, notably the gravity separator.

Follow-up: An essential management practice which involves "follow-up" checking to see that orders were followed and operations done properly.

Forklift: A special truck-like vehicle equipped with forks which are used to lift, hold and carry a loaded pallet or totebox. It is used in storage and handling of seed in bags or toteboxes.

Formal Seed Sector: Refers to the "organized" seed industry which operates under quality control systems for the express purpose of producing and supplying high-quality seed.

Foundation: The base on which a machine and/or its support stand or platform is installed. The foundation must provide a level surface which is sufficiently strong to control, absorb or prevent vibration or movement of the machine.

Friction Separator: An experimental machine which separates seed according to differences in their roughness or smoothness, which varies their flow pattern as they travel across a sloped moving belt.

Fumigation: Filling the air space within a seed lot or building with a toxic gas which will kill insects and pests present in the area. Fumigation can be efficiently and safely done only in a sealed, enclosed space, as the fumigant gas must remain at toxic levels for a period of time.

G

Genetic Quality: The degree of varietal (genetic) purity of a seed lot.

Good Seed: In seed conditioning, the fraction discharging from a cleaner or separator which contains the good crop seed (as opposed to other discharge components which are generally waste products removed from the good seed).

Grader: A generic term used to cover any machine which separates the different particles in a seed lot by using differences in a specific physical property of the seed and particles.

Grader (Width and/or Thickness): Term used for a separator which "grades" seed into different classes, such as different sizes of width and/or thickness.

Grate (in a Peanut Sheller): The screen-like bottom cover of the shelling compartment. After the shelling bar breaks open the peanut shells, the shelled peanut seed are small enough to fall through the openings in the grate, and exit the shelling compartment.

Gravity, Specific: The weight per unit of volume of seed; sometimes called density.

Gravity Separator: A specialized machine used to separate heavy (dense or high specific gravity) seed from light (less dense or low specific gravity) seed and/or particles.

H

Hand Screen: A small piece of the perforated or woven wire material used for screens in the air-screen cleaner. It is mounted within a frame, and is used to determine the perforation size required to make a desired separation.

Hard Seed: Seed (commonly of legumes) whose seed coat is impermeable to moisture or oxygen. As long as oxygen and/or moisture cannot penetrate through the seed coat, the seed will not germinate, even though they may be alive and possess the ability to germinate when the seed coats lose their impermeability due to scarification or aging.

Head, Elevator: The top section of a belt-and-bucket elevator. Here, the belt and loaded buckets pass over the head pulley and start back downward. At this point, the buckets empty their loads of seed into the head's discharge section.

Headroom: Clear height inside a building, below the roof support structures.

Health Quality: The quality of a seed lot as determined by freedom from diseases and pathogens, and/or presence of seed treatments, relating to both seed and seedlings.

Hood: A full or partial cover installed over or around a machine, to contain any dust generated by the machine, so that the dusty air can be exhausted and removed.

Hopper: (1) Primarily used for the feed intake of a machine, but also used for (2) a bin installed over a seed cleaner or separator, used to hold a supply of seed to feed the separator at a continuous and uniform rate. It also accumulates seed coming from the previous machine, and is often used to permit continuous operation of machines which have slightly different operating capacities.

Hopper Bin: A hopper or bin normally located over a machine, to hold seed and supply a uniform feed into the machine.

Horizontal Blower Separator: An air separator which blew seed in a horizontal direction, and separated heavy from light seed by differences in the distance they were blown.

Horizontal Disk Separator: An experimental machine which used a horizontal revolving disk to separate seed according to their ability to roll.

Hull: A husk or cover on seed which is not essential to maintaining the quality of the seed. Hulls are usually removed to make the seed less bulky and to improve their flow in conditioning.

Hulling: Removing hulls from seed such as small legume seed.

Huller: A machine used specifically for removing hulls from seed. With small legume seed, the same machine often scarifies the seed.

Hygroscopic: Able to gain or lose moisture to/from the surrounding atmosphere; *i.e.*, dry seed will gain moisture from humid surrounding air; moist seed will lose moisture to surrounding dry air (the concept which enables seed drying).

Hygrothermograph: An instrument which measures and records atmospheric humidity and temperature over a time period. It is useful in determining seed storage conditions.

I

Indented Cylinder Separator (Cylinder Separator): A machine which uses a pipe-like cylinder with indentations of a special size and shape it its wall, to separate seed of different lengths.

Identity (of a Seed Lot, Bag or Other Unit): Name and/or code which identifies each unit of a seed lot, and allows tracing the seed to its origin, operations and quality. Certain identity factors are required by law.

Initial (or Start-up or Starting) Adjustments: The adjustments used when a cleaner or separator is to be first started. Then, after the separator is operating, adjustments are fine-tuned or changed to get the desired separation at the most efficient capacity.

Install (or Installation): Putting a machine into position, with requisite power and seed flow and handling equipment, and adequate support, space and facilities for proper and safe operation and maintenance.

IQC (Internal Quality Control): Internal Quality Control in a seed enterprise, which is a special program which oversees/inspects/guides all activities so as to ensure high seed quality and efficient operations within the enterprise. IQC not only identifies seed quality; it guides operations so as to prevent losses, problems and low quality.

J

K

L

Lab Cleaners (Lab Separators): Model-size, low-capacity seed cleaners or separators. These are used to (1) determine settings and components required on high-capacity machines in the conditioning plant, or (2) to separate and clean small amounts of seed in an IQC lab, research program, or variety breeding program.

Label: A tag or other label on a seed container. Required by law on each container of seed, the label must include certain required information about the quality and origin of the seed. As a safeguard, the seed container usually bears a complete label plus a lot number separately stamped on the container.

Layout: The arrangement and/or sequence of (1) conditioning machines, or (2) conditioning operations, or (3) facilities/buildings within the conditioning facility.

Leg (or Legging): The long part(s) of a belt-and-bucket elevator between the boot and the head sections. The length of the leg determines the height of the elevator, and thus the vertical distance that the elevator will lift seed.

Let-Down Ladder: A device installed in a bin or under a discharge spout to let seed fall gently to a lower level without receiving impact force strong enough to cause mechanical injury. The "ladder" is commonly a series of closely-spaced steps or baffles which are padded with shock-absorbing rubber.

Liftings (in a Cylinder or Disk Separato*r)*: The shorter seed and particles which are lifted out of the seed mass by the indents in the cylinder wall or the pockets cast in the sides of the disks.

Liftings Trough (in a Cylinder Separator): The trough (whose position can be adjusted) inside the cylinder separator which receives the short lifted seed/particles and discharges them out a separate spout.

Limited Liability Company: Legal structure of an enterprise or corporation, in which the liabilities of the company are limited to the assets of the company, and do not include the personal assets of the owners.

Line (or Cleaning Line): The "line" or sequence of cleaners and separators installed and used in a seed conditioning plant.

Linter: Short fiber (not the fiber of textile use) left on cottonseed after ginning.

Log (Register): A summary record which contains summary entries of many different seed lots, operating days, etc. This is an overall record, and is also used to report activities or seed information.

Longevity: As used herein, refers to the period of time over which seed will remain viable (*i.e.*, alive and able to germinate).

Lot: A specific, separately-identified and handled quantity of seed. The entire lot is of uniform quality. Same as "seed lot."

Lot Number: The specific number or designation which identifies one specific lot of seed, and helps maintain its identity apart from all other lots. The lot number is usually constructed of several numbers and/or letters, so that it gives specific information about the lot.

Lower Air Liftings: In an air-screen cleaner, the light seed and particles separated by air after the seed have been cleaned and are leaving the air-screen cleaner.

M

Maintenance, Daily: Minor repairs, adjustments, lubrication, etc., which is done daily or regularly during operations.

Maintenance, Seasonal: Major repairs or replacement of machines, changes in layout or buildings, etc., which is done in the off-season when conditioning operations are not being done.

Magnetic Separator: A specialized separator for small legume seed. It adds moisture to the seed lot; only some kinds of seed hold the moisture. Then, a finely-ground iron powder is blended in; only the moist seed hold the iron powder. The seed lot is then passed over a strong magnet, which separates the seed which held the iron powder.

Mechanical Damage (or Injury): Damage to seed caused by the impact received when a hard moving part strikes the seed, or seed fall and impact against a hard surface. Types of mechanical damage include breaking, chipping, invisible internal damage, etc.

Middlings (or Middling) Fraction: The intermediate or "middle" fraction on the gravity separator deck, discharging between the heavy good seed and the light waste product.

Minimum Quality Standards: Minimum standards of quality of seed which may be offered for sale; usually specified in the Seed Law. However, many seed enterprises establish their own minimum quality standards which are higher than the minimum legally-required standards.

Mobile Conditioning Plant: A conditioning facility which is permanently mounted on wheels (truck and/or trailer) so that it can be moved from place to place.

Moisture Content: A measure of the water (moisture) contained in seed, expressed as a percentage by weight.

Moisture, Seed: The water (moisture) content of seed. Usually expressed as a percentage of either the wet or dry weight of the seed.

Multi-Story (Layout or Design of a Seed Plant): A plant designed and constructed so that cleaners and separators are installed on several different floor levels, in a vertical flow arrangement.

N

Needle Cylinder Separator: A separator which passed the seed (primarily pea) through a cylinder whose inner wall had many short needles. If a seed had a weevil hole, it was snagged on a needle and lifted and dropped into the liftings discharge trough.

Non-Vaporproof: Said of bag and package materials which allow the passage of air and humidity from the outside to the inside.

O

P

Pallet: A frame or structure on which seed bags can be piled. The pallet can be lifted by the forks of a forklift, so the entire pallet and its load can be moved or stacked as a unit, by forklift.

Palletized: Said of seed bags which have been stacked on a pallet, and are handled on the pallet by forklift.

Pearson Square: A simple method for determining the amounts of different sub-lots to blend together to get a final larger lot of the desired quality.

Pelletizing: Encrusting or encapsulating a seed in a material(s) which changes the size and shape of the seed, so that all seed are of the same size. The encrusting material may be inert or may carry nutritive and/or protective materials.

Perforation: An opening in a sheet-metal screen. The opening is of specific and precise shape, width and length, to make a specific separation. Except in the case of special dual-use screens, all perforations on a screen are of the same dimensions.

Physical Characters: The physical properties of seed which can be used by machines to separate different seed/particles. These characters of individual seed include size, weight (specific gravity), length, surface texture, shape (roundness), color, affinity for liquids, and electrical conductivity.

Physical Quality: The physical or mechanical purity of a seed lot, measuring its freedom from undesirable seed and particles such as other crop seed, weed seed, and inert.

Physiological Quality: The quality of the life and life processes of seed; generally measured by such tests as germination, vigor, TZ, and AA tests.

Picking Belt or Table: A device which slowly passes seed (or ear corn) over a belt or other well-lit flat surface, while operators visually examine the seed and remove all undesirable seed by hand.

Pitch (Slope) (of Screens): The slope, or angle from the horizontal, of a screen from the feed end to the discharge end. Degree of slope is used to control the speed at which seed flow over the screen.

Platform: The raised stand on which a machine is installed.

Plugging (of a Screen): Said of a perforation in a separating screen when a seed or other particle is stuck or jammed in the perforation so that the perforation is essentially plugged and not functioning.

Pneumatic Separator: An air blast separator which uses a positive-pressure air stream (as opposed to a negative-pressure) to make a separation of seed. The pneumatic separator has a fan at the air intake site, to draw in and force a positive-pressure air stream through the separating chamber.

Polisher: A machine which (1) brushed or (2) rubbed seed with a polishing compound, used to remove dust and/or external discolorations from seed (or food beans) to brighten their appearance.

Polishing Compound: A powdered compound used in a polisher to gently rub and polish seed. The compound also absorbed and held dust.

Portable (Cleaners): A separator or cleaner which can be readily moved from one place to another and set up again, to clean seed at the new location.

Pre-cleaning: Initial operations conducted on a seed lot so that later conditioning can be more effective or efficient. Pre-cleaning includes scalping, debearding, delinting, shelling/hulling, etc., but in common parlance, the scalper is mainly intended.

Pre-cleaner: Any machine used to pre-clean seed, but most commonly refers to the scalper.

Pre-conditioning: Refers to operations conducted before the principal operation of final conditioning the seed.

Pre-treating: Treating the seed lot in some manner in advance, to make it more suited to a particular operation; for example, pre-treating clover seed by applying a small amount of water, so that the seed can then receive iron powder and be separated on the magnetic separator.

Priming (Seed): Before planting, soaking seed in water (or a special solution in water) for a short time so as to initiate the early "pre-germination: processes and then re-drying the seed. The seed are then soon planted, and the priming results in faster germination and emergence in the field.

Producer (Seed): The farmer, grower, etc., who grows the crop which produces seed.

Production: The field operations involved in producing seed crops. Usually includes harvesting, but some programs are organized in such a manner that harvest is managed and handled separately from other field operations.

Protectant: A chemical seed treatment applied to seed, for the purpose of destroying pathogens and/or protecting the seed and/or seedling from harmful organisms.

Purity: The purity or freedom from foreign material of a seed lot, expressed as percentage by weight which is the pure crop seed.

Purity Test: The quality-control test used to determine the purity of a seed lot by testing a representative sample. In the purity test, the analyst examines each seed/particle in a given sample weight and separates it into categories, usually including (for IQC purposes) pure crop seed, other crop seed, weed seed, and inert material.

Q

Quality Achievement: A general term for the operations involved in ensuring that high seed quality is achieved and maintained.

Quality Assurance: General term covering the activities involved in testing and determining that seed quality has been achieved.

Quality Control: The processes involved in measuring the quality of seed and ensuring that it meets desired or established quality standards. This includes sampling, testing and similar operations.

Quality Control, External (EQC): A quality control agency or program "outside" the seed enterprise, which essentially establishes and/or enforces specified levels of quality in seed offered for sale. These include Seed Law

Enforcement Agencies, Certification, and similar programs. EQC applies mostly to seed already cleaned, and does not focus on achieving quality.

Quality Control, Internal (IQC): A quality control program within the seed enterprise which evaluates operations, samples and tests seed, and conducts other operations to ensure that the enterprise's seed meet the desired quality standards, and that operations are efficient and cost-effective. IQC focuses on achieving quality and minimizing costs and losses.

R

Rat Shield (Ratproof Lip or Ledge): A smooth-finished concrete projection, about 1 m long and 1 m above ground level, around the outside of a seed building. Its purpose is to prevent or discourage rats and mice from getting into the building.

Raw Seed: Harvested seed which have not been cleaned or separated to improve their quality.

Receiving: Generally used to indicate the receiving of raw unconditioned seed into the operations of a seed conditioning plant; the facilities and/or operations of bringing in raw seed and introducing it into the conditioning operations.

Re-clean: To clean or separate a seed lot again, in a second or third pass over the same machines, usually with changed adjustments.

Recycle: Re-clean, or pass seed through the system or machine a second time.

Rejects: Undesirable seed which have been "rejected" by a separator.

Relative Humidity (RH): A measure of the amount of moisture (as invisible vapor) which air contains. RH is expressed as a percentage of the maximum possible amount of moisture the air could hold at its current temperature.

Report: A report prepared for management or other persons who need overall information. In conditioning, reports summarize seed quality, quantity, and condition, and are periodically provided to management, marketing, etc.

Resilience Separator: An experimental machine which separated seed of different resilience or ability to bounce when dropped onto a sloped hard surface.

Retail Dealer (or Sales): Person or agency which sells seed directly to farmers who plant the seed, or the operations of selling to farmer-users.

Roguing: Purifying seed production fields by walking slowly through the field, looking at all plants, and then pulling up and carrying out of the field all offtype and undesirable plants.

Roll Mill: A seed separator which separates by surface texture differences. It is normally used for small legume seed, to separate the smooth-surfaced crop seed from rough-surfaced weed and other seed.

Rotary Air Lock (Airlift Elevators): Airtight chambers mounted on a slowly-revolving shaft. Used in an airlift type of elevator, as seed fall from the airstream into the revolving rotary air lock, seed are removed from the air system without permitting air to flow in an upset the conveying air flow system.

S

Sample: A small quantity of seed, taken from different parts of a seed lot, so that the quality of the sample is identical to the quality of the larger lot. If the sample is truly representative of the lot, tests made on the sample are valid representations of the quality of the entire lot.

Sampling: The process of "drawing" or taking a small sample from a seed lot in such as manner as to ensure that the sample's quality is identical to the quality of the larger lot.

Saw Linter (or Delinter): A machine similar to a gin which is used to remove linters (short fibers) from cottonseed.

Scalping: The operation of making a rough separation of undesirable materials from a seed lot. Scalping does not bring the seed to the finally-desired purity; it simply removes much of the undesirable material so that later final conditioning is more efficient.

Scalper: The machine used to scalp or pre-clean seed. It usually uses a combination of separating screens and air blasts.

Scalpings: Large undesirable particles separated from the good seed in a scalping operation (as a scalping screen).

Scarifier: A machine which scarifies the coats of hard seed so that the seed can germinate more uniformly and completely.

Scarification (or Scarifying): The process of scratching, abrading, or otherwise weakening the seed coats of "hard seed", so that moisture and/or oxygen can penetrate into the seed and allow germination processes to begin.

Screen: A separating surface in the air-screen cleaner, scalper, etc. Each screen has perforations over its surface, all of the same size and shape. Screens separating surfaces may be of perforated sheet metal or woven wire mesh; the surface may be flat, slightly grooved, or with perforations in recesses.

Screen Apron: The blank (not perforated) area at the feed end of a screen. Its purpose is to receive the seed flowing to the screen and direct it to the following perforated area. If the main seed flow were fed directly onto the perforated area, the perforations would largely be plugged, and would increase the time required to clean screens after use.

Screen Ball Tray System: A system to prevent plugging of screen perforations by installing a vibrating tray containing resilient rubber balls beneath each screen. As the tray moves, the balls bounce up against the bottom of the screen. The impacts of the balls dislodges stuck seed and particles.

Screen Brush System: A system to keep screen perforations from plugging up, by moving brushes from side-to-side beneath the screens to knock out jammed seed and particles.

Screen, Grading (Bottom): A screen with perforations small enough so that the good seed will not fall through, but seed and particles smaller than the good seed will fall through and be separated.

Screen, Scalping (Top): A screen with perforations large enough for the good crop seed (and smaller seed/particles) to fall through. However, larger undesirable materials such as trash cannot pass through the perforations, so they pass over the screen and are separated.

Screen Shake (or Oscillation or Vibration): The speed or intensity of vibration or shaking of a screen. Shake, combined with slope, is used to control the rate of seed flow over the screen.

Screw Conveyor: Auger conveyor.

Seed: In this publication, the true seed or botanical structure which can be handled as a seed, and is used to re-multiply a crop species.

Seed Conditioning: The sequence of operations in which seed are handled, moved, stored, etc., in the process of removing undesirable seed and particles, improving the quality, purity and condition of the seed, and making them more suitable for planting to produce a high-yielding/quality crop.

Seed Industry: The agri-industry which produces, conditions, handles, supplies and markets improved seed to farmers. Sometimes used to include variety development, testing and maintenance, but these activities are more realistically included in research and breeding programs.

Seed Law (Law, Seed): Law regulating the sale of seed, requiring certain conditions, and controlling the quality of seed offered for sale.

Seed Lot: A specific quantity of seed, usually of limited size, identified and handled as a separate unit. A lot is homogeneous in all of its parts.

Separating (or Separation): In seed conditioning, denotes using a machine or other method to separate one or more types of undesirable particles or seed from the crop seed lot.

Separating air Blast: An air blast or aspiration in a cleaner or separator which works on seed which has already been aspirated to remove dust. This air separation is intended to remove specific undesirable seed and particles.

Sequence: The order or sequence in which a series of conditioning machines are installed and/or used. Different crop seed often require a sequence of different separators.

Shell: (1) Of seed, to remove the shell from seed or to remove corn seed from the cobs; (2) of a machine such as the width and thickness grader, the cylindrical perforated "shells" used to separate seed.

Shelling: (1) Removing corn (maize) seed from the cobs, and also used for (2) the removal of seed of crops such as peanut (groundnut) from the shells.

Shelling Bar (in a Peanut Sheller): The bar which strikes the unhulled peanuts to crack open the shell in the shelling process.

Shoe (in Air-Screen Cleaner or Other Machine with Screens): The boxlike structure or frame which holds one or more screens in operating position in a cleaner.

Shutoff Gate (or Discharge Shutoff): An adjustable device at the discharge of a bin which allows controlling or shutting off the flow of seed from the bin. Most cleaners and separators also have a shutoff gate in their feed intake.

Single-Story (Design or Plant Layout): A seed conditioning plant designed to operate with all machines installed on a single floor level. Seed is moved from machine to machine by conveyors and elevators.

Singulated (Seed in a Color Sorter): Seed organized into a single row, so that the color sorter can view each seed individually.

Size-grading: Separating seed into different lots or "grades" according to their size (which may include width and thickness). Size-grading enables closer space-planting in the field. It is done primarily on corn seed, but is also used on other seed.

Sling Psychrometer: A hand-held device used to measure the humidity percentage of the air.

Sliptube: Used on the intake of an airlift or vacuum elevator, a sliptube is the intake pipe with a slightly larger but close-fitting pipe fitted over it. The outer pipe can slide or "slip" up and down to change the length of the intake pipe so it can reach seed at different points.

Slurry: A liquid formulation made by mixing a powder chemical with water.

Small Particle Waste: Undesirable waste materials smaller than the good seed, and removed by separations such as a bottom grading screen.

Small-Scale (Lab) Cleaning/Separating: Cleaning or separating small samples or lots, as in quality control, testing, research, etc.

Spiral Separator: A seed separator which separates round seed from flat-sided seed by allowing them to flow by gravity down spiraled flights. Round seed roll faster and move by centrifugal force further out, and drop into a separate discharge spout.

Splitter: A system used to divide the seed flow, so that each of the two split streams of seed flow to a different place. For example, in a cylinder separator installation, one cylinder splits or separates short from long seed and sends each to a different secondary cylinder for a closer separation. In an air-screen cleaner, a splitter usually placed after the first scalping screen splits the seed flow into two parts, with each part going to a separate but equal bottom grading screen; this increases the capacity of the cleaner.

Spout: A spout or pipe which discharges any separated fraction from any machine in a seed conditioning plant; it may discharge clean seed or any waste fraction, or even dust and light trash removed by the dust control system.

Stack: A vertical pile of bagged seed, pallets of seed bags, or toteboxes. Normally, only one lot is included in one stack, as a safeguard against contamination or errors in identification.

Stack Card: A sign placed on a stack or row of stacks of seed in storage. It identifies the lot, kind and amount of seed in the stack, by date.

Stoner: A special gravity separator designed to separate only one kind of seed/particles from the crop seed, usually separating heavy seed/particles from the lighter crop seed.

Storage: Used for (1) the period of time between seed harvest and planting, (2) a structure or facility used for storing seed, or (3) the process of holding seed until it is used for planting.

Storage, Conditioned: A storage structure or operation which holds seed in a storage where environmental conditions are artificially created. Normally, this involves refrigeration to provide lower-than-ambient temperature and dehumidification to provide lower-than-ambient relative humidity. Conditioned storage is usually used for long-term storage, and/or for valuable or special seed.

Storage, Non-Conditioned: Seed storage under ambient temperature and humidity conditions, which are modified only by using favorable construction aspects in the storage structure. This is normally used for short-term storage of seed for farmer planting.

Storage, Short-Term: Storage of seed from harvest until the next planting season. This is usually at most a few months.

Storage, Long-Term: Storage of seed for years, such as in a genetic materials bank or a Foundation seed program.

Storage, Medium-Term (or Intermediate-Term): Storage of seed beyond the first planting season following harvest, but for at most 1–2 or 3 years. A seed enterprise will often produce 5–10% more seed than it expects to sell, to provide a buffer in case more seed is needed. This seed will be kept in medium-term storage for sale and planting the following year.

Stratification: On the gravity separator and stoner, using an upward-flowing air stream to lift lighter seed while heavier seed remain at the bottom of the seed mass as it flows across the separating surface (deck). This "stratifies" the seed mass vertically, with successively lighter seed higher up.

Support (or Stand or Platform or Base): The structure on which a machine is installed so that it can be operated properly and safely.

Support Building/Facility: Building or facility for activities which are not directly involved in seed conditioning but are essential to supporting conditioning operations. This usually refers to shop facilities, personnel facilities, etc., which are required but should not be included in seed conditioning buildings for reasons such as fire hazard, water, or personnel movement.

Surface Texture Separation (or Separator): Seed separation by differences in their surface texture; *i.e.*, separating smooth-surfaced from rough-surfaced seed.

T

Table Separator: A machine which separates seed of differing specific gravity and resilience by their flow characteristics on a sloping, vibrating table surface, combined with bumping actions.

Tag: A label in the form of a tag which is affixed to a bag of seed. Information on the seed's identity and quality is shown on the tag.

Temper (or Tempering): In seed drying, denotes partial drying of the seed, then holding the seed without drying in a separate bin for several hours. Drying first removes moisture from surface layers of seed; tempering allows moisture to migrate from the center of seed to the surface, so as to allow more efficient drying. This also allows drying of seed such as rice, without cracking the seed due to extreme differences in moisture content of different layers.

Terminal Velocity (of a Seed): The maximum velocity a seed will reach in free-fall, due to the force of gravity and the flight characteristics of the seed. In a moving air stream, if the air speed is lower than the terminal velocity of the seed, the seed will fall downward. If the air speed is at the seed's terminal velocity, the seed will be suspended and held stationary in the air stream. If air speed exceeds the seed's terminal velocity, the seed will be blown upward.

Test Weight Apparatus: A special apparatus used to measure the density of a seed sample. Results may be expressed as pounds per bushel, or as kilograms per hectoliter, according to the apparatus used.

Testing: Conducting one or more of the standardize test procedures used to determine an aspect of the quality of a seed sample which represents a much larger seed lot.

Thresher: A machine or method used to remove seed from the crop plants at harvest time. May be a small machine, a large machine, or a moving combine thresher. The thresher uses one of several methods of beating to remove dry mature seed from the parent plants.

Tilt: As used in most seed separators, the slope or tilt of the separating mechanism from the feed end to the discharge end, to cause seed to move through the system by gravity flow.

Timothy Bumper Mill: A separator which used a rocking-and-bumping action to cause seed to roll or slide to effect a separation. Used primarily to remove weed seed from timothy grass seed, thus its name.

Totebox: A box or boxlike structure which holds bulk seed. With usual capacity of about 1 ton, toteboxes are handled, moved and stacked by forklift. Toteboxes can be filled from any spout, and can be dumped/emptied by turning them over with a forklift or special box dump.

Treater: A specialized machine which applies a specific dosage of a protective chemical solution to the surface of seed, uniformly applying the material to all seed.

Treating: Usually said of the process of uniformly applying a protective and/or disinfecting chemical to seed.

Treatment: Used for (1) the process of treating seed with protective chemical solution(s) or (2) the protective material(s) applied to seed.

Trier: A tubelike device, with side openings and an outer tube which can be turned to close the side openings, inserted into a bag or bulk quantity of seed, to take a sample.

Truck Scale (Weighbridge): A scale or weighing mechanism designed to weigh loaded or unloaded trucks which are driven onto the weighing platform.

U

Uncleaned Seed: "Raw" seed which has not been cleaned or conditioned.

V

Vapor Barrier: Any vaporproof material installed in any location, to provide an unbroken seal around a storage area or quantity (bag) of seed which prevents movement of air and humidity from the outside to the inside. In seed conditioning, this term is usually applied to a vaporproof layer installed in the floor of conditioning and storage buildings and/or in the walls and ceiling of a conditioned storage.

Vaporproof: Said of bags and packages which are constructed of vaporproof materials and the filling opening is sealed, so that no air and humidity from the outside can pass through the material and reach the inside.

Vaportight: A space such as inside a storage room, seed bag or seed package, which is completely sealed and enclosed by a vaportight material or vapor barrier. Outside air or vapor (moisture) cannot enter the vaportight space.

Vertically Integrated: Said of an enterprise which conducts several or all of the operations involved in producing, conditioning, distributing and marketing seed.

Vibrator Separator: A machine which separates seed by differences in their ability to roll when passed over a sloped vibrating surface. No air flow or stratification is used.

Visual Examination: The process of examining seed by eye, and using experience and training to identify seed condition and quality, and to determine conditioning or other needs. This is an important aspect of achieving high quality in seed conditioning.

W

Waste Products: Undesirable materials removed from seed lots in conditioning. This includes dust, trash, mudballs, small stones, broken seed, immature or low-quality crop seed, weed seed, other crop seed, etc.

Waterglass Test: A test using a diluted solution of liquid sodium silicate (waterglass) to identify differences in the specific gravity or density of seed, to determine if they are sufficiently different for the gravity separator to separate them efficiently.

Weighing: Using appropriate kinds of scales to measure the weight of seed. Weighing is done on bulk seed in trucks and toteboxes to determine the amount of seed contained, on bags and small packages to ensure that each bag/package contains the correct weight of seed.

Width or Thickness Separator: A specialized seed separator which uses recessed perforations in flat or cylindrical screens to separate seed by either width or thickness. One screen makes only one separation; for more than one separation, a different number of screens, each with the desired type of perforation and making only one separation, are used.

Workspace: Adequate space around a machine, on the floor or on its support stand, so that workers can safely operate, adjust, maintain, or clean it. Space must also permit safely removing/changing parts, such as screens in an air-screen cleaner or cylinders in a cylinder separator.

X

Y

Y-Valve: A two-way valve with a switching flap which allows sending seed to either of two pipes.

Z

CROP SEED CONDITIONING

ISTA (International Seed Testing Association) established and standardized many of the aspects described herein. ISTA also uses standardized definitions for many of the terms used to describe various aspects of seed. For uniformity, the following terms are reproduced from The International Rules for Seed Testing, published by ISTA.

A

Achene (Achenium): A dry indehiscent one-seeded fruit, formed from one free carpel (*e.g. Ranunculaceae, Geum*) with the seed coat distinct from the fruit coat; occasionally consisting of more than one carpel (*Asteraceae, Compositae*).

Aril, Arillus (plural Arilli): A fleshy, often colored covering or appendage of a seed growing out form the funicle or base of the ovule (see also caruncle, strophiole).

Awn, Arista: Slender, straight or bent bristle. In grasses, usually a continuation of the mid-nerve of lemmas or glumes.

B

Beak(ed): A long, pointed prolongation of a fruit.

Bristle: A stiff hair; sometimes applied to the upper part of an awn, when the awn is bent.

C

Caruncle: A small outgrowth of the micropylar region of a seed (see also aril, strophiole).

Caryopsis: A naked fruit of a grass, in which the testa is united with the pericarp.

Cluster: A densely crowded inflorescence or in *Beta*, a part of an inflorescence.

D

Drupe: Indehiscent, one-seeded fruit with stonyh endocarp and fleshy outer layers.

E

Embryo: The young plant which is enclosed in a seed.

F

Fascicle: A tuft of branches arising from a bout the same place.

Floret: The lemma and palea with enclosed pistil and stamens or the mature caryopsis in *Gramineae (Poaceae)*; for the purpose of the ISTA Rules, the term floret refers to the fertile floret with or without additional sterile lemmas.

G

Glume: One of the two usually sterile bracts at the base of a grass spikelet.

H

I

Indehiscent: Not opening, as fruits which do not open at maturity.

Integument: The envelope of an ovule which becomes the seed coat or testa (generally 2 integuments are present).

J

K

L

Lemma: the outer (lower) bract of a grass floret, sometimes referred to as the flowering glume or the lower or outer palea. Bract enclosing the caryopsis on the outer (dorsal) side.

Locule, Loculus (pl. Loculi): Compartment of the ovary containing the seed.

M

Mericarp: Part of schizocarp.

N

Nutlet: A small nut.

O

P

Palea: The upper (inner) bract of a grass floret, sometimes called the inner or upper palea. Bract enclosing the caryopsis on the inner (ventral) side.

Pappus: A ring of fine, sometimes feathery, hairs or scales, crowning an achene.

Pedicel: The stalk of each single flower in an inflorescence.

Pericarp (Fruit Coat): The wall of the mature ovary or fruit.

Pod: Dehiscent dry fruit, especially of Leguminosae.

Pyrene: Seed enclosed by the hard endocarp from a drupe (or similar structures from multi-seeded fruits).

Q

R

Rachilla: A secondary rachis, in particular, in grasses the axis that bears the floret.

Rachis (pl. Rachides): The main axis of an infloresncence.

S

Seed Unit: The commonly-found dispersal seed unit; *i.e.*, achenes and similar fruits, schizocarps, florets, etc., as defined for each genus or species in the Pure Seed Definitions.

Schizocarp: A dry fruit which separates into two or more units (mericarps) at maturity.

Siliqua: Dehiscent, dry, two-valved fruit derived from two carpels, *e.g.*, Cruciferae.

Spikelet: The unit of a grass inflorescence comprising one or more florets subtended by one or two sterile glumes. For the purposes of ISTA Rules, spikelet includes: as well as a fertile floret, either one or more additional fertile or completely infertile florets, or glumes.

Strophiole: Small aril, wartlike outgrowth (see also aril, caruncle).

T

Testa: Seed coat.

U

V

W

Wing: A flat membranous outgrowth from a fruit or seed.

X

Y

Z

Bibliography

Much information herein was derived from the following sources, from which further information, and leads to additional materials, can be obtained. Also, obtain descriptive literature from suppliers of equipment available in your area. Additional source material can be developed by maintaining detailed records of conditioning adjustments and machines used.

Usually not listed in this bibliography are personal communications and information supplied by persons listed in the Acknowledgements.

AEC corn sheller descriptive literature. 2006. AEC Enterprises, Inc. St. Charles, Iowa 50240.

AGRA industries products descriptive literature. 2005. AGRA Industries, Inc. Merrill, WI 54452.

Agrawal, P.K. and M. Dadlani. 1987. Techniques in seed science and technology. South Asian Publishers. New Delhi, India. 191 pp.

Anon. 1952. Manual for testing agricultural and vegetable seeds. USDA. Washington, DC. Handbook no. 30. 441 pp.

Anon. 1991. Rules for testing seeds. Assn. Of Official Seed Analysts. *J. Seed Tech.* 12(3). 122 pp.

Anon. 1993. International rules for seed testing. *Seed Science & Technology 21, supplement.* International Seed Testing Association. Zurich, Switzerland. 284 pp.

Bailey, L.H. 1949. Manual of cultivated plants. MacMillan. New York. 1116 pp.

Bainer, Roy 1942. Seed segmenting devices. *Proc. American Society of Sugar Beet Technologists* 3rd General Meeting pp. 216–219. Salt Lake City, Utah, Jan. 5–7.

Bainer, Roy. 1947. Precision planting equipment. *Agri. Engineering* 282: 49–57.

Bainer, Roy. 1948. Present status of sugar beet seed processing. *Proc. American Society of Sugar Beet Technologists,* 5th General Meeting pp. 120–124. San Francisco, CA. Jan. 12–15.

Bainer, Roy, and L.D. Leach. 1943. Processing sugar beet seed by decortication, burr reduction, and segmentation. *Proc. American Society of Sugar Beet Technologists.* Pp. 625–639.

Ball, Vic (ed.). 1975. Ball red book, 13th edition. Geo. J. Ball, Inc. West Chicago, IL. 502 pp.

Banton, R. 1946. Problems in cleaning seed. A.T. Ferrell & Co. Saginaw, Mich.

Billups, Gary. 2000. Personal communication. Crippen Intnl, Inc. Dallas, TX.

Billups, Gary L. and Hossam Lasheen. 1990–2005. Engineering designs from conditioning plant plans and machine installations. Crippen International. Dallas, TX.

Boller, Robert A. 1964. Studies on a commercial seed-blending operation. Agr. Mkt. Service, Publ. 368. US Dept. of Agri. Washington, DC.

Bunch, H.D. 1957. Separating seed by the use of electronics. Seedsmen's Digest. Feb.

Calkins seed treater descriptive literature. 1950's. Calkins Mfg. Co. Spokane, WA.

Cardboard totebox bins descriptive literature. 2005. SpaceKraft Inc. In *Seed Today*, 2nd quarter 2005.

Carpo electrostatic separators descriptive literature. (Undated). Carpco, Inc. Jacksonville, FL.

Cecka, M.C. 1962. Width and thickness separations with precision graders. *Proc., Workshop Conference for Seedsmen.* Inst. of Agri. Sciences., Wash. State Univ. Pullman.

Chalam, G.U., Amir Singh, and J.E. Douglas. 1967. Seed Testing Manual. I.M.H. Press Pvt Ltd. New Delhi, India. 267 pp.

Chantland Material Handling equipment descriptive literature. (Undated). Chantland Material Handling Solutions. Humboldt, IA 50548.

Ching, T.M., M.C. Parker, and D.D. Hill. 1959. Interaction of moisture and temperature on viability of forage seeds stored in hermetically-sealed cans. Agron. J. 51: 680–684.

Cimbria Delta Super air-screen cleaner descriptive literature. 2006. Cimbria Bratney, Des Moines, IA.

Cimbria Heid cleaners and separators descriptive literature. 2005. Cimbria Heid GmbH, Stockerau, Austria, and Cimbria Bratney Co., Des Moines, IA 50322.

Columbia Okura robotic palletizing systems descriptive literature. 2005. Columbia Okura LLC. Vancouver, WA 98661, and Bratney Companies, Des Moines, IA 50322.

Continental Eagle ginning, delinting, and acid-delinting equipment descriptive literature. (Undated). Continental Eagle Corp., Prattville, AL.

Continuous overlapping bucket lift descriptive literature. 2005. Anex Mfg., Burbank, CA 91510.

Convertible slide gates descriptive literature. 2005. Tom-Cin Metals, Inc., Hortonville, WI USA.

Coronatron electrostatic separators descriptive literature. 1966. Superior Separator Co. Hopkins, Minn.

Creamer Metal Products elevators, conveyors and handling equipment literature. 2007. London, Ohio.

Crippen cleaners and separators descriptive literature. 2006. Crippen International, Dallas, TX.

Clark, Don. 2005. Personal communications on gin equipment. Continental Eagle Inc. Prattville, AL.

Crippen cleaners and separators descriptive literature. 1990. Crippen Mfg. Co., Inc., Alma, Mich.

Crow, J.W. 1958. Electronic separator does the tough jobs. Western Feed and Seed. 13(3).

DAMAS seed cleaning and separating equipment descriptive literature. (Undated). DAMAS A/S, DK-5600 Faaborg, Denmark.

DAMAS Sigma vertical screen cleaner and other machines descriptive literature. 1995. DAMAS A/S, DK-5600 Faaborg, Denmark.

DAMAS SORLA table separator descriptive literature. 2006. Damas, DK-5600 Fasborg, Denmark.

Day dust control systems descriptive literature. (Undated). Day Company, Minneapolis, Minn.

Deamco Corp. elevators and bulk materials handling equipment information. (Undated). Deamco Corp., City of Commerce, CA 90040.

Delouche, James C. 1968a. Precepts for seed storage. *Proc. Short Course for Seedsmen*. Seed Techn. Lab., Miss. State Univ.

Delouche, James C. 1968b. Physiology of seed storage. *Proc. 23rd Corn and Sorghum Res. Conf.* Amer. Seed Trade Assn. 23: 83–90.

Delux continuous flow grain dryers descriptive literature. 2003. Delux Mfg. Co., Kearney, NB 68848.

Designs and engineering drafts for a complete seed conditioning plant. 1992–1993. Crippen International, Inc., Dallas, TX.

Dockins seed grader. (Undated). J.O. Dockins, Scott, AR.

Dodds, M.E., and J.W. White. 1961. Seed cleaners and separators. Canada Dept. of Agri. Research Publ. Ottawa.

Domingo, R., T. Castaneda, G. Yambao, JJr., B. Caballero, E. Sevilla, S. Manipon, C. Casayuran, F. Delmo, E. Rodirguez, R. Grimares, R. Malasaga, G. Pichay, H. Jeciel, S. Sison, I. Toyama, M. Aiyama, K. Kobayashi. 1981. Manual for operation and management of the seed plant. Bureau of Plant Industry. Manila, Philippines. 74 pp.

Donnell, William J. 2003. Arrow Corp. equipment descriptive literature. Arrow Corp., Winnipeg, Manitoba.

Dupont Seed Treating Manual. 1960. E.I. Dupont de Nemours & Co. Wilmington, Del. 30 pp.

Dustrol dust control systems descriptive literature. (Undated). Robbins & Myers, Inc. Springfield, Ohio.

Enviropac seed treaters and inoculators descriptive information. 2006. Enviropac, Inc., 229 State Route 251, Box 229, Compton, IL 61318.

Equipment and installations, various machines descriptive literature. 2005. Ken Bratney Co. Des Moines, IA 50322.

Equipment Catalog. 2006. *Grain Journal*, 3065 Pershing Ct., Decatur, IL 62526-1564.

ESM color sorting machine. 1969. Electric Sorting Machine Co. Houston, TX.

Esmueller bulk material handling equipment information. 2006. Essmueller Co., Laurel, MS 39441.

Ferrell, A.T., and Co., Inc. 1960's. Leaflets on Oxford bean polisher, Clipper cleaners, Clipper dust collectors, Clipper roll mills. Saginaw, Mich.

Fifty years of seed testing. 1958. Assn. of Official Seed Analysts. 101 pp.

Filcon dust filters descriptive literature. 2002. Filcon A/S. Centervej 26, 480 Soro, Denmark.

Filcon bag filters descriptive literature. 2003. Filcon A/S. Centervej 26, Soro, Denmark.

First aid kits. 2006. zeemedical.com. (On internet).

Fischbein bagging and weighing equipment descriptive literature. 2005. Fischbein Co., Statesville, NC 28625, and Bratney Companies, Des Moines, IA 50322.

Focus Industries literature on The Handler seed treaters. 2006. St. Albert, Alberta, Canada T8N 5C3.

Forsberg, Fred, & Son, Inc. (Undated). Forsberg gravity separators descriptive literature. Thief River Falls, Minn.

Forsberg gravity and vacuum gravity separators descriptive literature. 2003. Forsberg Inc., Inc. Thief River Falls, MN 56701.

Fernald, M.L. 1970. *Gray's manual of botany*, 8th ed. Van Nostrand Reinhold. New York. 1632 pp.

Garratt specific gravity separators descriptive literature. 1995. Garratt Industries Ltd. Milestone, SK SOG 3LO Canada.

Garratt stoners and gravity separators descriptive literature. 2003–2004. Garratt Industries, Ltd. Milestone, Saskatchewan SK S0G 3L0.

Gastel, A.J.G. van, M.A. Pagnotta, and E. Porceddu, (eds.). 1996. Seed science and technology. ICARDA. Aleppo, Syria. 311 pp.

Gompper magnetic seed separator descriptive literature. (Undated). Gompper-Maschinen-Gesellschaft KG. Buederich nr. Duesseldorf, Germany.

Grain conditioning technology. 2003. Behlen Mfg Co. Columbus, NB. 31 pp.

Grain dryers descriptive literature. 2002. Delux Mfg. Co. Kearney, NE 68848-1027.

Grain elevators of North America. 1942. *Grain & Feed Journals Consolidated*. Chicago, IL.

Grain Inspection Manual. Grain Branch, Production and Marketing Admin., USDA. Washington, DC.

Grain spreaders descriptive literature. 2002. NECO Division, TIC United Corp. Omaha, NE 68112.

Grainway belt conveyor descriptive literature. 2006. GrainWay LLC, Aplington, IA 50604.

Grain weighers and aspiration pre-cleaners descriptive literature. 2002. Maskinfabriken JE-MA A/S, DK-8850 Bjerringbro, Denmark.

Grass, L. and B. Gregg. 2000. Seed processing—a training manual. ICARDA. Aleppo, Syria. viii + 174 pp.

Greene, C.H. and G.B. Davis. 1959. Seed processing in the Willamette Valley. Misc. Paper 81. Agri. Exp. Sta., Oregon State College. Corvallis, OR.

Gregg, B.R. 1958. How to plan a seed plant. *Western Feed & Seed*. 13(6): 30.

Gregg, B.R. 1962. Processors plan, think competition is keen. *Seedsmen's Digest* 13(4).

Gregg, B.R. 1962. Getting maximum efficiency from the specific gravity separator. *Western Feed & Seed* 17(6).

Gregg, B.R. 1962. The specific gravity separation. *Proc. Workshop Conference for Seedsmen*. Inst. Of Agri. Sciences, Washington State Univ. Pullman.

Gregg, Billy R., Alvin G. Law, Sher S. Virdi, and John S. Balis. 1970. Seed processing. National Seeds Corporation. New Delhi, India. 396 pp.

Gregg, Bill. 1991. Safe seed storage. National Agri. Research Project, ARC, CAS. Cairo, Egypt. 54 pp.

Gregg, Bill. 1982. Vegetable seed data for Seed Division, Dept. of Agri. Extension, MOA. Bangkok. 75 pp.

Gregg, Bill. 1983. Seed conditioning, storage and marketing. Seed Club of Thailand. Bangkok. 98 pp.

Gregg, Bill. 1990. Seed conditioning equipment technical study manual. Unpublished work. 200 pp.

Gregg, Bill. 1995–2005. Powerpoint lectures on seed enterprise management, conditioning, storage and handling seed. Various. Personal materials. Starkville, MS.

Gregg, B.R., and A.J.G. van Gastel 2000. Seed production manual for the informal seed sector. IITA/GTZ/CSIR Project "Promotion of seed production and marketing in West Africa". Accra, Ghana. 46 pp.

Gregg, B.R., and A.J.G. van Gastel. 1997. Managing seed marketing. IITA/GTZ/CSIR Project "Promotion of seed production and marketing in West Africa". Accra, Ghana. 116 pp.

Gregg, B.R., F.A. Oresajo, W.D. Motsenbocker, and A.J.G. van Gastel. 2001. Starting and managing a successful seed enterprise. IITA/GTZ/CSIR Project "Promotion of seed production and marketing in West Africa". Accra, Ghana. 116 pp.

Gregg, B.R., A.J.G. van Gastel, A.W. Ebert, L.I. Delimini, and E.A. Asiedu. 1999. Preparing a realistic business plan for successful seed companies. IITA/GTZ/CSIR Project "Promotion of seed production and marketing in West Africa". Accra, Ghana. 53 pp.

Gregg, B.R., A.J.G. van Gastel, E. Asiedu, F. Donkoh, and R.W. White. 1999. Good seed, extension and farmers. IITA/GTZ/CSIR Project "Promotion of seed production and marketing in West Africa". Accra, Ghana. 102 pp.

Gregg, B.R., and A.J.G. van Gastel. 2000. Seed operations cost and income analyses sheets. IITA/GTZ/CSIR Project "Promotion of seed production and marketing in West Africa". Accra, Ghana. 36 pp.

Gregg, B.R., A.J.G. van Gastel, I. Djokoto, and E.A. Asiedu. 2001. The WASDU seedsaver drying-packaging system. IITA/GTZ/CSIR Project "Promotion of seed production and marketing in West Africa". Accra, Ghana. 45 pp.

Gregg, B., S. Abd El Wanis, Z. Bishaw and A.J.G. van Gastel. 1994. Safe seed storage. WANA Seed Network publication no. 5/94. ICARDA. Aleppo, Syria. 53 pp.

Gregg, B.R., *et al.* 1970. Seed processing. National Seeds Corp. and USAID. New Delhi, India. 396 pp.

Gregg, Bill. (Unpublished manuscript). Records and reports system for a seed enterprise. 200 pp.

Gregg, Bill. (Unpublished manuscripts). Crop seed conditioning requirements. 1,000 pp.

Grisez. Magnetic separator descriptive literature. (Undated, 1950's). J.F. Grisez, Crows Landing, CA.

Grisez, J.F. 1959. Magnetic seed cleaners. Western Feed & Seed 14(6).

Gunson, R.W. 1950's. Sortex electronic color separator descriptive literature. R.W. Gunson Ltd. London, England.

Gustafson and Mistomatic treaters descriptive literature. 1950's. Gustafson Manufacturing Co., Minneapolis, MN.

Hall, C.W. 1957. Drying farm crops. Edwards Bros., Inc. Ann Arbor, Mich. 336 pp.

Hance Manufacturing Co. 1950's. Brochures and other material on Vacaway cleaners. J.W. Hance. Westerville, Ohio.

Hansen, Edward L. 1961. Rodent control in seed processing plants. *Proc., Seed Processors Short Course*. Oregon State College. Corvallis.

Hardin, (ed.). 1961. Sampling technique. *Proc., Seed Processors Short Course*. Oregon State College. Corvallis.

Harmond, Jesse E. 1958. Pneumatic separators. Small Seed Harveing and Processing Section. Agri. Engineering Research Division, ARS, USDA. Corvallis, OR.

Harmond, Jesse E. 1961. A forward look at seed processing. *Seedsmen's Digest* 12(4): 20.

Harmond, Jesse E. 1961. A practical evaluation of pneumatic conveying. *Western Feed & Seed* Jan.

Harmond, Jesse E., and Norman Brandenburg. 1960's. A velvet-roll separator for seed testing. Mimeograph paper. Small Seed Harvesting and Processing Section, Agri. Engineering Research Division, USDA. Corvallis, OR.

Harmond, Jesse E, and Leonard M. Klein. 1968. Mechanical seed cleaning and handling. Agri. Handbook no. 354. ARS, USDA. Washington, DC.

Harrington, J.F. and J.E. Douglas. 1970. Seed storage and packaging applications for India. National Seeds Corp. and Rockefeller Foundation. New Delhi, India. 222 pp.

Hawthorn, L.R., and L.H. Pollard. 1954. Vegetable and flower seed production. Blakison Co., Inc. New York.

Hayes, Brian. 2005. Squeeze belt conveyor. *Seed Today*, 2nd quarter 2005, pp 84, and other information from Hutchinson-Mayrath Inc., Clay Center, KS.

Henderson, James. 1957. Fanning mill operation and screen nomenclature. *Seed Processors Short Course Proceedings*. Oregon State College. Corvallis.

Henderson, James. 1959. How to get the most from your seed cleaner. *Western Feed & Seed*. 14(12): 23.

High temperature rice drying. 2003. Rice quality workshop. University of California Davis. http://www,agronomy.ucdavis.edu/uccerice/quality/rqw2003/c-11heated_air_drying.pdf.

Hill, H.J. 1999. Advances in seed technology. *J New Seeds* 1(1). Haworth Press. (Summary in web site of Seed Dynamics).

Hill, Hank. 2006. Web site and online papers on seed coating and pelletizing. Seed Dynamics. http://www.seedquest. com/vegetables/lettuce/expo/seeddynamics/science.htm; http://www. seedquest.com/vegetables/lettuce/expo/seeddynamics/whyprimeseed.htm

Holzinger Bros. 1954. LiftMaster airlift elevators descriptive literature. Santa Fe Springs, CA.

Hoover, C. Dale. 1958. Annual Report, Agronomy Dept. Miss. State College. Miss. Farm Research 21(6): 4–5.

Hudson, Wilbur G. 1954. Why use pneumatic conveyors? *Chem. Engineering* 61: 191–195.

INCOTEC seed coatings descriptive information. 2005. INCOTEC Inc., Salinas, CA.

Industrial bulk bags descriptive literature. 2005. Greif, Inc. Omaha, NE. In *Seed Today*, 2nd quarter, 2005.

Installation and maintenance manual for Essmueller bucket elevators. 2006. Essmueller Co., Laurel, MS 39441-1966. 18 pp.

International rules for seed testing. 1996. International Seed Testing Association. *Seed Science & Technology* Vol. 24. Supplement. Zurich, Switzerland. 331 pp.

Justice, Oren L. and Louis N. Bass 1978. Principles and practices of seed storage. Agri. Handbook 506. USDA. Washington, D.C. 289 pp.

Kent, C.A. 1954. A study of the cleaning and blending of commercial seed lots. Assn. *Official Seed Analysts Proceedings* 44: 129–144.

Klein, Dr. Bob. Needle separator used in sorting pistachio nuts. Dec. 6, 2006. Needle separator use in sorting pistachio nuts. Personal communication. 1318 East Shaw Ave., Suite 420, Fresno, CA 93710-7912. web: www.pistachios.org.

Klein, L.M. 1958. Air-screen cleaner. Small Seed harvesting and Processing Section, Agri. Engineering Research Division, USDA. Corvallis, OR.

Krantz, G.W. 1961. Control of insects in seed cleaning plants. *Proceedings, Seed Processor's Short Course.* Oregon State College. Corvallis.

Kreyger, J. 1963. General considerations concerning the drying of seeds. *Intnl. Seed Testing Assoc. Proc.* 28: 753–784.

Kromer, O.W. 1965. Kromer seed treater descriptive materials. O.W. Kromer Co., Minneapolis, Minn.

Kubik, Keith 2006a. Putting pellets around seed—an art and a science. *Seed Techn. Newsletter* Vol. 6. Seed Quest. (On internet).

Kubic, Keith. 2006b. Pelleting seed in a rotary coater—a new trend. *Seed Techn. Newsletter* Vol. 18. Seed Quest. (On internet).

LeClerg, E.L. 1953. Seed-borne plant parasites. Plant Disease Reporter 37(9).

Madsen, Erik, and N.E. Langkilde. 1988. ISTA handbook for cleaning of agricultural and horticulturals seeds on small-scale machines. Part II. International Seed Testing Association. Zurich, Switzerland. 117 working sheets, 240 pp.

Madsen, Erik, and N.E. Langkilde (eds.). 1987. ISTA handbook for cleaning of agricultural and horticultural seeds on small-scale machines. Part I. International Seed Testing Association, Zurich. 240 pp.

Maguire, James D. 1962. Seed testing—aid in seed plant operations. *Proceedings, Conference Workshop for Seedsmen.* Institute of Agi. Sciences, Washington State Univ. Pullman.

Mahdavinader, Hooman. 2006. Magnetic separator information. AR Machine Co., Tehran, Iran.

Marincic, Jeff. 2005. Dryer burner maintenance. *Seed Today* 2nd quarter 2005. pp. 88–89. (jmarinci@maxoncorp.com).

Mascarenhas, J. 1961. An evaluation and comparison of new and standard methods of seed blending. Master's thesis. Miss. State Univ. State College, MS.

Maskinfabriken equipment descriptive literature. 2002. Maskinfabriken JE-MA A/S. Bjerringbro, Denmark.

Massie, Duward. 1962. Cleanout, keeping lots separate, and other problems of processing pure seed. *Proceedings, Conference Workshop for Seedsmen.* Institute of Agri. Sciences, Washington State Univ. Pullman.

Mepu dryers, screen and precleaner machines descriptive literature. 2003. Mepu Oy. Mynamaentie 59, Finland 21900 Ylane.

Meyer Machine Co. (Undated). Meyer pneumatic separators descriptive literature. San Antonio, TX.

Milling plants and layouts descriptive literature. 2002. Skiold Saeby A/S. DK-9300 Saeby, Denmark.

Miss. Crop Improvement Assn. 2003. Handbook of seed Certification regulations, Certification inspection forms. MS State, MS. 86 pp.

Model Mini 1218 air-screen cleaner descriptive information. 2006. Three Stars USA, Conroe, TX 77304.

Moder, J.J., Jr., and N.M. Penny. 1954. Industrial engineering and economic studies of peanut marketing. Engineering Exp. Sta., Georgia Institute of Technology. Bulletin 286. Atlanta.

Moffet, W.W., and W.B. Christian. 1955. A study of the cost of processing seed in Mississippi. MS Agri. Exp. Sta. Bul. 528. State College, MS.

Moore, C.E., H.D. Bunch, and G.B. Welch. 1958. Magnetic separation of seed. *MS Farm Research* 21(11): 3. Nov.

Benefits of modern seed treatment and other materials on Panogen treaters. 1950's. Morton Chemical Co. Chicago, IL.

Mrozinski, David. 2003–5. Information on acid-delinting plants, gins and machine delinters. Continental Eagle Inc. Prattville, AL.

Nema, N.P. 1986. Principles of seed certification and testing. Allied Publishers, New Delhi, India. 196 pp.

Nielsen, Jens. April 1900. Illustreret priskurant fra Jens Nielsen maskinfabriken "Vester-Aaby". Banestation Peirup pr. Vesterskjerninge, Danmark. 50 pp.

Ochse, J.J. and R.C. Bakhuizen van den Brink. 1980. Vegetables of the Dutch East Indies. A. Asher & Co. B.V. Amsterdam. 1006 pp.

OECD. 1966. Scheme for varietal certification of herbage seed moving in international trade. 1966. Publ. No. 76. Orgn. For Economic Cooperation and Development. May.

Oliver gravity separators and stoners descriptive literature. 1996. Oliver Mfg.Co. Rocky Ford, CO81067.

Oliver gravity separators descriptive literature. 2004. Oliver Mfg. Co., Inc. Rocky Ford, CO 81067.

Oliver gravity separators manuals and information. 1989–1992. Oliver Mfg.Co. Rocky Ford, CO.

Oresajo, Felix. 2001. Personal communication. Premier Seed Co. Zaria, Nigeria.

Oresajo, F.A., B.R. Gregg and A.J.G. van Gastel. 2001. Financial management for seed managers. IITA/GTZ/CSIR Project "Promotion of seed production and marketing in West Africa". Accra, Ghana. 46 pp.

Pal, B.P. 1966. Wheat. ICAR. New Delhi, India. 369 pp.

Pegasus model of gravity separator descriptive literature. 2006. Arrow Corp., Winnipeg, Manitoba Canada R3H0V5.

Petkus Agri Engineering Inc. 1993. Information on Petkus equipment. Des Moines, IA.

Petkus equipment descriptive literature. 2003. Petkus Agri Engineering. Des Moines, IA.

Pike, Patric. 2006. Personal communications on electronic color sorter. Satake America. Houston, TX.

Planning and installation of seed equipment and plants. 2005. Bratney Companies. Des Moines, IA 50322.

Plant installations and controls descriptive literature. 2003. Agra Construction Co. Merrill, WI.

Pouchpack seed packeting machine descriptive literature. 2005. O/K International Corp., Marlborough, MA.

Principles of Pest Control. 1960. Bureau of Correspondence Courses, General Extension Service, Washington State Univ. Pullman.

Proposed standards and procedures for seed storage installations used for long-term conservation of base collections. 1974. Report by a working group of the FAO panel of experts on plant exploration and introduction. Rome. May.

Quillen, C.S. 1954. Mixing—the universal operation. *Chem. Engineering* 61: 177–224.

Rat, your enemy No. 1. 1967. Directorate, Storage and Inspection, Ministry of Food and Agriculture, GOI. New Delhi, India.

Rice Symposium. 1964. The major insect pests of the rice plant. Internl. Rice Research Institute. John Hopkins Press. Baltimore, MD. 729 pp.

Rice Symposium. 1967. The virus diseases of the rice plant. Intnl. Rice Research Institute. John Hopkins Press. Baltimore, MD. 354 pp.

Rice, William A. 2005. Information and personal communications on the roll mill. Jerseyville, IL 62052.

Rollin, S.F. 1956. How to stay out of trouble with seed control officials. *Proceedings, Short Course for Seedsmen.* Miss. State College. State College, MS.

Rollin, S.F. 1961. A word to the wise on seed laws. *Proceedings, Short Course for Seed Processors.* Oregon State College. Corvallis.

Rollin, S.F. 1961. Seedsmen's records and the Federal Seed Act. *Seedsmen's Digest* 12(7): 8.

Rules and Regulations Under the Federal Seed Act. 1965. USDA, Agri. Marketing Service & Regulatory Announcement No. 156. Washington, DC.

Rules for Testing Seeds. 1965. *Proceedings, Assn of Official Seed Analysts*, no. 54. Agri. Marketing Service, USDA. Beltsville, MD.

S & S buckhorn separator descriptive literature. 1965. Sutton, Steele & Steele, Inc. Dallas, TX.

ScanMaster color sorters descriptive literature. 2005. Satake USA, Stafford, TX 77477, and Bratney Companies, Des Moines, IA 50322.

Schule table (paddy) separator descriptive literature. 2006. F.H. Schule Mohlenbau GmbH, Muhlenbau, Austria.

Schutte Buffalo hammermills descriptive literature. 2006. Schutte Buffalo Pulverizer Co., Inc. Buffalo, NY 14240-0546.

Seed box inverter descriptive literature. 2005. DBH LLC, Leroy, IL.

Seedburo catalog. 2005. Seedburo Equipment Co. West Chicago, IL. 144 pp.

Seed and grain cleaners descriptive literature. 2002. Arrow Corp. Alma, Michigan 48801.

Seed Equipment. 1970. *First All-India Symposium on Improved Seed Equipment.* National Seeds Corp. & USAID/MSU. New Delhi, India. 176 pp.

Simatek universal bucket elevators descriptive literature. 2003. Simatek Bulk Systems A/S. DK-4270 Denmark.

Seed coating. 2005. Landec Ag, Inc. Oxford, IN.

Seed cleaners, screens and components descriptive literature. 2005. Cimbria Bratney Co. Des Moines, IA 50322.

Seed Processing, Storage and Distribution Manual. 1967. Mimeograph by National Seeds Corp. and Rockefeller Foundation. New Delhi, India.

Seed Review Team, India. 1968. Seed Review Team Report. GOI/NSC/USAID. New Delhi.

Seed technology newsletter 18: pelleting seed in a rotary coater, a new trend. 2006. Integrated Technology Research. On web.

Seeds. 1961. Yearbook of Agriculture. USDA. Washington, DC. 591 pp.

Segler, G. 1960's. Pneumatic grain conveying. National Institute of Agri. Engineering. Silsoe, Bedfordshire, England. Published by Pfro. Dr. Ing. G. Segler, Braunschweig, Germany.

S. Howes Machinery Co. 1950's. Brochures and other material on Eureka cleaners. Silver Creek, NY.

Screw Conveyor Corporation equipment descriptive literature. 2007. Screw Conveyor Corp., Hammond, Indiana 46327.

Simon-Carter, Inc. 1950's. Brochures and other materials on Carter precision graders. Minneapolis, Minn.

Singh, K.N., B.B. Mohanty and M.R. Lokhande. 1968. Towards a Dynamic Agriculture. Navchetan Press Litd. Naya Bazar, New Delhi, India.

Sittisroung, Prasoot. 1967. Storage of rice *Oryza sativa* and cowpea *Vigna sinensis* seed, M.S. Thesis. Mississippi State University.

Strube, Harry L. 1954. Conveyors and elevators. Chem. Eng. 61: 195–212.

Stutler, Darin. 2006. Personal communications on machines, processes and designs of seed conditioning plants and operations. Bratney Companies. Des Moines, IA.

Sudenga equipment technical information. 2006. Sudenga Industries, Inc. George, Iowa 51237.

Sukprakarn, Sutevee, Sunanta Juntakool, and Rukui Huang. 2005. Saving your own vegetable seeds. AVRDC. 25 pp. Shanhua, Tainan, Taiwan.

Superior Rock-it corn grader descriptive literature. 1970. Superior Separator Division of the Daffin Corp. Hopkins, Minn.

Taylor packaging and container handling equipment descriptive literature. 2005. Taylor Products division of Magnum Systems, Parsons, KS 67357, and Bratney Companies, Des Moines, IA 50322.

Testing Agricultural and Vegetable Seeds. 1952. USDA. Handbook 30. Washington, DC. 440 pp.

Thomas, Jim. 1999. An introduction to gravity separators. Reprint from Powder and Bulk Engineering. 3 pp.

Thompson, Leonard E. 1959. How to choose conveyors and elevators. Western Feed & Seed.

Vaughan, Charles E., and C. Hunter Andrews. 1961. Let's clean up. Seed Technol. Lab., Miss State Univ. State College, MS.

Vaughan, Charles E., B.R. Gregg, and James C. Delouche. 1968. Seed processing and handling. Handbook No. 1. Seed Technol. Lab. Miss. State Univ. State College. 295 pp.

Virdi, S.S., B.R. Gregg, V. Sankaran, R. Kurmanaikulu, S.S. Rekhi, P.B. Young and Y.R. Mehta. 1972. Seed processing plant operator's handbook. National Seeds Corp. New Delhi, India. 510 pp.

Welch, G. Burns 1954. Seed processing equipment. *Miss. Agri. Exp. Sta. Bul.* 520. State College, MS.

Welch, G. Burns and James C. Delouche. 1968. Environmental and structural requirements for seed storage. *Miss. Agri. Exp. Sta. Journal Paper* 1607.

Wheeler, W.A., and D.D. Hill. 1957. Grassland Seeds. D. Van Nostrand. Princeton, NJ. 967 pp.

Wiese, M.V. 1987. Compendium of wheat diseases. 2nd edition.

Wiesner, Loren (ed.). 1991. Rules for testing seeds. Association of Official Seed Analysts. *Journal of Seed Technology* 12(3). 122 pp.

Index

For Product Safety Concerns and Information please contact our EU representative GPSR@taylorandfrancis.com Taylor & Francis Verlag GmbH, Kaufingerstraße 24, 80331 München, Germany

Batch number: 10725935

Printed by Printforce, the Netherlands